Semiconductor Lasers I

Fundamentals

OPTICS AND PHOTONICS
(formerly Quantum Electronics)

SERIES EDITORS
PAUL L. KELLEY

Tufts University
Medford, Massachusetts

IVAN P. KAMINOW

AT&T Bell Laboratories
Holmdel, New Jersey

GOVIND P. AGRAWAL

University of Rochester
Rochester, New York

CONTRIBUTORS

Alfred Adams
G. Björk
H. John E. Bowers
I. Heitmann
J. Inoue
Eli Kapon
F. Matinaga
Radhakrishnan Nagarajan
Eoin P. O'Reilly
Mark Silver
Amnon Yariv
Y. Yumamoto
Bin Zhao

A complete list of titles in this series appears at the end of this volume.

Semiconductor Lasers I
Fundamentals

Edited by Eli Kapon
Institute of Micro and Optoelectronics
Department of Physics
Swiss Federal Institute of Technology, Lausanne

OPTICS AND PHOTONICS

ACADEMIC PRESS

San Diego   London   Boston
New York   Sydney   Tokyo   Toronto

This book is printed on acid-free paper. ∞

Copyright © 1999 by Academic Press

All Rights Reserved.
No part of this publication may be reproduced or
transmitted in any form or by any means, electronic
or mechanical, including photocopy, recording, or
any information storage and retrieval system, without
permission in writing from the publisher.

ACADEMIC PRESS
a division of Harcourt Brace & Company
525 B Street, Suite 1900, San Diego, CA 92101-4495, USA
http://www.apnet.com

ACADEMIC PRESS
24–28 Oval Road, London NW1 7DX, UK
http://www.hbuk.co.uk/ap/

Library of Congress Cataloging-in-Publication Data
Semiconductor lasers : optics and photonics / edited by Eli Kapon.
    p. cm.
  Includes indexes.
  ISBN 0-12-397630-8 (v. 1). — ISBN 0-12-397631-6 (v. 2)
  1. Semiconductor lasers.   I. Kapon, Eli.
TA1700.S453 1998
621.36'6—dc21   98-18270
                                                        CIP

Printed in the United States of America
98 99 00 01 02 BB 9 8 7 6 5 4 3 2 1

Contents

Preface   ix

Chapter 1   Quantum Well Semiconductor Lasers
1.1   Introduction   1
1.2   Carriers and photons in semiconductor structures   8
    1.2.1   Electronic states in a semiconductor structure   8
    1.2.2   Carrier distribution functions and induced polarization   12
    1.2.3   Optical transitions and gain coefficients   16
1.3   Basics of quantum well lasers   21
    1.3.1   Transition matrix elements   22
    1.3.2   Density of states for QW structures   34
    1.3.3   Rate equations for quantum well laser structures   37
    1.3.4   General description of statics and dynamics   41
1.4   State filling in quantum well lasers   44
    1.4.1   Gain spectrum and sublinear gain relationship   45
    1.4.2   State filling on threshold current   49
    1.4.3   A puzzle in high-speed modulation of QW lasers   50
    1.4.4   State filling on differential gain of QW lasers   54
    1.4.5   State filling on spectral dynamics   63
1.5   Reduction of state filling in QW lasers   67
    1.5.1   Multiple quantum well structures   69
    1.5.2   Quantum well barrier height   70
    1.5.3   Separate confinement structures   74
    1.5.4   Strained quantum well structures   78
    1.5.5   Substrate orientation   81
    1.5.6   Bandgap offset at QW heterojunctions   81
1.6   Some performance characteristics of QW lasers   84
    1.6.1   Submilliampere threshold current   84

	1.6.2	High-speed modulation at low operation current	95
	1.6.3	Amplitude-phase coupling and spectral linewidth	98
	1.6.4	Wavelength tunability and switching	103
1.7	Conclusion and outlook		108
	References		109

Chapter 2 Strained Quantum Well Lasers

2.1	Introduction		123
2.2	Strained layer structures		127
	2.2.1	Elastic properties	127
	2.2.2	Critical layer thickness	129
2.3	Electronic structure and gain		131
	2.3.1	Requirements for efficient lasers	131
	2.3.2	Strained-layer band structure	133
	2.3.3	Strained valence band Hamiltonian	136
	2.3.4	Laser gain	141
	2.3.5	Strained layers on non-[001] substrates	145
2.4	Visible lasers		147
2.5	Long-wavelength lasers		152
	2.5.1	Introduction	152
	2.5.2	The loss mechanisms of Auger recombination and intervalence band adsorption	152
	2.5.3	Influence of strain on loss mechanisms	155
	2.5.4	The influence of strain on temperature sensitivity	161
2.6	Linewidth, chirp, and high-speed modulation		167
2.7	Strained laser amplifiers		169
2.8	Conclusions		170
	Acknowledgments		171
	References		171

Chapter 3 High-Speed Lasers

3.1	Introduction		177
3.2	Laser dynamics		179
	3.2.1	Rate equations	180
	3.2.2	Small-signal amplitude modulation	182
	3.2.3	Relative intensity noise	187
	3.2.4	Frequency modulation and chirping	190
	3.2.5	Carrier transport times	193
3.3	High-speed laser design		202

	3.3.1	Differential gain	204
	3.3.2	Optimization of carrier transport parameters	217
	3.3.3	Nonlinear gain	240
	3.3.4	Photon density	247
	3.3.5	Device operating conditions	248
	3.3.6	Device structures with low parasitics	250
	3.3.7	Device size and microwave propagation effects	256
3.4	Large-signal modulation		262
	3.4.1	Gain switching	263
	3.4.2	Modulation without prebias	266
	3.4.3	Mode locking	269
3.5	Conclusions and outlook		278
	Acknowledgments		280
	References		280

Chapter 4   Quantum Wire and Quantum Dot Lasers

4.1	Introduction		291
4.2	Principles of QWR and QD lasers		294
	4.2.1	Density of states	295
	4.2.2	Optical gain	299
	4.2.3	Threshold current	301
	4.2.4	High-speed modulation	303
	4.2.5	Spectral control	305
4.3	Quantum wire lasers		307
	4.3.1	Semiconductor lasers in high magnetic fields	307
	4.3.2	QWR lasers fabricated by etching and regrowth	309
	4.3.3	QWR lasers made by cleaved-edge overgrowth	312
	4.3.4	QWR lasers grown on vicinal substrates	315
	4.3.5	QWR lasers made by strained-induced self-ordering	319
	4.3.6	QWR lasers grown on nonplanar substrates	323
4.4	Quantum dot lasers		339
	4.4.1	QD lasers fabricated by etching and regrowth	342
	4.4.2	QD lasers made by self-organized growth	342
4.5	Conclusions and outlook		352
	Acknowledgments		353
	References		353

Chapter 5   Quantum Optics Effects in Semiconductor Lasers

5.1	Introduction	361

5.2	Squeezing in semiconductor lasers		
	5.2.1	Brief review of squeezed states	362
	5.2.2	Theory of squeezed-state generation in semiconductor lasers	366
	5.2.3	Experimental results	383
	5.2.4	Squeezed vacuum state generation	390
5.3	Controlled spontaneous emission in semiconductor lasers		414
	5.3.1	Brief review of cavity quantum electrodynamics	414
	5.3.2	Rate-equation analysis of microcavity lasers	417
	5.3.3	Semiconductor microcavity lasers: experiments	420
5.4	Conclusion		437
	References		437

Index   **443**

Preface

More than three decades have passed since lasing in semiconductors was first observed in several laboratories in 1962 (Hall *et al.*, 1962; Holonyak, Jr. *et al.*, 1962; Nathan *et al.*, 1962; Quist *et al.*, 1962). Although it was one of the first lasers to be demonstrated, the semiconductor laser had to await several important developments, both technological and those related to the understanding of its device physics, before it became fit for applications. Most notably, it was the introduction of heterostructures for achieving charge carrier and photon confinement in the late sixties and the understanding of device degradation mechanisms in the seventies that made possible the fabrication of reliable diode lasers operating with sufficiently low currents at room temperature. In parallel, progress in the technology of low loss optical fibers for optical communication applications has boosted the development of diode lasers for use in such systems. Several unique features of these devices, namely the low power consumption, the possibility of direct output modulation and the compatibility with mass production that they offer, have played a key role in this development. In addition the prospects for integration of diode lasers with other optical and electronic elements in optoelectronic integrated circuits (OEICs) served as a longer term motivation for their advancement.

The next developments that made semiconductor lasers truly ubiquitous took place during the eighties and the early nineties. In the eighties, applications of diode lasers in compact disc players and bar-code readers have benefited from their mass-production capabilities and drastically reduced the prices of their simplest versions. In parallel, more sophisticated devices were developed as the technology matured. Important examples are high power lasers exhibiting very high electrical to optical power conversion efficiency, most notably for solid state laser pumping and medical applications, and high modulation speed, single frequency, distributed feedback lasers for use in long-haul optical communication systems.

Moreover, progress in engineering of new diode laser materials covering emission wavelengths from the blue to the mid-infrared has motivated the replacement of many types of gas and solid state lasers by these compact and efficient devices in many applications.

The early nineties witnessed the maturing of yet another important diode laser technology, namely that utilizing quantum well heterostructures. Diode lasers incorporating quantum well active regions, particularly strained structures, made possible still higher efficiencies and further reduction in threshold currents. Quantum well diode lasers operating with sub-mA threshold currents have been demonstrated in many laboratories. Better understanding of the gain mechanisms in these lasers has also made possible their application in lasers with multi-GHz modulations speeds. Vertical cavity surface emitting lasers, utilizing a cavity configuration totally different than the traditional cleaved cavity, compatible with wafer-level production and high coupling efficiency with single mode optical fibers, have progressed significantly owing to continuous refinements in epitaxial technologies. Advances in cavity schemes for frequency control and linewidth reduction have yielded lasers with extremely low (kHz) linewidths and wide tuning ranges. Many of these recent developments have been driven by the information revolution we are experiencing. A major role in this revolution is likely to be played by dense arrays of high speed, low power diode lasers serving as light sources in computer data links and other mass-information transmission systems. Tunable diode lasers are developed mainly for use in wavelength division multiplexing communication systems in local area networks.

In spite of being a well established commercial device already used in many applications, the diode laser is still a subject for intensive research and development efforts in many laboratories. The development efforts are driven by the need to improve almost all characteristics of these devices in order to make them useful in new applications. The more basic research activities are also drive by the desire to better understand the fundamental mechanisms of lasing in semiconductors and by attempts to seek the ultimate limits of laser operation. An important current topic concerns the control of photon and carrier states and their interaction using micro- and nano-structures such as microcavities, photonic bandgap crystals, quantum wires, and quantum dots. Laser structures incorporating such novel cavity and heterostructure configurations are expected to show improved noise and high speed modulation properties and higher efficiency. Novel diode laser structures based on intersubband quantum-cascade transitions are explored for achieving efficient lasing in the mid infrared range. And new III nitride compounds are developed for extending the emission wavelength range to the blue and ultraviolet regime.

The increasing importance of semiconductor lasers as useful, mature device technology and, at the same time, the vitality of the research field related to these devices, make an up-to-date summary of their science and technology highly desirable. The purpose of this volume, and its companion volume *Semiconductor Lasers: Materials and Structures*, is to bring such a summary to the broad audience of students, teachers, engineers, and researchers working with or on semiconductor lasers.

The present volume concentrates on the fundamental mechanisms of semiconductor lasers, relating the basic carrier and photo states to the important laser parameters such as optical gain, emission spectra, modulation speed, and noise. Besides treating the more well established quantum well heterostructure and "large," cleaved optical cavities, the volume also introduces the fundamentals of novel structures such as quantum wires, quantum dots, and microcavities, and their potential application in improved diode laser devices. The companion volume deals with the more technological aspects of diode lasers related to the control of their emission wavelength, achievement of high output power, and surface emission configurations. Both volumes are organized in a way that facilitates the introduction of readers without a background in semiconductor lasers to this field. This is attempted by devoting the first section (or sections) in each chapter to a basic introduction to one of the aspects of the physics and technology of these devices. Subsequent sections deal with details of the topics under consideration.

Chapter 1 of the present volume, by Bin Zhao and Amnon Yariv, treats the fundamentals of quantum well lasers. It introduces the reader to the effect of quantum confinement on the electronic states, the transition selection rules, and the optical gain spectra. Several practical quantum well configurations and their impact on laser performance are discussed.

In Chapter 2, Alfred R. Adams, Eoin P. O'Reilly, and Mark Silver summarize the impact of strain on the properties of quantum well lasers. The effect of both compressive and tensile strain on the semiconductor band structure and optical gain are analyzed in detail. The evolution of threshold current density with the degree and sign of strain are examined, and model predictions are compared to reported experimental results.

The fundamentals and engineering of high speed diode lasers are discussed in Chapter 3, by Radhakrishnan Nagarajan and John E. Bowers.

Rate equations describing the carrier and photo dynamics are developed and solved. Fundamental limits on the modulation speed are reviewed, with special attention to carrier transport effects in quantum well structures. Short pulse generation techniques are also discussed.

Chapter 4, by Eli Kapon, describes the effects of lateral quantum confinement on the electronic states and the optical gain spectra. The potential improvement in static and dynamic laser properties by introducing two or three dimensional quantum confinement in quantum wire or quantum dot lasers are analyzed and recent performance results are compared.

Finally, Chapter 5, by Y. Yamamoto, S. Inoue, G. Björk, H. Heitmann, and F. Matinaga, discusses quantum optics effects in diode lasers employing novel current sources and microcavities. The generation of squeezed states of photons using semiconductor lasers is treated theoretically and experimental results are described and analyzed. The control of spontaneous emission using microcavity configurations is discussed. The possibility of achieving thresholdless laser operation in such structures is also examined.

While it is difficult to include all aspects of this very broad field in two volumes, we have attempted to include contributions by experienced persons in this area that cover the most important basic and practical facets of these fascinating devices. We hope that the readers will find this book useful.

References

Hall, R. N., Fenner, G. E., Kingsley, J. D., Soltys, T. J., and Carlson, R. O. (1962). *Phys. Rev. Lett.*, **9**, 366.

Holonyak, N. Jr., and Bevacqua, S. F. (1962). *Appl. Phys. Lett.*, **1**, 82.

Nathan, M. I., Dumke, W. P., Burns, G., Dill, F. H. Jr., and Lasher, G. (1962). *Appl. Phys. Lett.*, **1**, 62.

Quist, T. M., Rediker, R. II, Keyes, R. J., Krag, W. E., Lax, B., McWhorter, A. L., and Zeigler, H. J. (1962). *Appl. Phys. Lett.*, **1**, 91.

Chapter 1

Quantum Well Semiconductor Lasers

Bin Zhao
Rockwell Semiconductor Systems, Newport Beach, CA

Amnon Yariv
California Institute of Technology, Pasadena, CA

1.1 Introduction

Semiconductor lasers have assumed an important technological role since their invention in the early 1960s (Basov *et al.*, 1961; Bernard and Duraffourg, 1961; Hall *et al.*, 1962; Nathan *et al.*, 1962). Judged by economic impact, semiconductor lasers have become the most important class of lasers. They are now used in applications such as cable TV signal transmission, telephone and image transmission, computer interconnects and networks, compact disc (CD) players, bar-code readers, laser printers, and many military applications. They are now figuring in new applications ranging from two-dimensional display panels to erasable optical data and image storage. They are also invading new domains such as medical, welding, and spectroscopic applications that are now the captives of solid-state and dye lasers.

The main reasons behind this major surge in the role played by semiconductor lasers are their continued performance improvements especially in low-threshold current, high-speed direct current modulation, ultrashort optical pulse generation, narrow spectral linewidth, broad linewidth range, high optical output power, low cost, low electrical power consumption and high wall plug efficiency. Many of these achievements were based on joint progress in material growth technologies and theoretical understanding of a new generation of semiconductor lasers — the quantum well (QW) lasers. The pioneering work using molecular beam epitaxy (MBE) (Cho, 1971; Cho et al., 1976; Tsang, 1978; Tsang et al., 1979) and metal organic chemical vapor deposition (MOCVD) (Dupuis and Dapkus, 1977; Dupuis et al., 1978, 1979a, 1979b) to grow ultrathin semiconductor layers, on the order of ten atomic layers, had paved the way for the development of this new type of semiconductor laser. The early theoretical understanding and experimental investigations in the properties of QW lasers had helped speed up the development work (van der Ziel et al., 1975; Holonyak et al., 1980; Dutta, 1982; Burt, 1983; Asada et al., 1984; Arakawa et al., 1984; Arakawa and Yariv, 1985.)

As shown in Fig. 1.1, a semiconductor laser is basically a p-i-n diode. When it is forward-biased, electrons in the conduction band and holes in the valence band are injected into the intrinsic region (also called the *active region*) from the n-type doped and the p-type doped regions, respectively. The electrons and the holes accumulate in the active region and are induced to recombine by the lasing optical field present in the same region. The energy released by this process (a photon for each electron-hole recombination) is added coherently to the optical field (laser action). In conventional bulk semiconductor lasers, as shown in Fig. 1.1, a double heterostructure (DH) is usually used to confine the injected carriers and the optical field to the same spatial region, thus enhancing the interaction of the charge carriers with the optical field.

In order for optical radiation at frequency ν to experience gain (amplification) rather than loss in a semiconductor medium, the separation between the Fermi energies of electrons and holes in the medium must exceed the photon energy $h\nu$ (Basov et al., 1961; Bernard and Duraffourg, 1961). To achieve this state of affairs for lasing, a certain minimum value of injected carrier density N_{tr} (transparency carrier density) is required. This transparency carrier density is maintained by a (transparency) current in a semiconductor laser, which is usually the major component of the threshold current and can be written as

$$I_{tr} = J_{tr}wL \qquad (1.1)$$

1.1 Introduction

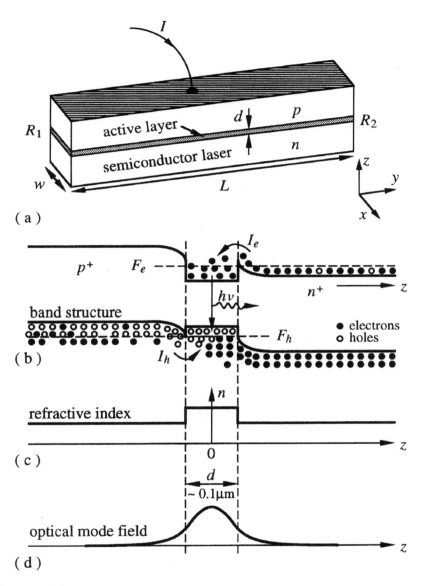

Figure 1.1: A schematic description of a semiconductor laser diode: (a) the laser device geometry; (b) the energy band structure of a forward-biased double heterostructure laser diode; (c) the spatial profile of the refractive index that is responsible for the dielectric waveguiding of the optical field; (d) the intensity profile of the fundamental optical mode.

where w is the laser diode width and L is the laser cavity length. J_{tr} is the transparency current density, which can be written as

$$J_{tr} = \frac{eN_{tr}d}{\tau_c} \tag{1.2}$$

where e is the fundamental electron charge, d is the active layer thickness, and τ_c is the carrier lifetime related to spontaneous electron-hole recombination and other carrier loss mechanism at injection carrier density N_{tr}. Equations (1.1) and (1.2) suggest the strategies to minimize the threshold current of a semiconductor laser: (1) to reduce the dimensions of the laser active region (w, L, d), (2) to reduce the necessary inversion carrier density N_{tr} for the required Fermi energy separation, and (3) to reduce the carriers' spontaneous recombination and other loss mechanism (increase τ_c). Each of these strategies has stimulated exciting research activities in semiconductor lasers. For example, pursuing strategy (1) has resulted in the generation of quantum well, quantum wire, quantum dot, and micro cavity semiconductor lasers. Pursuing strategy (2) has resulted in the electronic band engineering for semiconductor lasers, such as the reduction in valence band effective mass and increase in subband separation caused by addition of strain to the QW region. Pursuing strategy (3) has led to the development of various fantastic semiconductor laser structures and materials to reduce leakage current and to suppress the Auger recombination. It also has stimulated the interesting research in squeezing the spontaneous emission in micro cavity for thresholdless semiconductor lasers (see Chap. 5). In addition to threshold current, other important performance characteristics of semiconductor lasers have been improved by these and other related research and development activities, which include the modulation speed, optical output power, laser reliability, etc.

Figure 1.2 shows the schematic structures for three-dimensional (3D) bulk, two-dimensional (2D) quantum well, one-dimensional (1D) quantum wire, and zero-dimensional (0D) quantum dot and their corresponding carrier density of states (DOS). The electronic and optical properties of a semiconductor structure are strongly dependent on its DOS for the carriers. The use of these different structures as active regions in semiconductor lasers results in different performance characteristics because of the differences in their DOS as shown in Figure 1.2.

Equation (1.2) shows that a reduction in the active layer thickness d will lead to a reduction in the transparency current density, which is usually the major component of the threshold current density. As the

1.1 Introduction

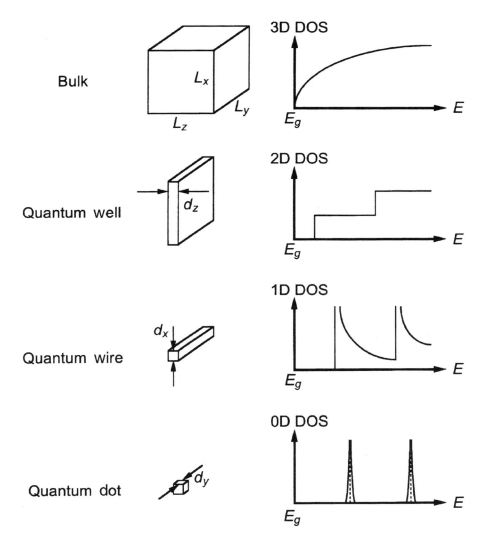

Figure 1.2: Schematic structures and corresponding carrier density of states (DOS) for three-dimensional (3D) bulk, two-dimensional (2D) quantum well, one-dimensional (1D) quantum wire, and zero-dimensional (0D) quantum dot semiconductor lasers.

active layer thickness d is reduced from ~ 1000 Å in conventional DH lasers by an order of magnitude to ~ 100 Å, the threshold current density, and hence the threshold current, should be reduced by roughly the same order of magnitude. However, as d approaches the 100-Å region, the DH structure shown in Fig. 1.3(a) cannot confine the optical field any more. To effectively confine a photon or an electron, the feature size of the confinement structure needs to be comparable with their wavelengths. Thus a separate confinement heterostructure (SCH) as shown in Fig. 1.3(b) is needed. In an SCH structure, the injected carriers are confined in the active region of quantum size, a size comparable to the material wavelength of electrons and holes, in the direction perpendicular to the active layer, while the optical field is confined in a region with size comparable with its wavelength. The active layer is a so-called quantum well, and the lasers are called *quantum well (QW) lasers*. The electrons and the holes in the quantum well display quantum effects evidenced mostly by the modification in the carrier DOS. The quantum effects greatly influence the laser performance features such as radiation polarization, modulation, spectral purity, ultrashort optical pulse generation, as well as lasing wavelength tuning and switching.

This chapter is devoted mainly to a general description of QW lasers. Extensive discussions on QW lasers were given by many experts in a book edited by Zory (1993). Various discussions on this subject also can be found in other books (e.g., Weisbuch and Vinter, 1991; Agrawal and Dutta, 1993; Chow et al., 1994; Coldren and Corzine, 1995; Coleman, 1995). In this chapter, efforts have been made to discuss QW lasers from different perspectives whenever it is possible. We start with a discussion of the fundamental issues for understanding the properties of semiconductor lasers, such as the interaction between injected carriers and optical field in a semiconductor medium. A universal optical gain theory is described, which generally can be applied to various semiconductor lasers of bulk, quantum well, quantum wire, or quantum dot structures. As the first chapter in this book, we hope these discussions are informative and entertaining. The following discussion on optical gain of QW lasers shows how the simple and widely used decoupled valence band approximation is derived from a more rigorous and more complicated valence band theory for the optical gain calculation. We then address a specific phenomenon, state filling or band filling, related to QW laser structures and discuss its influence on laser performance. Finally, we review some recent performance achievements of QW lasers, which include sub-microampere

1.1 Introduction

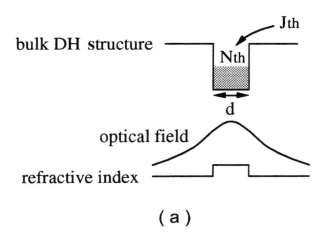

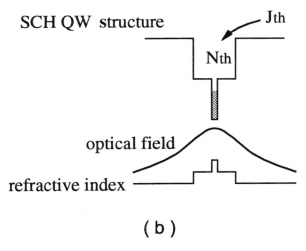

Figure 1.3: Schematic structures for (a) bulk double heterostructure (DH) semiconductor lasers; (b) separate confinement heterostructure (SCH) quantum well (QW) semiconductor lasers.

threshold currents, high-speed modulation at low operation currents, spectral linewidth behavior, and wide lasing wavelength tunability.

1.2 Carriers and photons in semiconductor structures

Before dealing with the optical properties of a semiconductor medium or a semiconductor laser structure, it is essential to understand the behavior of injected carriers in a semiconductor medium and how these injected carriers interact with an optical field. In this section we first review some elementary analysis on electronic band structures, i.e., the electronic states available for the injected carriers to occupy in a semiconductor medium. Detailed treatment of this topic can be found in numerous references dealing with the wave mechanics of solids (e.g., Luttinger and Kohn, 1955; Luttinger, 1956; Kane, 1966; Ashcroft and Mermin, 1976; Altarelli, 1985; Kittel, 1987). Then we discuss how these electronic states are occupied by the injected carriers, i.e., the carrier distribution functions under the presence of an optical field. Finally, we describe the optical transitions induced by the interaction between the injected carriers and the optical field. The discussion in this section is made as general as possible so that it is applicable to various semiconductor laser structures, i.e., bulk, quantum well, quantum wire, or quantum dot structures.

1.2.1 Electronic states in a semiconductor structure

In a semiconductor medium, the wavefunction of an electron in a given energy band is a solution of the Schrödinger equation

$$H_{\text{crystal}}\Psi_n(\mathbf{r}) = \left[\frac{\mathbf{p}^2}{2m_0} + U_p(\mathbf{r}) + U(\mathbf{r})\right]\Psi_n(\mathbf{r}) = E_n\Psi_n(\mathbf{r}) \quad (1.3)$$

where $\mathbf{p} = -i\hbar\nabla$ is the momentum operator, m_0 is the free electron mass, $U_p(\mathbf{r})$ is the periodic potential of the bulk semiconductor crystal, $U(\mathbf{r})$ is the additional potential (e.g., the potential caused by the material variation in a QW structure), and n designates the corresponding band. If $U(\mathbf{r}) = 0$, the solution to Eq. (1.3) is the Bloch function:

$$\Psi_{n,\mathbf{k}}(\mathbf{r}) = u_{n,\mathbf{k}}(\mathbf{r})\frac{1}{\sqrt{V}}e^{i\mathbf{k}\cdot\mathbf{r}} \quad (1.4)$$

1.2 Carriers and photons in semiconductor structures

where $u_{n,\mathbf{k}}(\mathbf{r})$ has the periodicity of the crystalline lattice, $\mathbf{k} = k_x\hat{\mathbf{x}} + k_y\hat{\mathbf{y}} + k_z\hat{\mathbf{z}}$ is the wavevector of the electron, k_q is quantized as

$$k_q = j\frac{2\pi}{L_q} \tag{1.5}$$

j is an integer, L_q is the length of the crystal in the q direction ($q = x, y, z$), and $V = L_x \cdot L_y \cdot L_z$. The functions $\Psi_{n,\mathbf{k}}(\mathbf{r})$ form a complete basis set. For $U(\mathbf{r}) \neq 0$, the solution of Eq. (1.3) can be written as an expansion of the basis set $\Psi_{n,\mathbf{k}}(\mathbf{r})$

$$\Psi_n(\mathbf{r}) = \sum_{m,\mathbf{k}} F_m(\mathbf{k}) u_{m,\mathbf{k}}(\mathbf{r}) \frac{1}{\sqrt{V}} e^{i\mathbf{k}\cdot\mathbf{r}} \tag{1.6}$$

Note that any periodic function can be expanded by the Bloch functions at the band edge ($\mathbf{k} = 0$), which form a complete basis set for the periodic functions:

$$u_{m,\mathbf{k}}(\mathbf{r}) = \sum_l c_{ml}(\mathbf{k}) u_{l,0}(\mathbf{r}) \tag{1.7}$$

Therefore, Eqs. (1.6) and (1.7) lead to

$$\Psi_n(\mathbf{r}) = \sum_l \Phi_l(\mathbf{r}) u_{l,0}(\mathbf{r}) \equiv \sum_l \Phi_l(\mathbf{r}) u_l(\mathbf{r}) \tag{1.8}$$

where

$$\Phi_l(\mathbf{r}) = \sum_{m,\mathbf{k}} F_m(\mathbf{k}) c_{ml}(\mathbf{k}) \frac{1}{\sqrt{V}} e^{i\mathbf{k}\cdot\mathbf{r}} \tag{1.9}$$

is the envelope function. $u_l(\mathbf{r}) \equiv u_{l,0}(\mathbf{r})$ satisfies

$$\langle u_l(\mathbf{r})|u_{l'}(\mathbf{r})\rangle = \frac{1}{\Omega}\int_{\text{cell}} u_l^*(\mathbf{r}) u_{l'}(\mathbf{r})\, d\mathbf{r} = \delta_{ll'} \tag{1.10}$$

where Ω is the volume of the unit cell. In the envelope function approximation, which has assumed that $\Phi_l(\mathbf{r})$ varies slowly within one unit cell of the crystal, a $k \cdot p$ approximation up to the second order of the wave vector showed that the $\Phi_l(\mathbf{r})$ values are governed by a set of coupled Schrödinger equations (Luttinger and Kohn, 1955)

$$\sum_{l'}\left\{\sum_{\alpha\alpha'}[D_{ll'}^{\alpha\alpha'}(-i\partial_\alpha)(-i\partial_{\alpha'})] + \delta_{ll'}U(\mathbf{r})\right\}\Phi_{l'}(\mathbf{r}) = [E_n - E_{l0}]\Phi_l(\mathbf{r}) \tag{1.11}$$

where $\partial_\alpha = \frac{\partial}{\partial_\alpha}$, $\partial_{\alpha'} = \frac{\partial}{\partial_{\alpha'}}$, $\{\alpha, \alpha'\} = \{x, y, z\}$, and E_{l0} is the energy of a *free* electron in the semiconductor medium at the band edge, i.e., the energy of an electron for $U(\mathbf{r}) = 0$ (free electron) and $\mathbf{k} = 0$ (at the band edge). $D_{ll'}^{\alpha\alpha'}$ are a set of constants depending on the crystal symmetry, the choice of coordinate system and corresponding basis functions $\{u_l\}$. The symmetry of the crystal, proper choice of the coordinate system and the corresponding $\{u_l\}$ will significantly simplify Eq. (1.11) because many of the $D_{ll'}^{\alpha\alpha'}$ coefficients vanish.

If $\partial U/\partial q = 0$ but $\partial U/\partial s \neq 0$ (i.e., there is no material variation in q-direction(s) and there is material variation in s-direction(s)), where q and s denote one or more of the x, y, z, respectively (i.e., $\{q, s\} = \{x, y, z\}$), the envelope functions can be solved by the variable separation method, and they can be written as

$$\Phi_l(\mathbf{r}) = \Phi_{l_s,k_q}(s) \frac{1}{\sqrt{L_q}} e^{ik_q q} \quad (1.12)$$

$\Phi_{l_s,k_q}(s)$ satisfies

$$\sum_{l'} \left\{ \sum_{\alpha\alpha'} [D_{ll'}^{\alpha\alpha'}(-i\partial_\alpha)(-i\partial_{\alpha'})] + \delta_{ll'} U(s) \right\} \Phi_{l'_s,k_q}(s) = [E_n(k_q) - E_{l0}] \Phi_{l_s,k_q}(s) \quad (1.13)$$

with $-i\partial_q$ replaced by k_q.

The preceding argument is general in the sense that it applies to different semiconductor structures. Table 1.1 lists the corresponding coordinates, wavefunctions, etc. for bulk, quantum well (Q-well), quantum wire (Q-wire) and quantum dot (Q-dot) structures that were shown schematically in Fig. 1.2. $E_n(k_q)$ can be obtained by solving Eq. (1.13). Essentially, the $E_n(k_q)$ relation gives the description of the electronic states for a given band, n in a semiconductor structure. From this relation, various electronic and optical properties of a semiconductor structure can be calculated and predicted.

If the band-to-band coupling is weak and negligible, as is the case for electrons in the conduction band, the wavefunction can be written as a single term:

$$\Psi_n(\mathbf{r}) = u_n(\mathbf{r}) \Phi_{n_s, k_q}(s) \frac{1}{\sqrt{L_q}} e^{ik_q q} \quad (1.14)$$

1.2 Carriers and photons in semiconductor structures

Structure	q	Unconfined dimension L_q	$\frac{1}{\sqrt{L_q}}e^{ik_q q}$	Confined dimension s	d_s	$\frac{\partial^2}{\partial^2 s}$
Bulk	$x\ y\ z$	$L_x\ L_y\ L_z$	$\frac{1}{\sqrt{L_x L_y L_z}}e^{i(k_x x + k_y y + k_z z)}$	—	—	—
Q-well	$x\ y$	$L_x\ L_y$	$\frac{1}{\sqrt{L_x L_y}}e^{i(k_x x + k_y y)}$	z	d_z	$\frac{\partial^2}{\partial^2 z}$
Q-wire	y	L_y	$\frac{1}{\sqrt{L_y}}e^{ik_y y}$	$x\ z$	$d_x\ d_z$	$\frac{\partial^2}{\partial^2 x}+\frac{\partial^2}{\partial^2 z}$
Q-dot	—	—	—	$x\ y\ z$	$d_x\ d_y\ d_z$	$\frac{\partial^2}{\partial^2 x}+\frac{\partial^2}{\partial^2 y}+\frac{\partial^2}{\partial^2 z}$

Table 1.1: The coordinates, dimensions, related wavefunctions, and differential in the wave equations for the analysis of energy band structures in bulk, quantum well, quantum wire, and quantum dot semiconductor structures.

where $\Phi_{n_s,k_q}(s)$ satisfies the Schrödinger equation

$$\left[-\frac{\hbar^2}{2m_n}\frac{\partial^2}{\partial s^2} + \frac{\hbar^2}{2m_n}k_q^2 + U(s)\right]\Phi_{n_s,k_q}(s) = [E_n(k_q) - E_{n0}]\Phi_{n_s,k_q}(s) \quad (1.15)$$

or

$$\left[-\frac{\hbar^2}{2m_n}\frac{\partial^2}{\partial s^2} + U(s)\right]\Phi_{n_s,k_q}(s) = E_{n_s}\Phi_{n_s,k_q}(s) \quad (1.16)$$

The electron energy is

$$E_n(k_q) = E_{n_s} + E_{n0} + \frac{\hbar^2}{2m_n}k_q^2 \quad (1.17)$$

and E_{n_s} is the quantized energy due to the confinement potential $U(s)$. Equation (1.17) shows that $E_n(k_q)$ is parabolic in k_q. The constant m_n is the effective mass of the electrons in the n band. In this case, the electronic band structure [$E_n(k_q)$ relation] is characterized by the constants of effective mass m_n and energy $E_{n0} + E_{n_s}$. This is the well-known parabolic band approximation.

If the band-to-band coupling (such as the case in the degenerate valence bands) needs to be taken into account, the $E_n(k_q)$ relation in these bands can be obtained by solving the coupled Schrödinger Eqs. (1.11) or

(1.13). Now the $E_n(k_q)$ is no longer parabolic, and the band structures are more complicated in description. More detailed discussion on the valence band structures will be given later on.

1.2.2 Carrier distribution functions and induced polarization

In the preceding subsection we discussed the electronic states in a semiconductor structure. Now we discuss the interaction between an optical field and the carriers occupying these electronic states.

In the presence of an optical field in a semiconductor medium, the Hamiltonian in the Schrödinger equation (1.3), H_{crystal}, changes to

$$H = \frac{[\mathbf{p} + e\mathbf{A}(\mathbf{r}, t)]^2}{2m_0} + U_p(\mathbf{r}) + U(\mathbf{r}) = H_{\text{crystal}} + H' \quad (1.18)$$

where $\mathbf{A}(\mathbf{r}, t)$ is the vector potential of the optical field [$\nabla \cdot \mathbf{A}(\mathbf{r}, t) = 0$, Coulomb gauge], and the interaction Hamiltonian is

$$H' = \frac{e}{m_0} \mathbf{A}(\mathbf{r}, t) \cdot \mathbf{p} + \frac{e^2}{2m_0} |\mathbf{A}(\mathbf{r}, t)|^2 \approx \frac{e}{m_0} \mathbf{A}(\mathbf{r}, t) \cdot \mathbf{p} \quad (1.19)$$

The $|\mathbf{A}(\mathbf{r}, t)|^2$ term approximately yields zero matrix elements for the interested interband transitions because $\langle u_l | u_{l'} \rangle = 0$ ($l \neq l'$ for interband transition) and $\mathbf{A}(\mathbf{r}, t)$ varies slowly within one unit cell of the crystal.

It was shown that the interaction Hamiltonian H' also can be written as (Sargent et al., 1974; Yariv, 1989)

$$H' = -(-e\mathbf{r}) \cdot \mathbf{E}(\mathbf{r}, t) = \hat{\mu} \cdot \mathbf{E}(\mathbf{r}, t) \quad (1.20)$$

where $\mathbf{E}(\mathbf{r}, t) = -\partial \mathbf{A}(\mathbf{r}, t)/\partial t = \mathcal{E}(\mathbf{r}, t)\hat{\mathbf{a}}$ is the optical field, and $\hat{\mathbf{a}}$ is the unit vector along the direction of the optical field polarization.

A semiclassical theory can be used to treat the interaction between the semiconductor medium and the optical field. The carriers in the semiconductor medium are described quantum mechanically by the Schrödinger equation, while the optical field is described classically by the Maxwell equations. The transition matrix element for optical field induced transition of an electron from the conduction band to the valence band, or in reverse, is

$$\begin{aligned} H'_{vc} &= \langle \Psi_v | \hat{\mu} | \Psi_c \rangle \cdot \mathbf{E}(\mathbf{r}, t) \\ &= e \int u_v^*(\mathbf{r}) \Phi_{nsv,k_{qv}}^*(s) \frac{1}{\sqrt{L_q}} e^{-ik_{qv}q} (\hat{\mathbf{a}} \cdot \mathbf{r}) u_c(\mathbf{r}) \Phi_{nsc,k_{qc}}(s) \frac{1}{\sqrt{L_q}} e^{ik_{qc}q} \, d\mathbf{r} \\ &\times \mathcal{E}(\mathbf{r}, t) \end{aligned}$$

$$(1.21)$$

1.2 Carriers and photons in semiconductor structures

Notice that since $\Phi_{n_{sl},k_{ql}}(s)$ and $e^{ik_{ql}q}$ ($l = c, v$) vary slowly within one unit cell of the crystal and $\langle u_v(\mathbf{r}) | u_c(\mathbf{r}) \rangle = 0$, we have

$$H'_{vc} = e \langle u_v(\mathbf{r}) | \hat{\mathbf{a}} \cdot \mathbf{r} | u_c(\mathbf{r}) \rangle \\ \times \int \Phi^*_{n_{sv},k_{qv}}(s) \frac{1}{\sqrt{L_q}} e^{-ik_{qv}q} \Phi_{n_{sc},k_{qc}}(s) \frac{1}{\sqrt{L_q}} e^{ik_{qc}q} \, d\mathbf{r} \times \mathscr{E}(\mathbf{r},t) \quad (1.22)$$

where

$$\langle u_v(\mathbf{r}) | \hat{\mathbf{a}} \cdot \mathbf{r} | u_c(\mathbf{r}) \rangle = \frac{1}{\Omega} \int_{\text{cell}} u_v^*(\mathbf{r}) (\hat{\mathbf{a}} \cdot \mathbf{r}) u_c(\mathbf{r}) \, d\mathbf{r} \quad (1.23)$$

Equation (1.22) indicates that $H'_{vc} = 0$ unless

$$k_{qv} = k_{qc} \equiv k_q \quad (1.24)$$

Equation (1.24) is a necessary condition for the band-to-band optical transition in a semiconductor structure and is called the *k-selection rule*. Under this selection rule, the transition matrix element becomes

$$H'_{vc} = \tilde{\mu}(\tilde{\alpha}) \mathscr{E}(\mathbf{r}, t) \quad (1.25)$$

where

$$\tilde{\mu}(\tilde{\alpha}) = e \langle u_v(\mathbf{r}) | \hat{\mathbf{a}} \cdot \mathbf{r} | u_c(\mathbf{r}) \rangle \int \Phi^*_{n_{sv},k_q}(s) \Phi_{n_{sc},k_q}(s) \, ds \quad (1.26)$$

and $\tilde{\alpha}$ represents a series of quantum numbers $\{k_q, n_{sv}, n_{sc}\}$. n_{sv} and n_{sc} are the quantum numbers associated with the wavefunctions $\Phi_{n_{sv},k_q}(s)$ and $\Phi_{n_{sc},k_q}(s)$ in the s direction(s), respectively, which are determined by the confinement potential $U(s)$ [see Eq. (1.13)].

Assuming a single-mode monochromatic optical field propagating along the y direction,

$$\mathscr{E}(\mathbf{r}, t) = \tfrac{1}{2} \mathscr{E}_0 e^{i(\beta y - \omega t)} + c.c. \quad (1.27)$$

then

$$H'_{vc} = \tfrac{1}{2} \tilde{\mu}(\tilde{\alpha})[\mathscr{E}_0 e^{i(\beta y - \omega t)} + c.c.] \quad (1.28)$$

where β is the propagation constant and ω is the angular optical frequency.

The density-matrix formalism can be used to obtain the carrier distribution functions — the probabilities that carriers occupy these states in a semiconductor medium in the presence of an optical field. One can use the density-matrix formalism for a two-level system and treat the semiconductor active medium as an ensemble of two-level systems with quantum number n_{sc} and n_{sv} and a rigorous k-selection rule applied to the recombining electron-hole pairs, i.e., $k_{qc} = k_{qv} = k_q$. The density-matrix equations can be rewritten in terms of the distribution (occupation) functions for electrons $\rho_{ee}(\tilde{\alpha})$ and holes $\rho_{hh}(\tilde{\alpha})$ in the presence of an optical field (Sargent et al., 1974; Yariv, 1989; Zhao et al., 1992f):

$$\frac{d}{dt}[\rho_{ee}(\tilde{\alpha})] = \frac{i}{\hbar}[H'_{vc}\,\rho_{eh}(\tilde{\alpha}) - c.c.] - \frac{\rho_{ee}(\tilde{\alpha}) - f_e}{\tau_e} \qquad (1.29)$$

$$\frac{d}{dt}[\rho_{hh}(\tilde{\alpha})] = \frac{i}{\hbar}[H'_{vc}\,\rho_{eh}(\tilde{\alpha}) - c.c.] - \frac{\rho_{hh}(\tilde{\alpha}) - f_h}{\tau_h} \qquad (1.30)$$

$$\frac{d}{dt}[\rho_{eh}(\tilde{\alpha})] = \frac{i}{\hbar}H'^{*}_{vc}[\rho_{ee}(\tilde{\alpha}) + \rho_{hh}(\tilde{\alpha}) - 1] - \frac{i}{\hbar}E_{\tilde{\alpha}}\rho_{eh}(\tilde{\alpha}) - \frac{\rho_{eh}(\tilde{\alpha})}{T_2} \qquad (1.31)$$

where

$$f_e = f_e(k_q, n_{sc}, F_e) = \frac{1}{\exp\left(\dfrac{E_e(k_q, n_{sc}) - F_e}{k_B T}\right) + 1} \qquad (1.32)$$

and

$$f_h = f_h(k_q, n_{sv}, F_h) = \frac{1}{\exp\left(\dfrac{E_h(k_q, n_{sv}) - F_h}{k_B T}\right) + 1} \qquad (1.33)$$

are the quasi-Fermi distributions that the electrons and the holes tend to relax to, $E_e(k_q, n_{sc})$ and $E_h(k_q, n_{sv})$ are the energy of an electron and a hole with wavevector k_q and quantum numbers n_{sc} and n_{sv} respectively, F_e and F_h are the quasi-Fermi energy levels for electrons and holes respectively, $\rho_{eh}(\tilde{\alpha})$ is the off-diagonal (electron-hole) density matrix element, $E_{\tilde{\alpha}}$ is the transition energy of an electron-hole pair with wavevector k_q and quantum numbers n_{sc} and n_{sv}, τ_e and τ_h are the quasi-equilibrium relaxation times for electrons in the conduction band and holes in the valence band, respectively, and T_2 is the interband dephasing time.

The times involved in Eqs. (1.29) to (1.31) are the stimulated emission time [related to the first term on the right hand side in Eqs. (1.29) and (1.30)] and the carrier intraband relaxation and interband dephasing times (τ_e, τ_h, T_2). They are usually in the sub-picosecond range and much smaller than the time scale related to laser performance. For example, the time scale for 100-GHz modulation is $\sim$10 ps. Thus quasi-equilibrium solutions ($d/dt = 0$) of Eqs. (1.29) to (1.31) can be used to study various dynamic and static properties related to semiconductor laser performance.

1.2 Carriers and photons in semiconductor structures

The quasi-equilibrium solutions of $\rho_{ee}(\tilde{\alpha})$ and $\rho_{hh}(\tilde{\alpha})$ can be obtained from Eqs. (1.29) to (1.31) as

$$\rho_{ee}(\tilde{\alpha}) = f_e - \frac{\tau_e}{\tau_e + \tau_h}[f_e + f_h - 1]\frac{\mathscr{L}(E - E_{\tilde{\alpha}})\frac{P}{\tilde{P}_s}}{1 + \mathscr{L}(E - E_{\tilde{\alpha}})\frac{P}{\tilde{P}_s}} \quad (1.34)$$

$$\rho_{hh}(\tilde{\alpha}) = f_h - \frac{\tau_h}{\tau_e + \tau_h}[f_e + f_h - 1]\frac{\mathscr{L}(E - E_{\tilde{\alpha}})\frac{P}{\tilde{P}_s}}{1 + \mathscr{L}(E - E_{\tilde{\alpha}})\frac{P}{\tilde{P}_s}} \quad (1.35)$$

where

$$\mathscr{L}(E - E_{\tilde{\alpha}}) = \frac{E_{T_2}^2}{E_{T_2}^2 + (E - E_{\tilde{\alpha}})^2} \quad (1.36)$$

$E_{T_2} = \hbar/T_2$, $E = \hbar\omega$ is the photon energy of the optical field, $P = \frac{1}{2}\epsilon_o n_r^2 |\mathscr{E}_0|^2/E$ is the photon density, $\tilde{P}_s = \hbar^2 \epsilon_o n_r^2/[|\tilde{\mu}(\tilde{\alpha})|^2 E(\tau_e + \tau_h)T_2]$ is the saturation photon density, ϵ_o is the electric permeability of the vacuum, and n_r is the refractive index for the optical field. Equations (1.34) and (1.35) show that the presence of an optical field causes spectral "holes" in the Fermi-like distributions of the electrons and holes. The spectral "holes" are represented by the terms involving P. This is the so-called *spectral hole burning* in a semiconductor medium (Yamada and Suematsu, 1981; Yamada, 1983; Asada and Suematsu, 1985; Chow et al., 1987; Agrawal, 1987, 1988).

The quasi-equilibrium solution for $\rho_{eh}(\tilde{\alpha})$ is

$$\rho_{eh}(\tilde{\alpha}) = -\frac{1}{2}\tilde{\mu}^*(\tilde{\alpha})[\rho_{ee}(\tilde{\alpha}) + \rho_{hh}(\tilde{\alpha}) - 1]\frac{1}{(E - E_{\tilde{\alpha}}) + iE_{T_2}}\mathscr{E}_0 e^{i(\beta y - \omega t)} \quad (1.37)$$

The induced polarization can be written as

$$\begin{aligned}\mathscr{P}_{\text{in}}(\mathbf{r}, t) &= \tfrac{1}{2}\mathscr{P}_{\text{in},0}e^{i(\beta y - \omega t)} + \text{c.c.} \\ &= \text{Tr}[\hat{\rho}(-\hat{\mu}/V)] = -\sum_{\tilde{\alpha}}\frac{\zeta(\mathbf{r},\tilde{\alpha})}{V(\tilde{\alpha})}[\rho_{eh}(\tilde{\alpha})\tilde{\mu}(\tilde{\alpha}) + \text{c.c.}] \\ &= \frac{1}{2}\sum_{\tilde{\alpha}}\frac{\zeta(\mathbf{r},\tilde{\alpha})}{V(\tilde{\alpha})}|\tilde{\mu}(\tilde{\alpha})|^2[\rho_{ee}(\tilde{\alpha}) + \rho_{hh}(\tilde{\alpha}) - 1] \quad (1.38) \\ &\quad \times \frac{1}{(E - E_{\tilde{\alpha}}) + iE_{T_2}}\mathscr{E}_0 e^{i(\beta y - \omega t)} + \text{c.c.} \\ &= \tfrac{1}{2}\epsilon_o(\chi_r + i\chi_i)\mathscr{E}_0 e^{i(\beta y - \omega t)} + \text{c.c.}\end{aligned}$$

where $V(\tilde{\alpha})$ is the confinement volume of electrons and holes and

$$\zeta(\mathbf{r}, \tilde{\alpha}) = \begin{cases} 1 & [\mathbf{r} \text{ inside } V(\tilde{\alpha})] \\ 0 & [\mathbf{r} \text{ outside } V(\tilde{\alpha})] \end{cases} \quad (1.39)$$

From Eq. (1.38), the susceptibility $\chi = \chi_r + i\chi_i$ can be obtained. The real part (χ_r) and imaginary part (χ_i) of the susceptibility can be written as

$$\chi_r(E) = -\sum_{\tilde{\alpha}} \frac{\zeta(\mathbf{r}, \tilde{\alpha})}{\epsilon_o V(\tilde{\alpha})} |\tilde{\mu}(\tilde{\alpha})|^2 [\rho_{ee}(\tilde{\alpha}) + \rho_{hh}(\tilde{\alpha}) - 1] \mathscr{L}_r(E_{\tilde{\alpha}} - E) \quad (1.40)$$

$$\chi_i(E) = -\sum_{\tilde{\alpha}} \frac{\zeta(\mathbf{r}, \tilde{\alpha})}{\epsilon_o V(\tilde{\alpha})} |\tilde{\mu}(\tilde{\alpha})|^2 [\rho_{ee}(\tilde{\alpha}) + \rho_{hh}(\tilde{\alpha}) - 1] \mathscr{L}_i(E_{\tilde{\alpha}} - E) \quad (1.41)$$

where

$$\mathscr{L}_r(E_{\tilde{\alpha}} - E) = \frac{E_{\tilde{\alpha}} - E}{E_{T_2}^2 + (E_{\tilde{\alpha}} - E)^2} \quad (1.42)$$

$$\mathscr{L}_i(E_{\tilde{\alpha}} - E) = \frac{E_{T_2}}{E_{T_2}^2 + (E_{\tilde{\alpha}} - E)^2} \quad (1.43)$$

1.2.3 Optical transitions and gain coefficients

The obtained polarization $\mathscr{P}_{\text{in}}$ [Eq. (1.38)] is due to the optical field–induced transitions (or optical field–induced electron-hole recombination in the active region). Equations (1.38) and (1.39) indicate that the active region is the region where electron wavefunctions and hole wavefunctions coexist or overlap. In the active region, the electron density and hole density can be related by a quasi-neutrality condition

$$\frac{1}{V_{MD}} \sum_{k_q, n_{cs}} f_e(k_q, n_{cs}, F_e) = \frac{1}{V_{MD}} \sum_{k_q, n_{vs}} f_h(k_q, n_{vs}, F_h) = N_{MD} \quad (1.44)$$

where V_{MD} represents the *dimensional volume* of a semiconductor laser structure in which the injected carriers are regarded as *M*-dimensional, e.g.,

$$V_{MD} = \begin{cases} V = L_x \cdot L_y \cdot L_z & (MD = 3D, \text{ bulk}) \\ S = L_x \cdot L_y & (MD = 2D, \text{ quantum well}) \\ L = L_y & (MD = 1D, \text{ quantum wire}) \\ 1 = 1 & (MD = 0D, \text{ quantum dot}) \end{cases} \quad (1.45)$$

and N_{MD} is the *M*-dimensional carrier density ($M = 3, 2, 1, 0$). Using the unperturbed quasi-Fermi distributions [Eqs. (1.32) and (1.33)] to calculate the carrier density is a very good approximation. The relative error due to the spectral hole burning is typically less than 1% for the carrier density and photon density values of interest.

Next we analyze how the optical field–induced transitions in turn contribute to the intensity of the optical field. We start with the wave equation derived from the Maxwell equations:

$$\nabla^2 \mathcal{E}(\mathbf{r}, t) - \mu_o \epsilon(\mathbf{r}) \frac{\partial^2}{\partial t^2} \mathcal{E}(\mathbf{r}, t) = \mu_o \frac{\partial^2}{\partial t^2} \mathcal{P}_{\text{in}}(\mathbf{r}, t) \tag{1.46}$$

where μ_o and $\epsilon(\mathbf{r})$ are the magnetic permeability and electric permeability, respectively. The variation in $\epsilon(\mathbf{r}) = \epsilon(s')$ (where $s' = \{x, z\}$) manifests the lateral and transverse optical confinement of the laser waveguide. We have already assumed that the optical wave propagates in the y-direction. Assume that the optical field $\mathcal{E}_0$ in Eq. (1.27) can be written as

$$\mathcal{E}_0 = A_0(t) E_0(s') \tag{1.47}$$

where A_0 is a complex number including both the amplitude and the phase of the optical field and $E_0(s')$ is the eigenmode function of the optical confinement structures with

$$\left(\frac{\partial^2}{\partial s'^2} - \beta^2 \right) E_0(s') + \omega^2 \mu_o \epsilon(s') E_0(s') = 0, \quad \frac{\partial^2}{\partial s'^2} = \frac{\partial^2}{\partial x^2} + \frac{\partial^2}{\partial z^2} \tag{1.48}$$

For most semiconductor laser waveguides, the lasing action occurs in the fundamental optical eigenmode. Substitution of Eqs. (1.27), (1.38), (1.47), and (1.48) into Eq. (1.46) leads to

$$2i\epsilon(s') E_0(s') \frac{dA_0}{dt} = -\omega \sum_{\tilde{\alpha}} \frac{\zeta(\mathbf{r}, \tilde{\alpha})}{V(\tilde{\alpha})} |\tilde{\mu}(\tilde{\alpha})|^2 \left[\rho_{ee}(\tilde{\alpha}) + \rho_{hh}(\tilde{\alpha}) - 1 \right]$$
$$\times \frac{1}{(E - E_{\tilde{\alpha}}) + iE_{T_2}} E_0(s') A_0 \tag{1.49}$$

where slow wave variation

$$\left| \frac{d^2 A_0}{dt^2} \right| << \omega \left| \frac{dA_0}{dt} \right| \tag{1.50}$$

has been used. We take the product of Eq. (1.49) with $E_0^*(s')$ and integrate over $s' = \{x, z\}$ from $-\infty$ to $+\infty$. The result is

$$\frac{dA_0}{dt} = \frac{i\omega}{2} A_0 \sum_{\tilde{\alpha}} \frac{1}{V(\tilde{\alpha})} \frac{\int \zeta(\mathbf{r}, \tilde{\alpha}) |E_0(s')|^2 ds'}{\int \epsilon(s') |E_0(s')|^2 ds'}$$
$$\times |\tilde{\mu}(\tilde{\alpha})|^2 \left[\rho_{ee}(\tilde{\alpha}) + \rho_{hh}(\tilde{\alpha}) - 1 \right] \frac{1}{(E - E_{\tilde{\alpha}}) + iE_{T_2}} \tag{1.51}$$

For different laser structures (e.g., quantum well, quantum wire), usually there is no potential energy variation along certain direction or directions. Using the definition in Sect. 1.2.1 and Table 1.1 for a specific laser structure, there is no additional potential variation in the q-direction(s) for the injected carriers. It is convenient to take the "dimensional volume" V_{MD} [see Eq. (1.45)] out of $V(\tilde{a})$ in Eq. (1.51) and define

$$v(n_{sc}, n_{sv}) = V(\tilde{a})/V_{MD} \tag{1.52}$$

$v(n_{sc}, n_{sv})$ can be regarded as the *dimensional volume* related to the s-direction(s) (where $\partial U/\partial s \neq 0$). It is the *volume* of these quantized dimension(s). Assume that the photons in the laser waveguides have a normalized distribution function in the directions perpendicular to the longitudinal optical wave propagation direction y, which is given by $\Theta(s') = \Theta(x, z)$; then

$$\Theta(s') = \epsilon(s') |E_0(s')|^2 / [\int \epsilon(s') |E_0(s')|^2 ds'] \tag{1.53}$$

Notice that since the carrier confinement is usually much tighter than the optical field confinement, $\Theta(s')$ in the active region can be approximated by its value at the center of the active region ($s' = 0$). Then, Eq. (1.51) can be written as

$$\frac{dA_0}{dt} = \frac{i\omega}{2\epsilon_o n_r^2} A_0 \sum_{\tilde{\alpha}} \Gamma_{MD} \frac{1}{V_{MD}} |\tilde{\mu}(\tilde{\alpha})|^2 [\rho_{ee}(\tilde{\alpha}) + \rho_{hh}(\tilde{\alpha}) - 1]$$
$$\times \frac{1}{(E - E_{\tilde{\alpha}}) + iE_{T_2}} \tag{1.54}$$

where n_r is the refractive index of the active region and

$$\Gamma_{MD} = \Theta(0) \frac{S_{ac}}{v(n_{sc}, n_{sv})} \tag{1.55}$$

is the *dimensional coupling factor* reflecting how the injected carriers in different semiconductor structures (which might be 3D bulk, 2D quantum well, or 1D quantum wire structures) interact with the photons that are confined by the waveguides. S_{ac} is the cross-sectional area of the active region the in xz plane that is normal to the optical mode propagation direction. In Table 1.2 the dimensional coupling factor and other related parameters are compared for bulk, quantum well (Q-well), and quantum wire (Q-wire) laser structures. For bulk and quantum well structures,

1.2 Carriers and photons in semiconductor structures

Device structure	Unconfined dimensional "volume" V_{MD}	Confined dimensional "volume" $v(n_{sc}, n_{sv})$	Active region cross-sectional area S_{ac}	Dimensional coupling factor Γ_{MD}
Bulk	$V = L_x \times L_y \times L_z$	1	$w \times L_z$	$\dfrac{L_z}{t_z}$
Q-well	$S = L_x \times L_y$	d_z	$w \times d_z$	$\dfrac{1}{t_z}$
Q-wire	$L = L_y$	$d_x \times d_z$	$d_x \times d_z$	$\dfrac{1}{A_{xz}}$

Note: For the bulk and quantum well structures, w is the laser strip width along the x direction, and t_z is the effective optical mode width along the z direction. For the quantum wire structure, d_x is the quantum wire width, and A_{xz} is the effective optical mode cross-sectional area in the xz plane.

Table 1.2: The coupling factors and other related parameters in semiconductor bulk, quantum well, and quantum wire structures.

usually the laser strip width w is large (see Fig. 1.1), so the optical field variation along the x-direction can be ignored. For the 3D bulk laser structures, the coupling factor is just the conventional confinement factor. We will discuss the coupling factor of 2D quantum well laser structures later in Sect. 1.3.3. For the 1D quantum wire laser structure, $A_{xz} = 1/\Theta(0)$ can be considered as the effective optical mode cross-sectional area, where the 0 in $\Theta(0)$ represents the position of the quantum wire in the x-z plane.

Equation (1.54) can be rewritten as

$$\frac{dA_0}{dt} = \frac{1}{2} v_g (\Gamma_{MD} G - i\Gamma_{MD} N_r) A_0 = \frac{1}{2} v_g (g - i\vartheta) A_0 \qquad (1.56)$$

where $v_g = c/n_r$, c is the speed of light in vacuum, $g = \Gamma_{MD} G(E)$ is the so-called modal (exponential) gain coefficient,

$$G(E) = \frac{1}{V_{MD}} \sum_{\tilde{\alpha}} \frac{|\tilde{\mu}(\tilde{\alpha})|^2}{\hbar c \epsilon_o n_r} E \left[\rho_{ee}(\tilde{\alpha}) + \rho_{hh}(\tilde{\alpha}) - 1\right] \mathscr{L}_i(E_{\tilde{\alpha}} - E)$$

$$= \frac{1}{V_{MD}} \sum_{\tilde{\alpha}} \frac{|\tilde{\mu}(\tilde{\alpha})|^2}{\hbar c \epsilon_o n_r} \frac{E}{E_{T2}} \left[\rho_{ee}(\tilde{\alpha}) + \rho_{hh}(\tilde{\alpha}) - 1\right] \mathscr{L}(E - E_{\tilde{\alpha}}) \qquad (1.57)$$

$$N_r(E) = \frac{1}{V_{MD}} \sum_{\tilde{\alpha}} \frac{|\tilde{\mu}(\tilde{\alpha})|^2}{\hbar c \,\epsilon_o n_r} E\, [\rho_{ee}(\tilde{\alpha}) + \rho_{hh}(\tilde{\alpha}) - 1]\, \mathscr{L}_r(E_{\tilde{\alpha}} - E) \qquad (1.58)$$

and $\vartheta = \Gamma_{MD} N_r(E)$. If the photon density inside the active region is denoted by P and

$$P = \tfrac{1}{2}\epsilon_o n_r^2 \,|\mathscr{E}_0(t, s' = 0)|^2/E = \tfrac{1}{2}\epsilon_o n_r^2\, |A_0 E_0(0)|^2/E \qquad (1.59)$$

using Eq. (1.56), we get

$$\frac{dP}{dt} = v_g g P = v_g \Gamma_{MD} G P \equiv \frac{dP}{dt}\Big|_{\text{stim}} \qquad (1.60)$$

which is the photon density increase rate caused by the stimulated optical transitions of the carriers in the active region.

On the other hand, the total optical power generated by the stimulated optical transitions in the active region can be written as

$$-\int \langle \mathscr{E}(\mathbf{r}, t) \frac{d\mathscr{P}_{\text{in}}(\mathbf{r}, t)}{dt}\rangle_t \, d\mathbf{r} = -\int \tfrac{1}{2} \text{Re}[\mathscr{E}_0(-i\omega \mathscr{P}_{\text{in},0})^*]\, d\mathbf{r} \qquad (1.61)$$

$$= -\hbar \omega V_{MD} \frac{dN_{MD}}{dt}\Big|_{\text{stim}}$$

where the spatial integral is over the active region and $\langle\ \rangle_t$ denotes a time average. $-V_{MD}(dN_{MD}/dt)|_{\text{stim}}$ is the total stimulated recombination rate of the carriers in the active region. Equations (1.61), (1.38), (1.57), and (1.59) lead to

$$\frac{dN_{MD}}{dt}\Big|_{\text{stim}} = -v_g G(E) P \qquad (1.62)$$

Equations (1.60) and (1.62) are the two very important relations that describe the interaction between the injected carriers and the photons in various semiconductor laser structures. From Eqs. (1.60) and (1.62), rate equations for photon density and carrier density can be obtained to describe the static and dynamic properties in semiconductor lasers of different structures.

Using Eqs. (1.57), (1.34), and (1.35) and assuming $P/\tilde{P}_s \ll 1$, the *dimensional gain coefficient* $G(E)$ can be approximately written as

$$G(E) = G_0(E) - G_1(E)\frac{P}{\tilde{P}_s} \qquad (1.63)$$

where

$$G_0(E) = \frac{1}{V_{MD}} \sum_{\tilde{\alpha}} \frac{|\tilde{\mu}(\tilde{\alpha})|^2}{\hbar c \epsilon_o n_r} \frac{E}{E_{T_2}} [f_e(\tilde{\alpha}) + f_h(\tilde{\alpha}) - 1]\, \mathscr{L}(E - E_{\tilde{\alpha}}) \quad (1.64)$$

$$G_1(E) = \frac{1}{V_{MD}} \sum_{\tilde{\alpha}} \frac{|\tilde{\mu}(\tilde{\alpha})|^2}{\hbar c \epsilon_o n_r} \frac{E}{E_{T_2}} [f_e(\tilde{\alpha}) + f_h(\tilde{\alpha}) - 1]\, \mathscr{L}^2(E - E_{\tilde{\alpha}}) \quad (1.65)$$

G_0 is the linear gain coefficient, and G_1 is the first-order nonlinear gain coefficient. $G_1 \neq 0$ describes the gain decrease at photon energy E because of the presence of the optical field, i.e., the gain suppression by spectral hole burning.

The definition for the *dimensional gain coefficient* $G(E)$ in Eq. (1.57) is not superfluous. It shows how to define the *material gain* in the low-dimensional world as the injected carriers become 2D, 1D, or 0D in the quantum structures. For quantum structures, the low-dimensional carrier density change rate is related to the 3D photon density by the low-dimensional gain coefficient. The 3D photon density change rate is related to the low-dimensional gain coefficient and the coupling between the low-dimensional carriers and the 3D photons (the dimensional coupling factor). The *differential gain dG/dN* has a universal unit for all the different structures where G is the *dimensional gain* and N is the *dimensional carrier density*. This facilitates comparison of the dynamic properties between the different lasers structures. Many dynamic properties in semiconductor lasers are influenced by the differential gain.

1.3 Basics of quantum well lasers

In the preceding section, analysis of optical transitions, carrier distribution functions, and gain coefficients was given for different semiconductor laser structures. Two things must be done to evaluate the optical properties in these semiconductor laser structures. First, we need to obtain the transition matrix elements $|\tilde{\mu}(\tilde{\alpha})|$. Second, we must accurately and efficiently make the summation $(1/V_{MD})\Sigma_{\tilde{\alpha}}$ of the interesting parameters over the quantum numbers $\tilde{\alpha} = \{k_q, n_{sv}, n_{sc}\}$. This can be accomplished by obtaining the density of states of the carriers and then making integrals of the interesting parameters with the density of states in the energy space. Both the transition matrix elements and the density of states are dependent on the structures of semiconductor active medium. In this section we concentrate our analysis on the transition matrix elements and the density of states of quantum well (QW) laser structures. The rate equation analysis on QW lasers in the last part of this section is still applicable to other semiconductor laser structures with proper terminology changes in the rate equations.

1.3.1 Transition matrix elements

The expressions for $G_0(E)$, $G_1(E)$, and $\tilde{P}_s$ depend on the transition matrix element $|\tilde{\mu}(\tilde{\alpha})|$. $\tilde{\mu}(\tilde{\alpha})$ is given in Eq. (1.26) for the single-band model. The transition matrix elements are quite different for different semiconductor laser structures due to the different spatial symmetry. In the following, we look at the transition matrix elements in semiconductor bulk and QW structures, respectively.

As we discussed in Sect. 1.2.1, to obtain the solutions for the electronic states, one needs to solve the Schrödinger equation or the coupled Schrödinger equations. Usually, the coordinate system is chosen such that the x, y, and z axes are along the crystalline axes. The coupled Schrödinger equations can thus be significantly simplified because many $D_{ll'}^{\alpha\alpha'}$ parameters are zero due to the symmetry of the crystal. This also facilitates the treatment of the interaction between the optical field and the semiconductor medium. Usually, the polarization of optical field is also along one of the crystalline axes.

In a zinc blende crystal with large direct band gap, the conduction band structure can be well described by the single-band model [Sect. 1.2.1 and Eqs. (1.14) to (1.17)]. From Eqs. (1.12) and (1.14), the wavefunction for the electrons in the conduction band is

$$\Psi_e(\mathbf{r}) = u_c(\mathbf{r})\Phi_e(\mathbf{r}) \tag{1.66}$$

The Bloch functions at the conduction band edge can be written as

$$u_c(\mathbf{r}) = \begin{cases} |S> \cdot |\uparrow> \\ |S> \cdot |\downarrow> \end{cases} \tag{1.67}$$

where $|S>$ is the S-like isotropic spatial function, and $|\uparrow>$ and $|\downarrow>$ are the spin wavefunctions with the spin along the $\mathbf{z}$ and $-\mathbf{z}$ directions,

1.3 Basics of quantum well lasers

respectively. The envelope function $\Phi_e(\mathbf{r})$ and the conduction band structure $E_e(k_q, n_{sc})$ can be obtained by solving

$$\frac{\hbar^2}{2m_e}[\hat{k}_x^2 + \hat{k}_y^2 + \hat{k}_z^2 + U_c(\mathbf{r})]\Phi_e(\mathbf{r}) = E_e\Phi_e(\mathbf{r}) \tag{1.68}$$

where m_e is the effective mass of the electrons in the conduction bands, $\hat{k}_r = -i\partial_r = -i\partial/\partial_r$ ($r = x, y, z$), and $U_c(\mathbf{r})$ is the additional energy potential due to the quantum confinement of the electrons in the conduction band for quantum structures. The electron energy E_e is measured upward from the bottom conduction band edge of the bulk structures [where $U_c(\mathbf{r}) = 0$]; i.e., $E_{c0} = 0$ is assumed.

For the zinc blende semiconductors of direct band gap and large spin-orbit energy separation, the valence bands have a fourfold degeneracy (including the twofold spin degeneracy) at the band edge for bulk structures. The Bloch functions at the valence band edge can be written as (Luttinger and Kohn, 1955)

$$u_v(\mathbf{r}) = |j\rangle = \begin{cases} \left|\frac{3}{2}\right\rangle = \frac{1}{\sqrt{2}}(|X\rangle + i|Y\rangle)\cdot|\uparrow\rangle \\ \left|\frac{1}{2}\right\rangle = \frac{i}{\sqrt{6}}[(|X\rangle + i|Y\rangle)\cdot|\downarrow\rangle - 2|Z\rangle\cdot|\uparrow\rangle] \\ \left|-\frac{1}{2}\right\rangle = \frac{1}{\sqrt{6}}[(|X\rangle - i|Y\rangle)\cdot|\uparrow\rangle + 2|Z\rangle\cdot|\downarrow\rangle] \\ \left|-\frac{3}{2}\right\rangle = \frac{i}{\sqrt{2}}(|X\rangle - i|Y\rangle)\cdot|\downarrow\rangle \end{cases} \tag{1.69}$$

where $|X\rangle, |Y\rangle$, and $|Z\rangle$ are the P-like spatial functions. $\left|\frac{3}{2}\right\rangle$ and $\left|-\frac{3}{2}\right\rangle$ correspond to the heavy hole valence bands; $\left|\frac{1}{2}\right\rangle$ and $\left|-\frac{1}{2}\right\rangle$ correspond to the light hole valence bands. The valence band structure can be obtained by solving the coupled Schrödinger equations [Eqs. (1.11) and (1.13)], which can be written explicitly as (Luttinger and Kohn, 1955; Luttinger, 1956)

$$\begin{bmatrix} P+Q & L & M & 0 \\ L^* & P-Q & 0 & M \\ M^* & 0 & P-Q & -L \\ 0 & M^* & -L^* & P+Q \end{bmatrix} \begin{bmatrix} \Phi_{h,\frac{3}{2}}(\mathbf{r}) \\ \Phi_{h,\frac{1}{2}}(\mathbf{r}) \\ \Phi_{h,-\frac{1}{2}}(\mathbf{r}) \\ \Phi_{h,-\frac{3}{2}}(\mathbf{r}) \end{bmatrix} = E_h \begin{bmatrix} \Phi_{h,\frac{3}{2}}(\mathbf{r}) \\ \Phi_{h,\frac{1}{2}}(\mathbf{r}) \\ \Phi_{h,-\frac{1}{2}}(\mathbf{r}) \\ \Phi_{h,-\frac{3}{2}}(\mathbf{r}) \end{bmatrix} \tag{1.70}$$

where

$$P = \frac{\hbar^2}{2m_0} \gamma_1 [\hat{k}_x^2 + \hat{k}_y^2 + \hat{k}_z^2 + U_v(\mathbf{r})] \tag{1.71}$$

$$Q = \frac{\hbar^2}{2m_0} \gamma_2 [\hat{k}_x^2 + \hat{k}_y^2 - 2\hat{k}_z^2] \tag{1.72}$$

$$L = -i\sqrt{12}\, \frac{\hbar^2}{2m_0} \gamma_3 [\hat{k}_x - i\hat{k}_y]\hat{k}_z \tag{1.73}$$

$$M = \sqrt{3}\, \frac{\hbar^2}{2m_0} \{\gamma_2 [\hat{k}_x^2 - \hat{k}_y^2] - i2\gamma_3 \hat{k}_x \hat{k}_y\} \tag{1.74}$$

γ_1, γ_2, and γ_3 are the Luttinger parameters, and $U_v(\mathbf{r})$ is the additional energy potential due to the quantum confinement of the holes in the valence band for the quantum structures. The valence band energy E_h is measured downward from the top valence band edge of the bulk structures [where $U_v(\mathbf{r}) = 0$]; i.e., $E_{v0} = 0$ and $E_h = -E_v \geq 0$ are assumed.

The wavefunction of the holes in the valence band is written as

$$\Psi_h(\mathbf{r}) = \sum_j |j\rangle\, \Phi_{h,j}(\mathbf{r}) \qquad (j = \tfrac{3}{2}, \tfrac{1}{2}, -\tfrac{1}{2}, -\tfrac{3}{2}) \tag{1.75}$$

Equations (1.69) to (1.74) were obtained by a $k \cdot p$ approximation up to the second order in the wave vector (k^2). There should be some linear k terms in Eqs. (1.70) to (1.74) due to the lack of inversion symmetry in the zinc blende structures in comparison with the diamond structures (such as the Si and Ge structures) (Luttinger and Kohn, 1955; Dresselhaus, 1955; Kane, 1957). However, these linear k terms are negligible for the zinc blende crystals with large spin-orbit energy separation (Dresselhaus, 1955; Kane, 1957; Broido and Sham, 1985) such as the GaAs/AlGaAs material system.

For the bulk semiconductor structure $U_c(\mathbf{r}) = 0$ and $U_v(\mathbf{r}) = 0$, the conduction band structure $E_e(\mathbf{k})$ and the valence band structure $E_h(\mathbf{k})$ can be obtained by taking the envelope functions $\Phi_l(\mathbf{r}) \propto \exp(i\mathbf{k} \cdot \mathbf{r})/\sqrt{V}$ ($l = e$ for electrons and $l = h, j$ for the holes). For the conduction band, Eq. (1.68) results in a parabolic relation $E_e(\mathbf{k}) = \hbar^2 k^2/(2m_e)$, where $k = |\mathbf{k}|$. For the valence band, using the spherical approximation (Baldereschi and Lipari, 1973) $\gamma_2 = \gamma_3 = \bar{\gamma} \equiv (2\gamma_2 + 3\gamma_3)/5$ in Eqs. (1.71) to (1.74), Eq. (1.70) can be diagonalized under a

1.3 Basics of quantum well lasers

unitary transformation (Luttinger, 1956; Kane, 1957). The new Bloch functions at the band edge under the unitary transformation are

$$u'_v(\mathbf{r}) = |j'> = \begin{cases} |\tfrac{3}{2}'> = \tfrac{1}{\sqrt{2}}[(\cos\phi\cos\theta - i\sin\phi)|X> \\ \qquad + (\sin\phi\cos\theta + i\cos\phi)|Y> - \sin\theta|Z>] \cdot |\uparrow'> \\ |\tfrac{1}{2}'> = \tfrac{i}{\sqrt{6}}\{[(\cos\phi\cos\theta - i\sin\phi)|X> \\ \qquad + (\sin\phi\cos\theta + i\cos\phi)|Y> - \sin\theta|Z>] \cdot |\downarrow'> \\ \qquad - 2[\cos\phi\sin\theta|X> + \sin\phi\sin\theta|Y> \\ \qquad + \cos\theta|Z>] \cdot |\uparrow'>\} \\ |-\tfrac{1}{2}'> = \tfrac{1}{\sqrt{6}}\{[(\cos\phi\cos\theta + i\sin\phi)|X> \\ \qquad + (\sin\phi\cos\theta - i\cos\phi)|Y> - \sin\theta|Z>] \cdot |\uparrow'> \\ \qquad + 2[\cos\phi\sin\theta|X> + \sin\phi\sin\theta|Y> \\ \qquad + \cos\theta|Z>] \cdot |\downarrow'>\} \\ |-\tfrac{3}{2}'> = \tfrac{i}{\sqrt{2}}[(\cos\phi\cos\theta + i\sin\phi)|X> \\ \qquad + (\sin\phi\cos\theta - i\cos\phi)|Y> - \sin\theta|Z>] \cdot |\downarrow'> \end{cases} \quad (1.76)$$

where ϕ and θ are the angles specifying the wave vector $\mathbf{k}$ in a bulk semiconductor structure as shown in Fig. 1.4, and $|\uparrow'>$ and $|\downarrow'>$ are the spin functions with the spin along the $\mathbf{k}$ and $-\mathbf{k}$ directions, respectively. The corresponding Bloch functions for the electrons in the conduction band under the unitary transformation are

$$u'_c(\mathbf{r}) = \begin{cases} |S> \cdot |\uparrow'> \\ |S> \cdot |\downarrow'> \end{cases} \quad (1.77)$$

Using the new Bloch functions, Eq. (1.70) becomes

$$\begin{bmatrix} P_{hh} & 0 & M & 0 \\ 0 & P_{hl} & 0 & 0 \\ 0 & 0 & P_{hl} & 0 \\ 0 & 0 & 0 & P_{hh} \end{bmatrix} \begin{bmatrix} \Phi_{h,\tfrac{3}{2}'}(\mathbf{r}) \\ \Phi_{h,\tfrac{1}{2}'}(\mathbf{r}) \\ \Phi_{h,-\tfrac{1}{2}'}(\mathbf{r}) \\ \Phi_{h,-\tfrac{3}{2}'}(\mathbf{r}) \end{bmatrix} = E_h \begin{bmatrix} \Phi_{h,\tfrac{3}{2}'}(\mathbf{r}) \\ \Phi_{h,\tfrac{1}{2}'}(\mathbf{r}) \\ \Phi_{h,-\tfrac{1}{2}'}(\mathbf{r}) \\ \Phi_{h,-\tfrac{3}{2}'}(\mathbf{r}) \end{bmatrix} \quad (1.78)$$

where

$$P_{hh} = \frac{\hbar^2}{2m_0}(\gamma_1 + 2\bar{\gamma})(\hat{k}_x^2 + \hat{k}_y^2 + \hat{k}_z^2) \quad (1.79)$$

$$P_{hl} = \frac{\hbar^2}{2m_0}(\gamma_1 + 2\bar{\gamma})(\hat{k}_x^2 + \hat{k}_y^2 + \hat{k}_z^2) \quad (1.80)$$

Equations (1.78) to (1.80) show that the Bloch functions $|\tfrac{3}{2}'>$ and $|-\tfrac{3}{2}'>$ correspond to the heavy hole valence bands with effective masses of m_{hh}

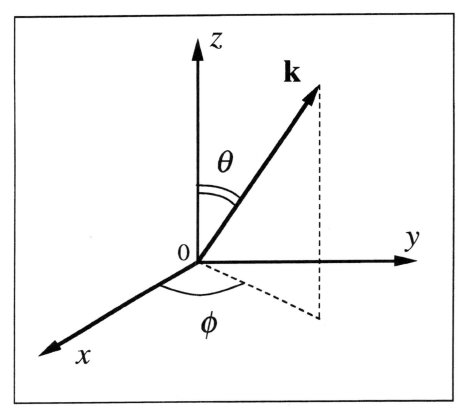

Figure 1.4: Representation of the wave vector for an electron or a hole in the crystalline coordinate system.

$= m_0/(\gamma_1 - 2\bar{\gamma})$ and $|\tfrac{1}{2}'>$ and $|-\tfrac{1}{2}'>$ correspond to the light hole valence bands with effective masses of $m_{hh} = m_0/(\gamma_1 + 2\bar{\gamma})$. The heavy hole valence bands and light hole valence bands are completely decoupled. Thus the valence bands can be treated by the single-band model, and the valence band structures are described by the parabolic relations with the corresponding effective masses, i.e., $E_{hh}(\mathbf{k}) = \hbar^2 k^2/(2m_{hh})$ and $E_{hl}(\mathbf{k}) = \hbar^2 k^2/(2m_{hl})$.

Assuming that the optical wave propagates along the y direction, the square of transition matrix elements $|\tilde{\mu}(k)|^2$ for the transitions between the different conduction bands and the different valence bands for the

1.3 Basics of quantum well lasers

TE mode ($\hat{\mathbf{a}} \parallel \hat{\mathbf{x}}$) and the TM mode ($\hat{\mathbf{a}} \parallel \hat{\mathbf{z}}$) are shown in Table 1.3. The $|\tilde{\mu}(k)|^2$ averages over ϕ and θ are also shown in Table 1.3, where $\langle \ \rangle_{\phi,\theta}$ represents the average over ϕ and θ. The square of the transition matrix elements shown in Table 1.3 are in units of μ^2, which is defined as

$$\mu^2 \equiv \frac{e^2}{3}|\langle S|x|X\rangle|^2 = \frac{e^2}{3}|\langle S|y|Y\rangle|^2 = \frac{e^2}{3}|\langle S|z|Z\rangle|^2 \qquad (1.81)$$

Table 1.3 can be summarized as that the electron-heavy hole transitions and the electron-light hole transitions have the same transition matrix element square μ^2 for both the TE and the TM modes in the bulk semiconductor structures.

In a QW semiconductor structure, as shown in Fig. 1.5, a thin layer of semiconductor material of smaller energy bandgap E_{g1} is bounded on either side by semiconductor materials of bigger energy bandgap E_{g2}. This

	TE Mode	
	$|S\rangle \cdot |\uparrow'\rangle$	$|S\rangle \cdot |\downarrow'\rangle$
$|\tfrac{3}{2}'\rangle$	$\langle \tfrac{3}{2}(\cos^2\phi \cos^2\theta + \sin^2\phi)\rangle_{\phi,\theta} = 1$	0
$|\tfrac{1}{2}'\rangle$	$\langle \tfrac{1}{2}(\cos^2\phi \cos^2\theta + \sin^2\phi)\rangle_{\phi,\theta} = \tfrac{1}{3}$	$\langle 2\cos^2\phi \sin^2\theta\rangle_{\phi,\theta} = \tfrac{2}{3}$
$|-\tfrac{1}{2}'\rangle$	$\langle 2\cos^2\phi \sin^2\theta\rangle_{\phi,\theta} = \tfrac{2}{3}$	$\langle \tfrac{1}{2}(\cos^2\phi \cos^2\theta + \sin^2\phi)\rangle_{\phi,\theta} = \tfrac{1}{3}$
$|-\tfrac{3}{2}'\rangle$	0	$\langle \tfrac{3}{2}(\cos^2\phi \cos^2\theta + \sin^2\phi)\rangle_{\phi,\theta} = 1$
	TM Mode	
	$|S\rangle \cdot |\uparrow'\rangle$	$|S\rangle \cdot |\downarrow'\rangle$
$|\tfrac{3}{2}'\rangle$	$\langle \tfrac{3}{2}\sin^2\theta\rangle_{\phi,\theta} = 1$	0
$|\tfrac{1}{2}'\rangle$	$\langle \tfrac{1}{2}\sin^2\theta\rangle_{\phi,\theta} = \tfrac{1}{3}$	$\langle 2\cos^2\theta\rangle_{\phi,\theta} = \tfrac{2}{3}$
$|-\tfrac{1}{2}'\rangle$	$\langle 2\cos^2\theta\rangle_{\phi,\theta} = \tfrac{2}{3}$	$\langle \tfrac{1}{2}\sin^2\theta\rangle_{\phi,\theta} = \tfrac{1}{3}$
$|-\tfrac{3}{2}'\rangle$	0	$\langle \tfrac{3}{2}\sin^2\theta\rangle_{\phi,\theta} = 1$

Note: See text for definition of μ^2.

Table 1.3: The transition matrix elements and their angular average for the TE and TM modes in a bulk semiconductor structure in units of μ^2.

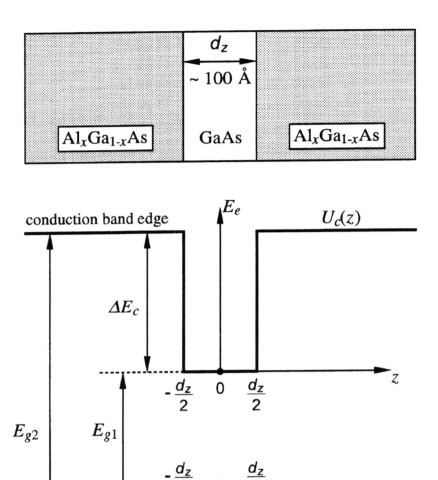

Figure 1.5: The layered structure and the conduction and valence band edges of a AlGaAs/GaAs/AlGaAs QW laser.

1.3 Basics of quantum well lasers

introduces the additional potential $U(\mathbf{r})$ for the electrons in the conduction band and for the holes in the valence band. Assume that the z direction is perpendicular to the E_{g1} layer; then $U(\mathbf{r}) = U(z)$; i.e., $s = z$ and $q = x, y$ for the formulation shown in Sect. 1.2. The thin E_{g1} layer acts as a trap for the electrons in the conduction band and for the holes in the valence band, respectively. If the thickness of the E_{g1} layer is ≈ 100 Å, the confined electrons and holes display quantum effects. As an example, consider the $Al_xGa_{1-x}As/GaAs/Al_xGa_{1-x}As$ QW structure as depicted in Fig. 1.5. The conduction band edge is lower by $\Delta E_c \approx 0.67 \cdot \Delta E_g$ in the GaAs inner region compared with the two sides $Al_xGa_{1-x}As$, where $\Delta E_g \equiv E_g(Al_xGa_{1-x}As) - E_g(GaAs) = x \cdot 1.27$ eV and x is the Al mole fraction in $Al_xGa_{1-x}As$. Similarly, there exists a discontinuity in the valence band edge $\Delta E_v \approx 0.33 \cdot \Delta E_g$. The conduction band edge and the valence band edge are the potential functions $U_c(z)$ and $U_v(z)$ for the electrons and holes, respectively

$$U_c(z) = \begin{cases} 0 & \left(|z| \leq \dfrac{d_z}{2}\right) \\ \Delta E_c & \left(|z| > \dfrac{d_z}{2}\right) \end{cases} \quad (1.82)$$

$$U_v(z) = \begin{cases} 0 & \left(|z| \leq \dfrac{d_z}{2}\right) \\ \Delta E_v & \left(|z| > \dfrac{d_z}{2}\right) \end{cases} \quad (1.83)$$

In QW laser structures, the wavefunction of an electron in the conduction band is given by Eq. (1.66), where the envelope function can be written as

$$\Phi_e(\mathbf{r}) = \Phi_{e,z}(z) \frac{1}{\sqrt{S}} e^{i\mathbf{k}_\| \cdot \mathbf{r}_\|} \quad (1.84)$$

where $\mathbf{r}_\|$ and $\mathbf{k}_\|$ are the position vector and the wave vector in the QW plane, respectively, and $S = L_x \times L_y$. Substituting Eq. (1.84) into Eq. (1.68) leads to

$$\left[-\frac{\hbar^2}{2m_e}\frac{\partial^2}{\partial z^2} + U_c(z)\right]\Phi_{e,z}(z) = E_{e,z}\Phi_{e,z}(z) \quad (1.85)$$

and the electron energy

$$E_e = \frac{\hbar^2 k^2}{2m_e} + E_{e,z} \qquad (1.86)$$

where $k = |\mathbf{k}_\parallel|$, $E_{e,z}$ are a series of quantized energy levels, and $\Phi_{e,z}(z)$ are the corresponding eigenfunctions

$$E_{e,z} = E_{e,z}(n_e), \quad \Phi_{e,z}(z) = \Phi_{e,z}(z, n_e) \qquad (n_e = 1, 2, \ldots) \qquad (1.87)$$

that are solved from Eq. (1.85) by using the boundary conditions for the eigenfunctions at the QW interfaces ($z = \pm d_z/2$). The QW thickness d_z and the QW potential energy depth ΔE_c directly influence the quantized energy levels and the eigenfunctions.

The wavefunctions for the holes in the valence band are given by Eq. (1.75), where, for the QW structures, the envelope functions $\Phi_{h,j}(\mathbf{r})$ can be written as

$$\Phi_{h,j}(\mathbf{r}) = \Phi_{hj,z}(z) \frac{1}{\sqrt{S}} e^{i\mathbf{k}_\parallel \cdot \mathbf{r}_\parallel} \qquad j = \frac{3}{2}, \frac{1}{2}, -\frac{1}{2}, -\frac{3}{2} \qquad (1.88)$$

In Eqs. (1.84) and (1.88), the wave vectors $\mathbf{k}_\parallel$ are not distinguished for the electrons and the holes because for interband transitions the wave vectors must be the same for the electrons and the holes under the k-selection rule. Substitution of Eq. (1.88) into Eq. (1.70) leads to a set of coupled differential equations similar to Eq. (1.70) with $\Phi_{h,j}(\mathbf{r})$ replaced by $\Phi_{hj,z}(z)$ and $\hat{k}_x$ and $\hat{k}_y$ replaced by k_x and k_y, respectively. The solutions to this set of coupled Schrödinger equations are a set of energy levels $E_h = E_h(n_h, k_x, k_y)$ and the corresponding eigenfunctions $\Phi_{hj,z}(z) = \Phi_{hj,z}(z, n_h, k_x, k_y)$ ($j = \frac{3}{2}, \frac{1}{2}, -\frac{1}{2}, -\frac{3}{2}$ and $n_h = 1, 2, \ldots$). They are determined by the QW thickness d_z and the QW potential energy depth ΔE_v through the boundary conditions for the eigenfunctions at the QW interfaces ($z = \pm d_z/2$).

For the QW semiconductor structures, the transition matrix elements are quite different from those for the bulk semiconductor structures because some of the crystal symmetries are not present due to the potential $U(z)$ in the QW structures. The transition matrix elements can be evaluated by following a similar derivation shown in Sect. 1.2.2 from Eq. (1.21) to Eq. (1.26) and using Eqs. (1.66) and (1.67) for the electrons and Eqs. (1.75) and (1.69) for the holes. The square of transition matrix elements in the QW structures generally can be written as

$$|\tilde{\mu}(\tilde{\alpha})|^2 = |\tilde{\mu}(\sigma, n_e, n_h, k_x, k_y)|^2 = \mu^2 \delta \qquad (1.89)$$

1.3 Basics of quantum well lasers

where δ is defined as the polarization modification factor, σ represents the spin states $|\uparrow\rangle$ or $|\downarrow\rangle$, and μ^2 is the square of the transition matrix element for the bulk structures given by Eq. (1.81). In Table 1.4 the polarization modification factor δ is shown for the TE and the TM modes, respectively, for the transitions between the two different spin states of the carriers where

$$\langle \Phi_{h,j}|\Phi_e\rangle = \int \Phi^*_{hj,z}(z, n_h, k_x, k_y)\Phi_{e,z}(z, n_e)dz \quad (1.90)$$

Table 1.4 and Eqs. (1.89) and (1.90) show how the transition matrix elements in the QW structures are obtained from the electron wavefunctions $\Phi_{e,z}(z, n_e, k_x, k_y)$ and the hole wavefunctions $\Phi_{hz,j}(z, n_h, k_x, k_y)$. For the differently polarized optical field (the TE or TM modes), the gain coefficients can be evaluated by using Eqs. (1.64) and (1.65) and making summations over k_x, k_y, n_e, n_h and σ (spin).

For the QW structures, the process of solving the coupled Schrödinger equations [Eq. (1.70)] for the holes in the valence band is much more complicated and difficult than that of solving the single Schrödinger equation [Eq. (1.85)] for the electrons in the conduction band. There have been several approximation methods used to simplify this process and to obtain the valence band structures, hole wavefunctions, and transition matrix elements more easily. First of all, Equation (1.70) can be block diagonalized

	TE Mode
$|\uparrow\rangle$	$\frac{3}{2}|\langle \Phi_{h,3/2}|\Phi_e\rangle|^2 + \frac{1}{2}|\langle \Phi_{h,-1/2}|\Phi_e\rangle|^2 + \sqrt{3}\,\text{Re}[\langle \Phi_{h,3/2}|\Phi_e\rangle\langle \Phi_e|\Phi_{h,-1/2}\rangle]$
$|\downarrow\rangle$	$\frac{3}{2}|\langle \Phi_{h,-3/2}|\Phi_e\rangle|^2 + \frac{1}{2}|\langle \Phi_{h,1/2}|\Phi_e\rangle|^2 + \sqrt{3}\,\text{Re}[\langle \Phi_{h,-3/2}|\Phi_e\rangle\langle \Phi_e|\Phi_{h,1/2}\rangle]$
	TM Mode
$|\uparrow\rangle$	$2\,|\langle \Phi_{h,1/2}|\Phi_e\rangle|^2$
$|\downarrow\rangle$	$2\,|\langle \Phi_{h,-1/2}|\Phi_e\rangle|^2$

Note: The definition of $\langle\Phi_{h,j}|\Phi_e\rangle$ is given in text.

Table 1.4: The polarization modification factors for the TE and TM modes with transitions between the two different spin states of the carriers in a QW structure.

into 2 × 2 blocks using a unitary transformation (Broido and Sham, 1985). Thus the number of the coupled equations can be reduced from four to two. This can be further simplified by the axial approximation. The axial approximation assumes that $\gamma_2 = \gamma_3 = \bar{\gamma} \equiv (\gamma_2 + \gamma_3)/2$ in the M term of Eq. (1.70) (Altarelli et al., 1985; Twardowski and Hermann, 1987; Ahn et al., 1988). This introduces a rotation symmetry to the valence band structures in the $k_x k_y$ plane.

The simplest method to obtain the transition matrix elements in the QW structures is the decoupled valence band approximation. It assumes that the electrons and the holes have a virtual wave vector k_z along the z direction that is related to their corresponding quantized energies in the z direction (Yamanishi and Suemune, 1984; Asada et al., 1984; Yamada et al., 1984, 1985). Similar to the treatment for bulk structures, Eq. (1.70) can be diagonalized under the spherical approximation and the same unitary transformation. The new Bloch functions are the same as given by Eqs. (1.76) and (1.77). The valence bands are then decoupled, and two equations similar to Eq. (1.68) can be obtained with corresponding heavy hole effective mass m_{hh} and light hole effective mass m_{hl}, respectively. The polarization modification factors δ for the TE and the TM modes are shown in Table 1.5 for the transitions between the different conduction bands and valence bands. The δ values shown in Table 1.5 have been averaged over φ, i.e., averaged over various k_x and k_y for a given value of $k = \sqrt{k_x^2 + k_x^2}$, and

$$\cos^2 \theta = \cos^2 \theta_j \approx \frac{E_{e,z} + E_{hj,z}}{E_e + E_{hj}} \qquad (j = h, l) \qquad (1.91)$$

In Eq. (1.91), E_e is the total electron energy including both the quantized energy $E_{e,z}$ related to the quantum confinement in the z direction and the transverse kinetic energy in the xy plane [see Eq. (1.86)]. Similarly, E_{hj} is the total hole energy including both the quantized energy $E_{hj,z}$ related to the quantum confinement in the z direction and the transverse kinetic energy in the xy plane, where $j = h$ denotes the heavy holes ($|\frac{3'}{2}\rangle, |-\frac{3'}{2}\rangle$) and $j = l$ denotes the light holes ($|\frac{1'}{2}>, |-\frac{1'}{2}>$).

The quantized energies $E_{e,z}$ for the electrons can be obtained by solving Eq. (1.85). The solutions can be found in many books on quantum mechanics. With m_e replaced by m_{hj} ($j = h, l$) and U_c replaced by U_v, similar results can be obtained for the holes under the decoupled valence

1.3 Basics of quantum well lasers

	TE Mode	
	$\|S\rangle \cdot \|\uparrow'\rangle$	$\|S\rangle \cdot \|\downarrow'\rangle$
$\|\frac{3'}{2}\rangle$	$\frac{3}{4}(\cos^2\theta + 1)$	0
$\|\frac{1'}{2}\rangle$	$\frac{1}{4}(\cos^2\theta + 1)$	$1 - \cos^2\theta$
$\|-\frac{1'}{2}\rangle$	$1 - \cos^2\theta$	$\frac{1}{4}(\cos^2\theta + 1)$
$\|-\frac{3'}{2}\rangle$	0	$\frac{3}{4}(\cos^2\theta + 1)$
	TM Mode	
	$\|S\rangle \cdot \|\uparrow'\rangle$	$\|S\rangle \cdot \|\downarrow'\rangle$
$\|\frac{3'}{2}\rangle$	$\frac{3}{2}(1 - \cos^2\theta)$	0
$\|\frac{1'}{2}\rangle$	$\frac{1}{2}(1 - \cos^2\theta)$	$2\cos^2\theta$
$\|-\frac{1'}{2}\rangle$	$2\cos^2\theta$	$\frac{1}{2}(1 - \cos^2\theta)$
$\|-\frac{3'}{2}\rangle$	0	$\frac{3}{2}(1 - \cos^2\theta)$

Table 1.5: The polarization modification factors for the TE and TM modes with the transitions between different spin states of electrons and holes in a QW structure under the decoupled valance band approximation.

band approximation. The quantized energy levels and corresponding wavefunctions for the holes in the valence bands are

$$E_{hj,z} = E_{hj,z}(n_{hj}), \quad \Phi_{hj,z}(z) = \Phi_{hj,z}(z, n_{hj}) \quad (j = h, l, \quad n_{hj} = 1, 2, \ldots) \tag{1.92}$$

The total energies for the holes in the valence bands are given by

$$E_{hj}(k, n_{hj}) = \frac{\hbar^2 k^2}{2m_{hj}} + E_{hj,z}(n_{hj}) \quad (j = h, l, \quad n_{hj} = 1, 2, \ldots) \tag{1.93}$$

As shown in Eq. (1.26), another factor determining the transition matrix elements is $\int \Phi_{e,z}^*(n_e, z) \Phi_{hj,z}(n_{hj}, z) \, dz$. For a deep and symmetric QW, one can find that

$$\int \Phi_{e,z}^*(n_e, z) \Phi_{hj,z}(n_{hj}, z) \, dz \approx \begin{cases} 1 & n_e = n_{hj} \\ 0 & n_e \neq n_{hj} \end{cases} \tag{1.94}$$

Equation (1.94) gives another selection rule

$$n_e = n_{hj} \quad (j = h, l) \tag{1.95}$$

for the interband transitions in the QW structures. For the interesting transitions near the band edges (i.e., $k \to 0$), $\cos^2 \theta \to 1$ in Eq. (1.91) due to the small transverse kinetic energy. Applying $\cos^2 \theta \to 1$ in Table 1.5, one can find that the electron-heavy hole transitions dominate the TE mode and the electron-light hole transitions dominate the TM mode.

1.3.2 Density of states for QW structures

The $(1/V_{MD}) \Sigma_{\tilde{\alpha}}$ summation of interesting parameters over the quantum numbers $\tilde{\alpha} = \{k_q, n_{sv}, n_{sc}\}$ can be done directly in the wavevector space. Sometimes, it is more convenient to obtain the density of states first and then to make the integral in the energy space. Under the decoupled valence band approximation, the density of states for QW structures is obtained as shown in the following paragraphs.

Since there is no quantum confinement in the x and y directions, the transverse wavevector has the values

$$k_x = m_x \frac{2\pi}{L_x} \quad m_x = 0, \pm 1, \pm 2, \ldots \tag{1.96}$$

$$k_y = m_y \frac{2\pi}{L_y} \quad m_y = 0, \pm 1, \pm 2, \ldots \tag{1.97}$$

Thus every electronic state occupies a "volume" of $4\pi^2/L_x L_y = 4\pi^2/S = 4\pi^2/V_{2D}$ in the $\mathbf{k}_\parallel$ space. The number of electron states with transverse wavevectors less than k ($k = |\mathbf{k}_\parallel|$) is

$$N(k) = 2 \times \frac{\pi k^2}{4\pi^2/S} = \frac{k^2 S}{2\pi} \tag{1.98}$$

where a factor of 2 was included to account for the spin degeneracy. The number of states between k and $k + dk$ is

$$\rho(k)\,dk = \frac{dN}{dk}\,dk = S\frac{k}{\pi}\,dk \tag{1.99}$$

1.3 Basics of quantum well lasers

For the electrons, the number of states with total energy between E_e and $E_e + dE_e$ is

$$D_e(E_e)\,dE_e = \frac{dN}{dE_e}\,dE_e = \frac{dN}{dk}\frac{dk}{dE_e}\,dE_e \qquad (1.100)$$

Equations (1.86) and (1.87) with $n_e = 1$ lead to

$$k = \sqrt{\frac{2m_e}{\hbar^2}[E_e - E_{e,z}(1)]} \qquad (1.101)$$

Thus the number of states between E_e and $E_e + dE_e$ per unit area is

$$\rho_e(E_e)\,dE_e = \frac{1}{S}D_e(E_e)\,dE_e = \frac{m_e}{\pi\hbar^2}\,dE_e \qquad (1.102)$$

In the derivation leading to Eq. (1.102), only one quantum state $n_e = 1$ was considered. Once $E > E_{e,z}(2)$, as an example, an electron of a given total energy E_e can be found in either $n_e = 1$ or $n_e = 2$ state so that the density of states doubles. The total density of states increases by $m_e/\pi\hbar^2$ at each of the energies $E_{e,z}(n_e)$; thus

$$\rho_e(E_e) = \sum_{n_e} \frac{m_e}{\pi\hbar^2} H[E_e - E_{e,z}(n_e)] \qquad (1.103)$$

where $H[x]$ is the Heaviside function, which is equal to unity when $x > 0$ and is zero when $x < 0$. Replacing E_e, m_e, $E_{e,z}(n_e)$, and n_e in Eq. (1.103) by E_{hj}, m_{hj}, $E_{hj,z}(n_{hj})$, and n_{hj}, respectively, the density of states for the holes in the valence bands $\rho_{hj}(E_{hj})$ ($j = h, l$) can be expressed similarly by Eq. (103). Thus

$$\frac{1}{V_{2D}}\sum_{\mathbf{k}_\|} = \frac{1}{S}\sum_{\mathbf{k}_\|} \rightarrow \frac{1}{S}\int dk D(k) = \int dE_i \rho_i(E_i) \qquad (i = e, hh, hl) \qquad (1.104)$$

The steplike density of states for the electrons in an infinitely deep QW structure is shown in Fig. 1.6. In Fig. 1.6 the 2D QW density of states for the electrons $\rho_{qw} \equiv \rho_{2D}$ is plotted against the 2D "bulk" density of states ρ_{bulk}. ρ_{bulk} is the 2D density of states for the electrons in the QW when the quantum effects are not considered. $\rho_{bulk} \equiv d_z \cdot \rho_{3D}$, where ρ_{3D} constitutes the 3D density of states for the electrons and d_z is the QW width. The value of 2D QW density of states ρ_{qw} equals the value of 2D "bulk" density of states ρ_{bulk} at each of the quantized energy levels for an infinitely deep QW.

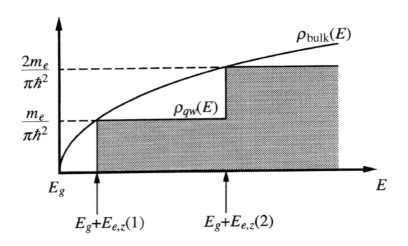

For infinitely deep QW, $E_{e,z}(1) = \dfrac{\pi^2 \hbar^2}{2 m_e d_z^2}$ & $E_{e,z}(2) = 4 E_{e,z}(1)$

Figure 1.6: The 2D QW density of states (ρ_{qw}) for electrons in an infinitely deep QW and the 2D "bulk" density of states (ρ_{bulk}). ρ_{bulk} is the 2D density of states for the electrons in the QW when the quantum effects are not considered. $\rho_{\text{bulk}} \equiv d_z \cdot \rho_{3D}$, where ρ_{3D} is the 3D density of states for the electrons, and d_z is the QW width.

Equation (1.104) can be used to calculate the carrier density N_{2D} by an integral of the density of states with the Fermi function over the electron energy or the hole energies

$$N_{2D} = \int dE_e \rho_e(E_e) f_e(E_e, F_e) = \sum_{j=h,l} \int dE_{hj} \rho_{hj}(E_{hj}) f_h(E_{hj}, F_h) \qquad (1.105)$$

In the expressions for the susceptibilities and the gain coefficients [Eqs. (1.40), (1.41), (1.64), and (1.65)], $\mathscr{L}_i(E_{\tilde{\alpha}} - E)$, $\mathscr{L}_r(E_{\tilde{\alpha}} - E)$, and $\mathscr{I}(E - E_{\tilde{\alpha}})$ are functions of the electron-hole transition energy $E_{\tilde{\alpha}}$. When calculating the susceptibilities or the gain coefficients, it is convenient to perform the integral over the transition energy $\mathscr{E}$ ($\mathscr{E} \equiv E_{\tilde{\alpha}}$) instead of the electron energy E_e, which are related by

$$\mathscr{E} = E_g + E_e + E_{hj} \qquad (j = h, l) \qquad (1.106)$$

1.3 Basics of quantum well lasers

Equations (1.86), (1.93), and (1.106) result in

$$E_e - E_{e,z}(n_e) = \frac{m_{rj}}{m_e}[\mathcal{E} - E_g - E_{e,z}(n_e) - E_{hj,z}(n_{hj})] \qquad (j = h, l) \qquad (1.107)$$

where the reduced mass m_{rj} is defined by $m_{rj}^{-1} = m_e^{-1} + m_{hj}^{-1}$ ($j = h, l$). Applying the transition selection rule $n_e = n_{hj} \equiv n$ in Eq. (1.107) leads to

$$E_e = E_{e,z}(n) + \frac{m_{rj}}{m_e}[\mathcal{E} - E_g - E_{e,z}(n) - E_{hj,z}(n)] \qquad (j = h, l) \qquad (1.108)$$

And similarly, for the holes,

$$E_{hj} = E_{hj,z}(n) + \frac{m_{rj}}{m_{hj}}[\mathcal{E} - E_g - E_{e,z}(n) - E_{hj,z}(n)] \qquad (j = h, l) \qquad (1.109)$$

Thus

$$\int dE_e \rho_e(E_e) = \int dE_e \sum_n \frac{m_e}{\pi \hbar^2} H[E_e - E_{e,z}(n)] \qquad (1.110)$$

$$= \int d\mathcal{E} \sum_n \frac{m_{rj}}{\pi \hbar^2} H[\mathcal{E} - E_g - E_{e,z}(n) - E_{hj,z}(n)] \equiv \int d\mathcal{E} \rho_{rj}(\mathcal{E})$$

$\rho_{rj}(\mathcal{E})$ is called the *reduced density of states*. Equations (1.108) and (1.109) can be used to express the Fermi distributions $f_e(E_e, F_e)$ and $f_h(E_h, F_h)$ as functions of the transition energy $\mathcal{E}$ instead of the electron energy or the hole energy. The polarization modification factor δ also can be expressed by the transition energy $\mathcal{E}$ by noticing that

$$\cos^2 \theta \approx \frac{E_{e,z}(n) + E_{hj,z}(n)}{\mathcal{E} - E_g} \qquad (1.111)$$

Using these results, the gain coefficients or the susceptibilities can be calculated by an integral over the transition energy $\mathcal{E}$.

1.3.3 Rate equations for quantum well laser structures

Along with the development of semiconductor lasers, a set of coupled rate equations, which involve injection current, injected carrier density, and photon density in the active region, has been found very useful in understanding the behaviors of semiconductor lasers (Channin, 1979; Lee et al., 1982; Koch and Bowers, 1984; Lau and Yariv, 1985a; Tucker, 1985; Bowers et al., 1986; Su and Lanzisera, 1986). All these rate equations treated the carrier density three-dimensionally. Bookkeeping analysis using 3D carrier density for the QW structures has been used widely. For QW structures, it is more appropriate to obtain and use the rate equations that involve the 2D carrier density, 2D gain coefficients, etc. The carrier density for QW structures is two-dimensional in nature due to the quantum confinement of the carriers in one direction. Here we introduce such a set of rate equations from the analysis and results shown above. Rate equations of the same forms can be obtained similarly to study other laser structures (e.g., 3D bulk structure) with proper terminology changes in the rate equations.

For a separate confinement heterostructure (SCH) QW structure as shown schematically in Fig. 1.7, from Eqs. (1.60), (1.62), and (1.63), a set of rate equations can be obtained to describe the QW laser behavior:

$$\frac{dP}{dt} = \Gamma G_0 v_g P - \Gamma G_1 v_g \frac{P^2}{P_s} - \frac{P}{\tau_p} + \xi \Gamma \frac{N}{\tau_n} \tag{1.112}$$

$$\frac{dN}{dt} = \frac{N'}{\tau_r} - G_0 v_g P + G_1 v_g \frac{P^2}{P_s} - \frac{N}{\tau_n} \tag{1.113}$$

$$\frac{dN'}{dt} = \frac{J}{e} - \frac{N'}{\tau_r} - \frac{N'}{\tau_n} \tag{1.114}$$

where P is the photon density at the QW active region, N and N' are, respectively, the 2D quasi-equilibrium carrier density and non-quasi-equilibrium carrier density in the SCH QW structure as shown schematically in Fig. 1.7, τ_n is the carrier lifetime, $\tau_p = (v_g \alpha_t)^{-1}$ is the photon lifetime and $\alpha_t = \alpha_i + (1/2L)\ln(1/R_1 R_2)$ is the total optical loss, L, R_1, and R_2 are the cavity length and the mirror reflectivity, respectively, J is the injection current density, ξ is the spontaneous emission coupling factor defined as the ratio of the spontaneous emission coupled into the lasing mode to the total spontaneous emission, and Γ is the 2D dimensional coupling factor we discussed in Sect. 1.2.3. In these rate equations, carrier relaxation time τ_r phenomenologically represents the carrier relaxation phenomena that cannot be described by the fast quasi-equilibrium relaxation times (τ_e, τ_h, and T_2 in Sect. 1.2.2). More discussion about τ_r can be found in Sect. 1.4.

1.3 Basics of quantum well lasers

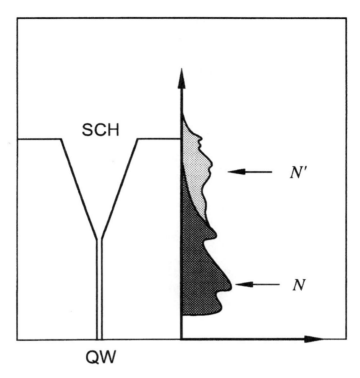

Figure 1.7: Schematic SCH QW structure and corresponding carrier densities. N and N' are, respectively, the 2D quasi-equilibrium carrier density in the SCH QW region and the non-quasi-equilibrium carrier density both inside and outside the SCH QW region.

From Eqs. (1.64) and (1.65) and the analysis in Sects. 1.3.1 and 1.3.2, the linear and nonlinear gain coefficients G_0 and G_1 for a QW laser structure can be written as

$$G_0(E) = \frac{\mu^2}{\hbar c \epsilon_o n_r} \frac{E}{E_{T_2}} \sum_{i=l,h} \int \delta_i \rho_{ri} (f_e + f_h - 1) \mathscr{L}(\mathscr{E} - E) \, d\mathscr{E} \quad (1.115)$$

$$G_1(E) = \frac{\mu^2}{\hbar c \epsilon_o n_r} \frac{E}{E_{T_2}} \sum_{i=l,h} \int \delta_i^2 \rho_{ri} (f_e + f_h - 1) \mathscr{L}^2(\mathscr{E} - E) \, d\mathscr{E} \quad (1.116)$$

where

$$\mathscr{L}(\mathscr{E} - E) = \frac{E_{T_2}^2}{E_{T_2}^2 + (\mathscr{E} - E)^2} \quad (1.117)$$

ρ_{ri} is the steplike 2D reduced density of states for the carriers in the SCH QW structure, i designates either light holes ($i = l$) or heavy holes ($i = h$), δ_i is the polarization modification factor of the transition matrix elements for light and heavy holes, and P_s is the saturation photon density, i.e.,

$$P_s = \frac{\hbar^2 \epsilon_o n_r^2}{\mu^2 E (\tau_e + \tau_h) T_2} \tag{1.118}$$

Equation (1.55) and Table 1.2 show that the coupling factor Γ for a QW structure is $\Gamma = 1/t_z$, where t_z is the effective optical mode width in the z direction. Assume that the photons have a normalized distribution function $\Theta_z(z)$ in the z-direction

$$\int_{-\infty}^{\infty} \Theta_z(z)\, dz = 1 \tag{1.119}$$

and $\Theta_z(z)$ has its maximum value at the QW active region ($z = 0$). Then we have $t_z = 1/\Theta_z(0)$. The 2D treatment of the carrier density N in a typical SCH structure is justified by noticing that the thickness of the quasi-equilibrium carrier population region of the SCH QW structure is usually much smaller than the carrier diffusion length.

The rate equations are basically space-averaged conservation equations for the photons and the injected carriers. They are approximate equations with applicable conditions. We are not going to discuss the validity conditions in detail here. However, we would like to point out two major limitations for the rate equations. First, the rate equations cannot be used where there is a large spatial variation in the photon density and carrier density along the laser cavity, i.e., the y-direction. Such an example is the case that the mirror reflectivity R is < 0.1 for the Fabry-Perot cavity semiconductor lasers. Second, the rate equations cannot be used to analyze the very fast optical phenomena with the time scale comparable with or shorter than cavity transit time of the photons. Such an example is the high-frequency modulation phenomena > 60 GHz for a 300-μm-long laser or the high-frequency modulation phenomena > 180 GHz for a 100-μm-long laser. More detailed discussion about the range of validity for rate equations can be found in the literature (Ulbrich and Pilkuhn, 1970; Hasuo and Ohmi, 1974; Moreno, 1977; Lau and Yariv, 1985b). The debate on distributed microwave effects of applied electrical signal (injection current) in semiconductor lasers can be found in the literature (Tauber et al., 1994; Tiwari, 1994; Wu et al., 1995).

1.3.4 General description of statics and dynamics

When the operating conditions of semiconductor lasers are within the validity range of the rate equation analysis (actually, this is the scenario in many of today's applications), the static and dynamic properties of semiconductor lasers can be analyzed by the rate equations. For example, the photon density at steady state can be obtained as (Zhao et al., 1992f)

$$P = \frac{(\Gamma G_0 - \alpha_t) + \sqrt{(\Gamma G_0 - \alpha_t)^2 + \xi \frac{N}{\tau_n} \frac{4\Gamma^2 G_1}{v_g P_s}}}{2\Gamma G_1} P_s \qquad (1.120)$$

If the spontaneous emission coupling is omitted, i.e., $\xi \to 0$, Eq. (1.120) becomes

$$P = \frac{(\Gamma G_0 - \alpha_t) + |\Gamma G_0 - \alpha_t|}{2\Gamma G_1} P_s = \begin{cases} 0 & (\Gamma G_0 < \alpha_t) \\ \frac{\Gamma G_0 - \alpha_t}{\Gamma G_1} P_s & (\Gamma G_0 > \alpha_t) \end{cases} \qquad (1.121)$$

From Eq. (1.121) it follows that the modal gain $g \equiv \Gamma G_0 > \alpha_t$ is a necessary condition for the lasing action (the coherent output power). The lasing threshold is determined by the linear gain coefficient G_0. $g_{th} \equiv \alpha_t$ is called the *threshold modal gain*. $\xi \neq 0$ will soften this threshold condition because part of the cavity loss is compensated by the spontaneous emission coupling.

To study the modulation dynamics, one needs to apply a small-signal analysis to the rate equations [Eqs. (1.112) to (1.114)]. The modulation response can be obtained as

$$R(f) = \frac{|p/i|^2(f)}{|p/i|^2(f \to 0)} = R_c(f) \cdot R_i(f) \qquad (1.122)$$

$$R_c(f) = \frac{1}{1 + (f/f_c)^2} \qquad (1.123)$$

$$R_i(f) = \frac{f_r^4}{(f^2 - f_r^2)^2 + (\gamma/2\pi)^2 f^2} \qquad (1.124)$$

where i is the small-signal amplitude of the sinusoidal current component and p is the corresponding modulation amplitude in the optical output power, and f is the modulation frequency. $R_c(f)$ is a response function similar to that for an RC low-frequency filter. $f_c = 1/(2\pi\tau_r)$ represents both the phenomena of slow carrier relaxation processes, which cannot be characterized by the fast quasi-equilibrium relaxation, and the laser device capacitive parasitics. All these effects are related to the carrier storage outside the carrier quasi-equilibrium in the SCH QW structures. We will discuss these effects more in the next section. The intrinsic response $R_i(f)$ is more fundamental in the sense that it is determined by the interaction between the quasi-equilibrium carriers and the photons at the active region.

In the expression for the intrinsic response $R_i(f)$, f_r is the relaxation resonance frequency, i.e.,

$$f_r^2 = \frac{v_g^2 \Gamma}{4\pi^2} \left[\left(G_0 - G_1 \frac{P_0}{P_s} \right) \left(G_0' - G_1' \frac{P_0}{P_s} \right) + \frac{G_1}{P_s} \frac{1}{v_g \tau_n} \right] P_0 \qquad (1.125)$$

and the damping rate γ is given by

$$\gamma = \frac{1}{\tau_n} + v_g \left[\left(G_0' - G_1' \frac{P_0}{P_s} \right) + \frac{\Gamma G_1}{P_s} \right] P_0 \qquad (1.126)$$

where P_0 is the stationary photon density in the QW active region, and $G_0' \equiv dG_0/dN$ and $G_1' \equiv dG_1/dN$ are the differential gain coefficients. Note that if we neglect gain suppression, i.e., $G_1 = 0$, the conventional result

$$f_r = \frac{1}{2\pi} \sqrt{v_g G_0' P_0 / \tau_p} \qquad (1.127)$$

can be obtained. The 3-dB modulation bandwidth of the intrinsic response $R_i(f)$ can be obtained as

$$f_{3\,dB} = \sqrt{f_r^2 - \frac{\gamma^2}{8\pi^2} + \sqrt{\left(f_r^2 - \frac{\gamma^2}{8\pi^2} \right)^2 + f_r^4}} \qquad (1.128)$$

Equations (1.125), (1.126), and (1.128) clearly indicate that the intrinsic 3-dB modulation bandwidth $f_{3\,dB}$ depends on the stationary (or the bias) photon density P_0 at the active region.

In the limit of low optical output power (low P_0) or low bias current, the damping term γ is negligible in $f_{3\,dB}$ so that

$$\begin{aligned} f_{3\,dB} &\approx \sqrt{1+\sqrt{2}}\, f_r \approx \frac{1.55 v_g}{2\pi} \sqrt{\Gamma G_0 G_0' P_0} \\ &= \frac{1.55}{2\pi} \sqrt{\frac{v_g G_0' P_0}{\tau_p}} \end{aligned} \qquad (1.129)$$

1.3 Basics of quantum well lasers

As P_0 increases, f_r increases, which leads to an increase in $f_{3\text{ dB}}$. However, as P_0 increases further, the damping term γ cannot be omitted. From Eq. (1.128), one can see that the presence of damping leads to a reduction in the $f_{3\text{ dB}}$. Notice that because $f_r^2 \propto P_0$ and $\gamma^2 \propto P_0^2$, there should exist a value of P_0 that maximizes $f_{3\text{ dB}}$. For P_0 larger than this value, an increase in P_0 leads to a decrease in $f_{3\text{ dB}}$.

For a better understanding of the modulation bandwidth optimization and limit, we consider the 0-dB modulation bandwidth ($f_{0\text{ dB}}$) for the intrinsic modulation response. As shown in Fig. 1.8, $f_{0\text{ dB}}$ is the frequency at which the intrinsic modulation response is equal to its low frequency limit. From Eq. (1.124), the $f_{0\text{ dB}}$ can be easily obtained as

$$f_{0\text{ dB}} = \sqrt{2\left(f_r^2 - \frac{\gamma^2}{8\pi^2}\right)} \qquad (1.130)$$

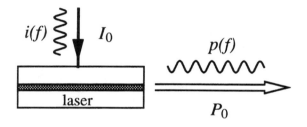

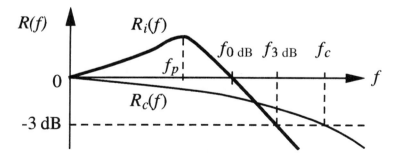

Figure 1.8: Schematic modulation response curves and bandwidths under the discussion in the text for a semiconductor laser biased at an injection current I_0 with a modulation current $i(f)$. The bandwidth $f_{3\text{dB}}$ and modulation peak frequency f_p shown here and discussed in the text are for the intrinsic modulation response $R_i(f)$. when we discuss the experimental results later in this chapter, $f_{3\text{dB}}$ and f_p are for the total modulation response $R(f) = R_c(f) \cdot R_i(f)$.

It is straightforward to show that $f_{0\,\text{dB}}$ has a maximum value that can be written explicitly as

$$f_{0\,\text{dB}}(\max) = \frac{v_g}{2\pi} \frac{\Gamma G_0}{G_0' + \Gamma G_1/P_s} G_0' \qquad (1.131)$$
$$\approx \frac{1}{2\pi} v_g P_s \frac{G_0}{G_1} G_0' \quad (\text{usually } G_0' \ll \Gamma G_1/P_s)$$

at a photon density

$$P_0 = P_{0m} = \frac{\Gamma G_0 G_0'}{(G_0' + \Gamma G_1/P_s)^2} \qquad (1.132)$$

Equations (1.129) and (1.131) show that the differential gain $G_0' = dG_0/dN$ directly influences the modulation bandwidth. A larger differential gain will lead to a larger modulation bandwidth at low-power condition. And a larger differential gain will lead to a larger maximum modulation bandwidth achievable in a specific laser device.

Finally, we would like to point out that the analysis and results in this subsection are not limited to QW lasers only. They are applicable to other types of semiconductor lasers such as the bulk and quantum wire lasers. This is so because rate equations of the same forms as Eqs. (1.112) to (1.114) can be obtained to describe other type lasers with only some terminology changes in the rate equations. The results in this subsection basically show how some of the static and dynamic properties of a semiconductor laser depend on the modal gain coefficients (ΓG_0, ΓG_1) and differential gain coefficients (G_0', G_1'). The terminology to describe the carrier density (N, N') and the dimensional gain (G_0, G_1) varies for these different laser structures due to the different carrier confinement. However, the modal gain coefficients and differential gain coefficients have a universal unit, respectively, for these different laser structures. Thus the lasers of different structures can be compared with each other. Examples comparing QW lasers with bulk lasers will be given in the next section.

1.4 State filling in quantum well lasers

Many optical properties of semiconductor lasers depend on the spectral gain $[G_0(E)]$ and its dependence on injected carrier density at active region. It is evident from Eq. (1.115) that the spectral gain strongly depends on

1.4. State filling in quantum well lasers

the density of states (DOS) of the injected carriers in the laser structures. The difference in energy distribution of the injected carriers, caused by the difference in DOS, leads to different performance features in various semiconductor lasers. According to the fundamental Fermi-Dirac statistics for the injected carriers, the injected electrons and holes not only populate the energy states corresponding to the lasing transitions (usually the energy states contributing the maximum optical gain) but also populate many other energy states not corresponding to the lasing transitions. This phenomenon is called *band filling* or *state filling*. In QW laser structures, the steplike DOS of the injected carriers in QW and the separate confinement structure for optical confinement make the consideration of band/state filling very important. QW lasers have been predicted to be superior to conventional bulk double-heterostructure (DH) lasers in nearly all the important device characteristics (Dutta, 1982; Burt, 1983, 1984; Asada et al., 1984; Arakawa et al., 1984; Arakawa and Yariv, 1985, 1986; Weisbuch and Nagle, 1987; Yariv, 1988). However, the consideration of state filling in QW laser structures changes some of the conclusions and leads to some new understanding on QW lasers. In this section we focus on carrier distribution and state filling in practical QW laser structures and discuss their influence on optical gain, threshold current, modulation, and spectral performance.

1.4.1 Gain spectrum and sublinear gain relationship

The gain spectrum is related to the reduced density of states and Fermi functions by Eq. (1.115) for a QW laser structure. In a typical GaAs QW laser, the lasing occurs in the TE mode and is dominated by the electron-heavy hole transitions. For the electron-heavy hole transitions, the reduced density of states is a steplike function [see $\rho_{rj}(\mathscr{E})$ in Eq. (1.110)] very similar to that shown in Fig. 1.6 with m_e replaced by $m_{rh} = m_e \cdot m_{hh}/(m_e + m_{hh})$, $E_{e,z}(1)$ replaced by $E_{e,z}(1) + E_{hh,z}(1)$, $E_{e,z}(2)$ replaced by $E_{e,z}(2) + E_{hh,z}(2)$, etc. As the injection level increases above transparency [i.e., the maximum of $G_0(E)$ spectrum > 0], the maximum optical gain in the spectral range near $E_g + E_{e,z}(1) + E_{hh,z}(1)$ experiences a large increase due to the step-jump in the reduced density of states at $E_g + E_{e,z}(1) + E_{hh,z}(1)$. However, the energy states available for carrier population are limited by the flat feature in the DOS right above $E_g + E_{e,z}(1) + E_{hh,z}(1)$. As the injection level increases further, the increase in carrier population near $E_g + E_{e,z}(1) + E_{hh,z}(1)$ starts to saturate. The excess injected carriers have to populate higher energy states away from $E_g + E_{e,z}(1) + E_{hh,z}(1)$ (state filling/carrier filling into higher energy states).

As a consequence of the preceding discussion, the gain available at any energy in the spectral range $E_g + E_{e,z}(1) + E_{hh,z}(1) < E < E_g + E_{e,z}(2) + E_{hh,z}(2)$ is limited by the flat feature of the DOS. The maximum optical gain tends to saturate due to the limited carrier population increase at the corresponding energy. This results in a sublinear dependence of the maximum gain on injected carrier density in QW laser structures. For comparison, a much more linear relationship exists for bulk semiconductor laser structures (Asada and Suematsu, 1985). Thus, in QW laser structures, the differential gain (dG_0/dN) at the gain maximum is not a constant and is strongly dependent on the injection levels. In Fig. 1.9 the modal gain and differential gain are plotted versus the injection current density for typical GaAs/AlGaAs QW structures with different number of QWs (Arakawa and Yariv, 1985). They were obtained by a bookkeeping analysis using the material gain, carrier density, and confinement factor for 3D bulk DH lasers.

Historically, it was natural and easy to analyze QW lasers by the bookkeeping approach. In the bookkeeping analysis for QW lasers, the 3D linear gain $G_0(3D)$ was obtained by an expression similar to Eq. (1.115) with ρ_{rj} replaced by ρ_{rj}/d_z so that $G_0(3D) = G_0/d_z$ where $G_0 \equiv G_0(2D)$ is the 2D gain coefficient and d_z is the QW width. Equation (1.115) relates the 2D gain coefficient G_0 to the 2D density of states ρ_{rj} [see Eq. (1.110)]. The optical confinement factor was $\Gamma(3D) = d_z/t$, where t is the effective optical mode width. The 3D carrier density N_{3D} was related to the 2D carrier density $(N \equiv N_{2D})$ by $N_{3D} = N/d_z$. The modal gain g and differential gain are actually

$$g = \Gamma(3D) \cdot G_0(3D) = \frac{d_z}{t} \cdot \frac{G_0}{d_z} = \Gamma G_0 \qquad (1.133)$$

$$\frac{dG_0(3D)}{dN_{3D}} = \frac{d(G_0/d_z)}{d(N/d_z)} = \frac{dG_0}{dN} \qquad (1.134)$$

where $\Gamma = 1/t$ is the coupling factor for QW structures, as discussed before. The modal gain and differential gain in the bookkeeping analysis are thus equivalent to those obtained by the 2D analysis given before for the QW laser structures.

The carrier state filling into the higher energy states also results in a wide gain spectrum in QW laser structures. At high injection levels,

1.4. State filling in quantum well lasers

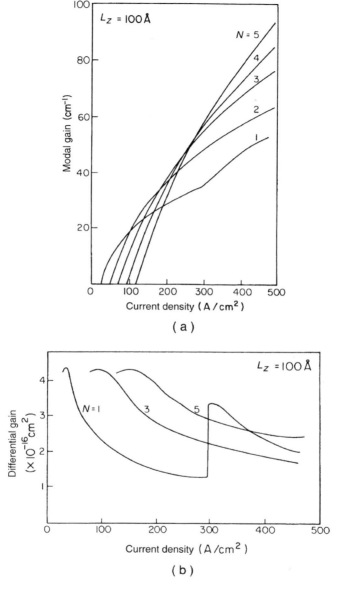

Figure 1.9: The modal gain (a) and the differential gain (b) as a function of injection current density for typical GaAs/AlGaAs QW structures with various numbers of quantum wells. The GaAs QW thickness is 100 Å. The N values shown in the figures are the number of quantum wells. [From Arakawa and Yariv (1985).]

the injected carriers start to populate the second quantized states. The step-jump in the reduced density of states at $E_g + E_{e,z}(2) + E_{hh,z}(2)$ leads to another peak in the gain spectra. Very wide and flat gain spectra are thus obtained because of the carrier state filling into the second quantized states. In Fig. 1.10 the calculated modal gain spectra of a GaAs/AlGaAs single QW laser structure is shown for different injection levels (Mittelstein *et al.*, 1989) for the TE mode. The gain maximum shifts from low to high photon energies (or from long to short wavelengths) as the injection level increases. By decreasing laser cavity length to increase the threshold modal gain, lasing wavelength switch and second quantized state lasing at shorter wavelengths were observed (Mittelstein *et al.*, 1986) in a single QW laser. Broad lasing wavelength range and light emission of wide spectrum can be obtained from QW structures due to the broad gain spectrum available.

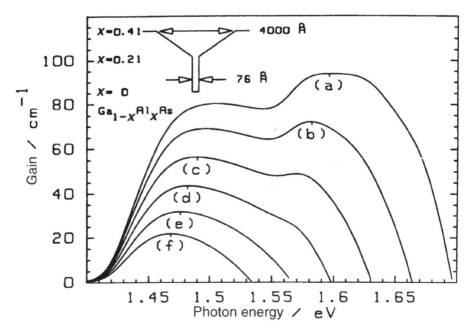

Figure 1.10: A theoretical plot of modal gain spectra at different injection levels for a GaAs/AlGaAs single QW laser structure. The injection current density is (a) 1830, (b) 1350, (c) 950, (d) 630, (e) 405, (f) 225 A/cm^2, respectively. [From Mittelstein *et al.* (1989).]

1.4.2 State filling on threshold current

It is obvious that the carriers populating the second quantized states are wasted when the lasing transitions occur between the first quantized states. These carriers do not contribute to the gain maximum, but they produce part of threshold current because of spontaneous recombination and other nonradiative carrier loss processes. In a practical QW laser structure, separate confinement heterostructure (SCH) is commonly used to achieve the optical confinement and high injection quantum efficiency. In a SCH QW structure, the QW active region is sandwiched between two optical confining layers (OCL), as shown schematically in Fig. 1.11. The OCL thickness is on the order of the optical wavelength of the laser field, i.e., a few thousands angstroms. In comparison with QW thickness, the large OCL thickness results in a large density of states in the OCL region. Obviously, at finite temperatures, the injected carriers populate not only the energy states of the QW but also the energy states of the

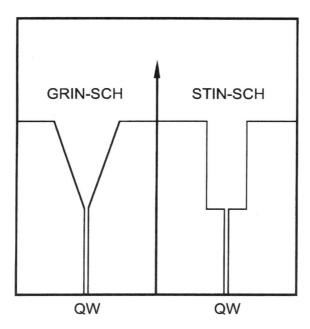

Figure 1.11: Schematic band diagram for graded-index (GRIN) separate confinement heterostructure (SCH) QW structure and step-index (STIN) separate confinement heterostructure (SCH) QW structure.

OCL. These carriers populating the OCL contribute to the threshold current, either radiatively or nonradiatively.

Nagle et al. (1986) first analyzed the effect of carrier population in the optical confining region on the threshold currents in SCH QW laser structures. It was shown that at the threshold, the carrier population in the OCL of a step-index SCH (STIN-SCH) QW structure could be comparable with or significantly larger than the carrier population in the QW as the QW thickness changes from 120 to 60 Å. In comparison, a graded-index SCH (GRIN-SCH) QW structure of the same optical confinement generated a much smaller carrier population in the OCL region at the threshold. The GRIN-SCH QW and STIN-SCH QW structures are shown schematically in Fig.1.11 for comparison. The smaller thickness at the bottom of OCL in GRIN-SCH structures leads to a smaller density of states at the bottom of the OCL, thus resulting in smaller state filling effects. More detailed analysis with additional consideration of other upper energy states, such as indirect L conduction band minima, were given by Chinn et al. (1988) for GRIN-SCH QW lasers and by Nagarajan et al. (1989) for STIN-SCH QW lasers. In addition to larger threshold current, the carrier population in the OCL region could lead to poor temperature characteristics in threshold current (small T_0).

The influences of OCL carrier population on other QW laser performance features were not noticed immediately after its influence on threshold current was analyzed. Despite possibly significant influence on threshold current in some SCH QW lasers, the impact of OCL carrier population was shadowed by the steady progress in reduction of threshold current density and threshold current of QW lasers (Tsang, 1982; Chen et al., 1987, 1990; Derry et al., 1988; Lau et al., 1988; Eng et al., 1989; Kapon et al., 1990; Choi and Wang, 1990; Chand et al., 1991; Williams et al., 1991). The achievements were made by improvement in crystal growth, modification of QW materials (e.g., addition of strain), optimization in optical confinement, improvement in current confinement, etc. The influence of OCL carrier population on high-speed modulation of QW lasers was noticed about 5 years after its impact on threshold current was first analyzed.

1.4.3 A puzzle in high-speed modulation of QW lasers

It had been predicted theoretically that the differential gain of QW lasers should be enhanced by a factor of 2 to 4 in comparison with that of bulk

DH lasers (Burt, 1983; Arakawa and Yariv, 1985, 1986). This differential gain enhancement would be independent of the number of quantum wells [e.g., see Fig. 1.9(b)]. The differential gain enhancement resulted from the quantum effects (or from the steplike density of states) of the two-dimensional injected carriers in the QW. As shown in Sect. 1.3.4, the enhancement of differential gain should lead directly to improvement in high-speed modulation bandwidth.

However, the high-speed modulation experiments by the end of 1992 showed results in contradiction with the theoretical predictions (Lau *et al.*, 1984; Derry *et al.*, 1988; Wolf *et al.*, 1989; Ralston *et al.*, 1991, 1992; Meland *et al.*, 1990; Morton *et al.*, 1992; Lealman *et al.*, 1991; Nagarajan *et al.*, 1991a; Lipsanen *et al.*, 1992; Chen *et al.*, 1992c; Lester *et al.*, 1992; Weisser *et al.*, 1992). These results indicated that no significant improvement in QW laser modulation bandwidth had been observed from the pure quantum effects in comparison with bulk DH lasers. Table 1.6 shows a summary of the representative achievements in modulation bandwidth for various semiconductor lasers by the end of 1992. There exist three definite trends in these results. First, in comparison with bulk laser counterparts, unstrained single quantum well (SQW) lasers do not show significant improvement in the modulation bandwidth. Second, the use of multiple quantum well (MQW) structures as active region leads to improvement in modulation bandwidth over SQW laser counterparts. Third, strained QW lasers always show improved modulation bandwidth in comparison with their unstrained QW laser counterparts. The last fact is due to the strain-induced valence band modification that will be discussed briefly later. Strained bulk semiconductor lasers are not available for comparison because of the limitation from the critical thickness of strained materials. The first two facts were contrary to the previous theoretical predictions on high-speed modulation of QW lasers. The anomalous behavior in high-speed modulation of QW lasers had drawn a lot of attention, and several theoretical models were proposed to explain this puzzle.

Rideout *et al.* (1991) first pointed out that the possible slow carrier capture by QW from the surrounding barriers would cause a spatial hole in carrier distribution at the QW called *well-barrier hole burning*. Similar to spectral hole burning, well-barrier hole burning would cause a damping effect to the high-speed modulation response. The enhanced damping by well-barrier hole burning would degrade the high-speed modulation performance of QW lasers. The carrier capture/escape dynamics between

Materials	Structures	Lasing wavelength	3dB modulation bandwidth	Reference
GaAs/AlGaAs	Bulk	~0.85 μm	11 GHz	Lau et al. (1984)
	SQW	10.85 μm	9 GHz	Chen et al. (1992c)
	3QW	~0.85 μm	15 GHz	Ralston et al. (1991)
InGaAs/AlGaAs	Compressive strained SQW	~0.98 μm	15 GHz	Nagarajan et al. (1991a)
	Compressive strained 4QW	~0.98 μm	28 GHz	Lester et al. (1992)
	Compressive strained p-doped 4QW	~0.98 μm	30 GHz	Weisser et al. (1992)
InGaAsP/InP	Bulk	~1.3 μm	24 GHz	Meland et al. (1990)
	p-Doped 16QW	~1.55 μm	17 GHz	Lealman et al. (1991)
	Compressive strained p-doped 7QW	~1.55 μm	25 GHz	Morton et al. (1992)

Table 1.6: A summary of the representative achievements in high-speed modulation bandwidth in various semiconductor lasers of Fabry-Perot cavity by the end of 1992.

QW and barrier and its influence on modulation response of QW lasers were immediately studied by Kan et al. (1992a, 1992b) and Tessler et al. (1992a, 1992b, 1993). An improved model of well-barrier hole burning further showed that the well-barrier hole burning would degrade the differential gain by a factor of $1 + \eta$, where η is the ratio of capture and escape times for carriers into and out of the QW (Vassell et al., 1992). The QW carrier capture time plays an important role in this model. However, there exists a large variation in its value both theoretically and

1.4. State filling in quantum well lasers

experimentally. For example, the carrier capture time by QW has been reported to be from several picoseconds (Kan et al., 1992a; Morin et al., 1991) to 0.1 to 0.2 ps (Tsai et al., 1994; Hirayama and Asada, 1994).

Nagarajan et al. (1991b, 1992a, 1992b, 1992c) proposed a carrier transport theory showing that the possible slow carrier transport across the SCH region could influence the modulation response significantly by introducing a low-frequency rolloff to the modulation response. The carrier transport can be part of the "slow" carrier relaxation process, which can be characterized by τ_r in Sect. 1.3.3. Equations (1.122) to (1.124) show that the slow carrier relaxation τ_r leads to a low-frequency rolloff to the modulation response. However, τ_r also represents the parasitic effect in the laser devices. For the low-frequency rolloff, the carrier transport and device parasitics are not distinguishable from each other because both effects are related to carrier storage outside the quasi-equilibrium carrier distribution in the SCH QW structures. In the carrier transport model, another impact of slow carrier transport is the reduction of differential gain by a factor of $1 + \eta$, where η is defined as carrier transport (across SCH) and capture (by QW) time devided by carrier escape time out of the QW.

A state-filling theory was developed for understanding the modulation dynamics of QW lasers (Zhao et al., 1991, 1992b–1992d). This theory showed that the carrier population in the SCH OCL region could significantly influence the modulation dynamics of QW lasers by its influence on the differential gain even if the well-barrier hole burning and carrier transport effects were not significant and could be ignored. In the first two models, if the carrier capture and carrier transport are much faster than the carrier escape or emission from the QW, no degradation in differential gain will occur because $\eta \to 0$. The state-filling theory showed that the differential gain could still be degraded significantly by the carrier population in the SCH OCL region.

In addition, it was shown theoretically that the hot carrier effects could have a significant impact on the high-speed modulation bandwidth of QW lasers (Willatzen et al., 1992; Lester and Ridley, 1992). In a semiconductor laser, the carriers are injected from high energy levels into the active region, and the carriers lose their energy, i.e., emit phonons, to reach a quasi-thermal equilibrium distribution in the active region. The emitted phonons make the carrier temperature and the lattice temperature increase. Moreover, if the injected carriers cannot lose their energy fast enough, in comparison with the stimulated emission process and the injected carrier modulation, a non-quasi-thermal equilibrium carrier distribution may result. Similar to the spectral hole burning, this may cause a wide range of spectral distortion to the quasi-equilibrium carrier distribution. These phenomena are called *hot carrier effects*. The temperature increase and the spectral distortion will lead to differential gain degradation and a large damping effect on the modulation response through a large nonlinear gain coefficient. Thus the high-speed modulation performance is degraded.

Each of these different theories captured certain important aspect related to the modulation dynamics of QW lasers. There is one thing common in these theories; i.e., the injected carriers outside the QW can significantly influence the modulation response. Understanding of these different theories would help optimize QW laser structures for better performance. After development of these theories and, of course, with the continuous improvement in crystal growth, SCH QW structure optimization, reduction in laser device parasitics, etc., 3-dB modulation bandwidths of 21 GHz in a GaAs (~0.85 μm wavelength) MQW laser (Dong *et al.*, 1996), 40 GHz in a strained InGaAs (~0.98 μm wavelength) MQW laser (Weisser *et al.*, 1996), 43 GHz in a tunneling injection-strained InGaAs (~0.98 μm wavelength) MQW laser (Zhang *et al.*, 1996), and 30 GHz in long wavelength (1.55 μm) strain-compensated InGaAsP MQW lasers (Matsui *et al.*, 1997; Kjebon *et al.*, 1997) have been achieved.

A review of barrier-hole burning, carrier transport, carrier capture, and escape dynamics has been given by Lau (1993). A thorough discussion on carrier transport theory can be found in Chap. 3 (by Nagarajan and Bowers) of this book. Thus in the following we focus on the state-filling theory to understand the dynamics of QW lasers. This theory can be used easily to explain the two experimental facts in the modulation bandwidth of QW lasers: (1) Unstrained SQW lasers do not show improvement in modulation bandwidth in comparison with their bulk counterparts, and (2) the use of MQW as active region leads to improvement in modulation bandwidth over SQW lasers. We will briefly revisit the hot carrier effects in Sect. 1.5.3.

1.4.4 State filling on differential gain of QW lasers

Intuitive analysis using a simplified model

First, let us examine a simple model that leads to a simple accounting for state-filling effects. The simple two-level model is shown schematically

1.4. State filling in quantum well lasers

in Fig. 1.12. The energy level E_0 represents the first quantized state ($n = 1$ band edge) of the carriers in the QW. The optical transition is assumed to occur between the first quantized states of the electrons and holes in the QW. The energy level E_1 represents the upper energy states in the OCL structure, and D is the effective number of these subbands. At room temperature, the effective number of states, i.e., the number of energy states within one kT from the top of the QW in the OCL structure, is on the order of 20 for a typical SCH QW structure, where T is the temperature and k is the Boltzmann constant. And these states are on the order of 80

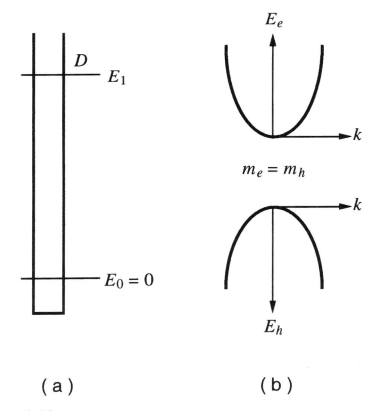

Figure 1.12: A simple two-level model accounting for state-filling effects: (a) the QW ground lasing state at E_0 and D number of upper energy states at E_1; (b) the electronic band structures for the electrons and the holes are assumed to be the same for simplicity.

meV above the first quantized state in a typical QW laser. For the sake of simplicity, we also assume that the electrons and holes have the same parabolic electronic band structure and effective mass ($m_e = m_h$) in this simple model.

Neglecting dephasing collision (T_2) broadening, the maximum optical gain (occurring at band edge $E = E_0$) can be written as

$$G_0 = A_0 \rho_r (2f_e - 1) \tag{1.135}$$

where A_0 is a material-dependent parameter, $\rho_r = \frac{1}{2}(m_e/\pi\hbar^2) = \frac{1}{2}\rho_e$ is the reduced density of states for the first quantized states, and the Fermi function is given by

$$f_e = \frac{1}{1 + \exp\left(-\dfrac{F}{kT}\right)} \tag{1.136}$$

F is the quasi-Fermi energy level of electrons measured from the first quantized state, i.e., assuming $E_0 = 0$. The 2D carrier density is obtained by integrating the density of states and the Fermi function:

$$N = \rho_e kT \left\{ \ln\left[1 + \exp\left(\frac{F}{kT}\right)\right] + D \ln\left[1 + \exp\left(\frac{F - E_1}{kT}\right)\right] \right\} \tag{1.137}$$

From Eqs. (1.135) to (1.137) it is straightforward to derive the differential gain:

$$\frac{dG}{dN} = \frac{(A_0/kT)/[1 + \exp(F/kT)]}{1 + D[1 + \exp(F/kT)]/[\exp(E_1/kT) + \exp(F/kT)]} \tag{1.138}$$

For a given value of gain G_0 (i.e., a given value of F), Eqs. (1.137) and (1.138) show clearly that the presence of (D) upper subbands increases the injected carrier density and reduces the differential gain. They also indicate that a large separation between the first quantized state and upper subbands (i.e., a large E_1 parameter) diminishes the influence of OCL states, resulting in a smaller carrier density and a larger differential gain. Certainly, reduction in the number of upper subbands (D) leads to the same conclusion.

More accurate theory and comparison with bulk lasers

Using the band and optical gain theory discussed in Sect. 1.3, the 2D gain coefficient and carrier density can be calculated more accurately for QW

1.4. State filling in quantum well lasers

laser structures. Figure 1.13 shows the calculated peak gain G_0 and the corresponding differential gain G_0' for a typical GaAs/AlGaAs SCH SQW structure using the more accurate theory. Results with and without inclusion of carrier population in the SCH OCL region are compared in Fig. 1.13. It is shown that the exclusion of carrier population in the SCH OCL region leads to an overestimation of the differential gain and an underestimation of the transparency carrier density. The maximum differential gain is overestimated by about a factor of 2 if the injected carrier population in the OCL region is omitted.

In the following we compare the differential gain in typical bulk and QW lasers with consideration of the state-filling effects in the QW laser structures. For 3D DH bulk lasers, their characteristics can be described by the same set of rate equations [Eqs. (1.112) to (1.114)] with J replaced by J/L_z, where L_z is the active layer thickness of the DH laser structures. N and N' are the corresponding 3D carrier densities in the bulk laser structures. The gain coefficients G_0 and G_1 are given by Eqs. (1.115) and (1.116) with $\delta_j = 1$ ($j = l, h$), and ρ_{rj} is the corresponding 3D reduced density of states:

$$\rho_{rj}(\mathcal{E}) = \frac{1}{2\pi^2} \left(\frac{2m_{rj}}{\hbar^2}\right)^{3/2} \sqrt{\mathcal{E} - E_g} \quad (1.139)$$

where m_{rj} are the reduced effective masses ($j = l, h$). The coupling factor for the bulk DH laser structures is the conventional optical confinement factor, which can be obtained from Eq. (1.55) and Table 1.2 as

$$\Gamma = \Gamma_{DH} = \Gamma_{3D} = \frac{1}{wt_z} \cdot \frac{wL_z}{1} = \frac{L_z}{t_z} \quad (1.140)$$

In Sect. 1.3.3 we showed how to evaluate the effective optical mode width t_z. After the appropriate terminology change shown above for bulk DH lasers, all the results obtained for threshold and modulation bandwidth of QW lasers in Sect. 1.3.4 are now valid for bulk DH lasers [Eqs. (1.120) to (1.132)]. In Eqs. (1.120) to (1.132), the modal gain ΓG_0 and ΓG_1 and the differential gain $G_0' = dG_0/dN$ have the same units, respectively, for 2D QW and 3D bulk DH lasers. Thus their performance characteristics can be compared with each other.

Figure 1.14 shows the calculated differential gain G_0' at the peak of gain spectrum $G_0(E)$ as a function of peak modal gain $g = \Gamma G_0$ for typical GaAs/AlGaAs bulk DH lasers and SCH QW lasers of single quantum well

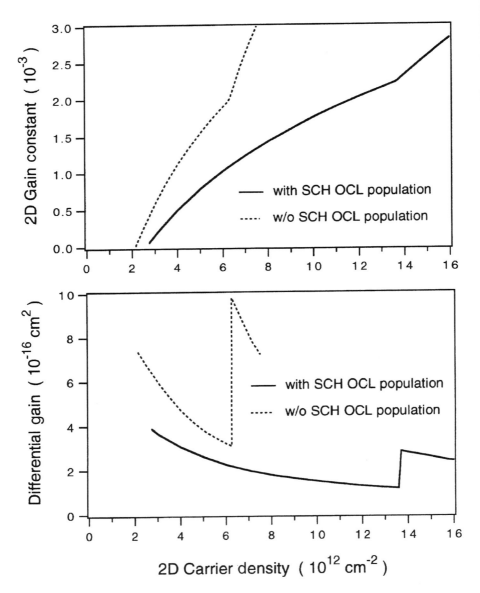

Figure 1.13: Calculated 2D peak gain G_0 (top) and corresponding differential gain G_0' (bottom) as a function of carrier density in a typical GaAs/AlGaAs SCH 100-Å single QW laser with and without consideration of the state-filling effects. The calculation was made by using the decoupled valence band and optical gain theory discussed in Sect. 1.3.

1.4. State filling in quantum well lasers

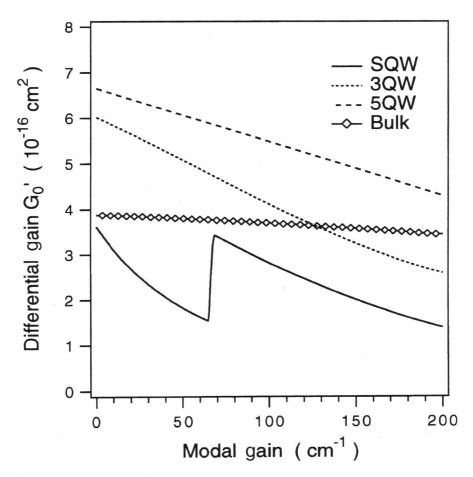

Figure 1.14: The differential gain G_0' at gain peak as a function of peak modal gain $g = \Gamma G_0$ for typical GaAs/AlGaAs bulk DH and QW laser structures with different numbers of quantum wells.

(SQW), three quantum wells (3QW), and five quantum wells (5QW). The calculations were made for the dominant TE modes in these structures at room temperature. The typical bulk DH laser structure is assumed to be a GaAs/Al$_{0.35}$Ga$_{0.65}$As DH structure with a GaAs active layer thickness of 0.1 μm. The confinement factor (or coupling factor) for the bulk DH laser structure was estimated as $\Gamma_{DH} = 0.33$. The typical SCH QW laser structures used for comparison consist of a pair of 2000-Å $Al_{0.5}Ga_{0.5}As/Al_{0.25}Ga_{0.75}As$ parabolic-shaped GRIN-SCH layers and 100-Å quantum well or quantum wells located near the center of the GRIN-SCH structure. The coupling factor for one QW was estimated as $\Gamma_{QW} = \Gamma_{2D} = 3.9 \times 10^4$ cm^{-1}.

Figure 1.14 shows that the differential gain of the SQW structure is comparable with that of the DH structure in the very low modal gain region. In the high modal gain region, the DH structure possesses higher differential gain. The decrease in G'_0 with increasing modal gain in the SQW structure results from the finite available gain per quantized state. The abrupt increase in G'_0 of the SQW structure is due to the onset of the second quantized state lasing.

The qualitative physical reason for the smaller differential gain in SQW structures is as follows: Due to state filling, an increasing fraction of the injected carriers goes into occupation of the large density of states in the SCH OCL region. These electrons (and holes) contribute very little to the peak gain because of the $(\mathscr{E} - E)^2$ term in the denominator of $\mathscr{L}(\mathscr{E} - E)$ in Eq. (1.115). On the other hand, the optical gain (at peak) is limited by the flat feature of the 2D steplike density of states of the QW structure. The result is a lowering of differential gain $G'_0 = dG_0/dN$ in the SQW structure. However, it was shown that in comparison with the DH structure, there is a differential gain enhancement in the SQW structure at low temperatures (Zhao et al., 1991). At lower temperatures, the Fermi-Dirac occupation factor for the energy states in the SCH OCL region is reduced considerably. The more abrupt cutoff of the Fermi-Dirac occupation functions at low temperatures take advantage of the step-jump 2D density of states profile in contributing to relatively large increase in G_0 with increasing injected carrier density N. Thus the differential gain enhancement due to the quantum effect of injected carriers in SQW laser structures is recovered at low temperatures.

Additional differential gain enhancement in MQW lasers

In QW structures, above transparency, the differential gain decreases as the carrier density (or the modal gain) increases because of the sublinearity in the gain versus carrier density dependence. This sublinearity is attributed to the flat feature of the 2D steplike density of states of QW structures. For a given value of modal gain, the differential gain increases as the number of quantum wells (N_{qw}) increases because the carrier density in one QW decreases with the increase of N_{qw}. This is the conven-

1.4. State filling in quantum well lasers

tional differential gain enhancement in MQW structures. Without consideration of the state-filling effect, it was predicted that larger differential gain should be achieved at lower injection levels and that the maximum attainable differential gain was independent of the number of quantum wells N_{qw} [see Fig. 1.9(b)]. Figure 1.14 indicates that the inclusion of OCL state filling considerably modifies this picture. It is shown that the use of multiple quantum wells (MQW) as active region actually leads to an additional differential gain enhancement. More specifically, the maximum attainable differential gain increases as the number of quantum wells increases. An $N_{qw} = 5$ QW laser with a threshold modal gain of 40 cm^{-1}, for example, will have a differential gain that is appreciably larger than the *maximum* attainable differential gain in an SQW laser and nearly three times as large as an SQW laser with the same threshold modal gain value. At the threshold modal gain of 40 cm^{-1}, the differential gain of an MQW laser with $N_{qw} = 5$ is about 60% larger than that in a DH laser.

The physical reason for the additional increase in differential gain in MQW structures is the following: In an SQW laser the need to obtain a sufficient optical gain (to overcome the losses) forces the Fermi energy F to rise toward the top of the quantum well as the pumping level is increased. The OCL states, say, at energy E_{ocl}, whose occupation is roughly determined by the factor $\exp[-(E_{ocl} - F)/kT]$ are thus more heavily populated. The increase in OCL carrier density contributes negligibly to the increase in peak gain because of the nonresonant nature of these carriers' transitions with the lasing transitions at the peak gain. This leads to reduced differential gain, since part of the injected carriers are wasted. The wasted carriers in the OCL region must be there due to the fundamental quasi-Fermi-Dirac statistics. In an MQW laser, there are several active quantum wells contributing to the gain. The necessary total gain is reached with a much lower population in any one well. This results in a much lower F, and consequently, the OCL states occupation factor $\exp[-(E_{ocl} - F)/kT]$ is smaller.

Using the simple two-level model accounting for the state filling, one can see the underlying physics more clearly. In the simple model, it is straightforward to get the differential gain for MQW lasers as

$$\frac{dG_0}{dN} = \frac{(A_0/kT)/[1 + \exp(F/kT)]}{1 + (D/N_{qw})[1 + \exp(F/kT)]/[\exp(E_1/kT) + \exp(F/kT)]} \quad (1.141)$$

Note that the increase in dG_0/dN with increasing number of wells (N_{qw}) is due to the explicit dependence on N_{qw} as well as to the implicit dependence on F, since F decreases with increasing N_{qw} for a given modal gain.

The explicit term D/N_{qw} in Eq. (1.141) shows that the penalty due to the OCL state filling is reduced by "distributing" it among the N_{qw} quantum wells in the MQW structures.

An analysis on the influence of coupling between the quantum wells on the differential gain was given by Akhtar and Xu (1995) for GaAs/AlGaAs MQW laser structures. It was shown that the influence was negligible for QW width of 100 Å. For narrow QW width of 30 Å, narrow barrier thickness (<50 Å) would lead to a reduction in differential gain. As the QW coupling increases with the decrease of the barrier thickness, the nearly degenerate ground state of the MQW structures becomes non-degenerate and splits into sub-energy states. For narrow QW and large QW coupling, the energy splitting of these sub-energy-states is large. For low-threshold modal gain, only the carriers populating the lowest sub-energy state contribute to the gain. In such a case, an MQW laser is equivalent to an SQW laser. Thus the differential gain is smaller due to the state filling effects.

Comparison with experimental achievements in modulation bandwidth

The preceding theoretical results on differential gain agree well with experimental observations. The nonenhancement of differential gain in unstrained GaAs SQW lasers explains why modulation bandwidth enhancement was not observed in the SQW lasers in comparison with the bulk DH lasers. A 3-dB modulation bandwidth of 11 GHz was demonstrated in GaAs/AlGaAs DH bulk lasers (Lau et al., 1984). However, for GaAs/AlGaAs SQW lasers, the typical modulation bandwidth was just about 6 GHz (Derry et al., 1988). Using higher QW barriers to reduce the state-filling effect, the modulation bandwidth was improved to 9 GHz in a GaAs SQW laser (Chen et al., 1992c). The additional differential gain enhancement in the MQW lasers explains the observed differential gain enhancement in MQW structures (Uomi et al., 1985, 1987; Takahashi et al., 1991) and the improved high-speed modulation performance in MQW lasers over their SQW and DH counterparts. A CW 3-dB modulation bandwidth of 15 GHz was reported previously in a GaAs/AlGaAs MQW laser (Ralston et al., 1991). Recently, a 3-dB modulation bandwidth of 21 GHz was demonstrated in a GaAs/AlGaAs MQW laser (Dong et al., 1996). The state-filling theory can be applied to other material systems, such as InGaAs and InGaAsP, to understand the similar experimental trends in modulation bandwidth of bulk DH, SQW, and MQW lasers.

1.4.5 State filling on spectral dynamics

In semiconductor lasers there is a fundamental spectral linewidth enhancement resulting from the coupling between amplitude and phase fluctuations of the optical field that were induced by the spontaneous emission. This leads to the modified Schawlow-Townes formula for the spectral linewidth of semiconductor lasers (Henry, 1982; Vahala and Yariv, 1983a, 1983b):

$$\Delta \nu = \frac{v_g^2 h \nu g \alpha_m n_{sp} (1 + \alpha^2)}{8 \pi P} \quad (1.142)$$

where $h\nu$ is the lasing energy, g is the modal gain, α_m is the mirror loss, P is the optical output power, n_{sp} is the spontaneous emission factor, and α is the *linewidth enhancement factor* that reflects the amplitude-phase coupling in the lasers. The α parameter is also called the *amplitude-phase coupling factor*. The name *linewidth enhancement factor* emphasizes that it leads to a spectral linewidth enhancement. The name *amplitude-phase coupling factor* emphasizes its physical origin.

The amplitude and phase coupling is also the major factor leading to undesired frequency modulation (FM) (i.e., wavelength or frequency chirping) caused by amplitude modulation (AM) of semiconductor lasers (Harder *et al.*, 1983; Koch and Bowers, 1984; Koch and Linke, 1986). A detailed review of the α parameter and its influence on various spectral properties of semiconductor lasers was given by Osiński and Buus (1987). Reduction of spectral linewidth and frequency chirping in semiconductor lasers is important for optical communications. Narrower spectral linewidth and smaller frequency chirping enable a larger number of communication channels by reducing the pulse spreading in digital systems and reducing signal distortion in analog systems, which are all influenced by the fiber dispersion. The α parameter is of fundamental importance in determining the spectral linewidth and frequency chirping and plays a crucial role in the efforts to control the spectral properties of semiconductor lasers. Vahala and Newkirk (1990) have shown a method to reduce the amplitude noise or phase noise (frequency noise) by decorrelation of the amplitude-phase coupling in semiconductor lasers.

It was predicted theoretically that QW lasers would be good candidates to achieve narrow spectral linewidth because of the predicted reduction in the α parameter compared with bulk lasers (Burt, 1984; Arakawa *et al.*, 1984; Arakawa and Yariv, 1986). In addition, the SQW lasers were expected to be superior to their MQW counterparts in spectral linewidth because of the smaller spontaneous emission factor n_{sp} in the SQW lasers (Arakawa and Yariv, 1985). The reduction in the α parameter in the QW lasers was attributed to the differential gain enhancement compared with bulk lasers. Since the differential gain is influenced by the state filling in QW lasers, the α parameter should be affected by the state filling as well (Zhao et al., 1993d). In the following we discuss the influence of state filling on the amplitude-phase coupling and spectral linewidth of QW lasers.

Instead of the Kramers-Kronig approach, the linewidth enhancement factor α can be obtained by direct calculation of the real and imaginary parts of the complex susceptibility (Vahala et al., 1984). The α factor is defined as the ratio of the real and imaginary parts of the susceptibility variations, i.e.,

$$\alpha = \frac{\partial \chi_{r0}}{\partial \chi_{i0}} = \frac{\partial N_{r0}}{\partial G_0} = \frac{\partial N_{r0}/\partial N}{\partial G_0/\partial N} \qquad (1.143)$$

where χ_{r0}, χ_{i0}, G_0, and N_{r0} can be obtained from Eqs. (1.40), (1.41), (1.57), and (1.58) with ρ_{ii} replaced by the Fermi functions f_i ($i = e, h$) to account for the linear term only. N is the injected carrier density. $\partial N_{r0}/\partial N$ can be called the *differential index*. Using the same derivation as in Eq. (1.57) to Eq. (1.115) for G_0, an expression for N_{r0} can be obtained from Eq. (1.58):

$$N_{r0}(E) = \frac{\mu^2}{\hbar c \epsilon_o n_r} \frac{E}{E_{T_2}} \sum_{i=l,h} \int \delta_i \rho_{ri}(f_e + f_h - 1) \mathscr{L}'(\mathscr{E} - E) \, d\mathscr{E} \qquad (1.144)$$

where

$$\mathscr{L}'(\mathscr{E} - E) = \frac{E_{T_2}(\mathscr{E} - E)}{E_{T_2}^2 + (\mathscr{E} - E)^2} \qquad (1.145)$$

Using Eqs. (1.143), (1.115) and (1.144), the α parameter can be calculated for different laser structures. Figure 1.15(a) shows the calculated α parameter at gain peak as a function of modal gain for typical GaAs/AlGaAs QW and bulk DH lasers. The typical QW and bulk DH structures are the same ones used previously in comparing the differential gain (Sect. 1.4.4). The calculation is made also for the dominant TE mode at room temperature. Figure 1.15(a) shows that there is no reduction of α parameter in

1.4. State filling in quantum well lasers

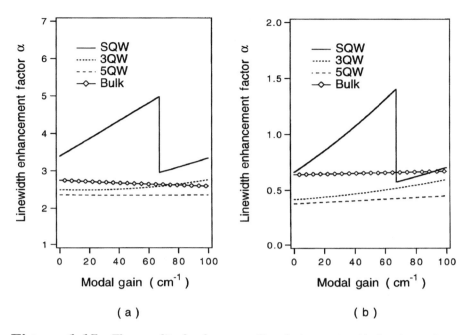

Figure 1.15: The amplitude-phase coupling factor α at optical gain peak as a function of the modal gain in bulk DH and QW lasers with different numbers of quantum wells: (a) the interband transition component α_{tr} and (b) the free-carrier component α_{ca}.

the SQW structure. On the contrary, the α parameter in the SQW structure is much larger than that in the bulk DH structure.

The larger α parameter in the SQW structure is due to the carrier state filling in the OCL region. The carrier fluctuation caused by spontaneous emission leads to a corresponding change in quasi-Fermi levels. The change in quasi-Fermi levels, in turn, results in a quite different carrier population change in the QW and the OCL regions. At reasonably high injection levels, the carrier population change in the QW is small due to the flat feature of 2D density of states. The changes in G_0 and N_{r0} caused by the carrier population change in the QW are thus small. On the other hand, the carrier population change in the SCH OCL region is relatively large because of the large number of states. The carrier population change in the OCL region contributes relatively little to the change in peak G_0 because $\mathscr{L}(\mathscr{E} - E)$ in G_0 [Eq. (1.115)] introduces a nonresonance between the OCL energy states and the lasing transitions at the peak G_0. The resulting change in the denominator of Eq. (1.143) is thus small. However, the carrier population change in the OCL region contributes a relatively large amount to the change of N_{r0} because of the reduced nonresonance between the OCL states and the lasing transitions in N_{r0}. For N_{r0} [Eq. (1.144)], the linear term $(\mathcal{E} - E)$ in the numerator of $\mathscr{L}'(\mathcal{E} - E)$ reduces the nonresonance, and $\mathscr{L}'(\mathcal{E} - E)$ has a long tail at high energy [see Eq. (1.145)]. Thus a large α parameter results. Apparently, a large carrier population in the OCL region will enhance this effect. The abrupt drop in the α curve for the SQW structure in Fig. 1.15(a) is due to the onset of second quantized state lasing, which leads to a differential gain enhancement. Figure 1.15(a) also shows that the use of MQW leads to reduction in the α parameter. This is attributed to the reduction in state filling in the OCL region, as discussed before, which comes along with an additional differential gain enhancement in the MQW lasers. However, in comparison with the bulk DH structure, the reduction in α in the unstrained MQW structures is not significant.

In semiconductor lasers, the amplitude-phase coupling factor α consists of an interband-transition component α_{tr}, as discussed above and shown in Fig. 1.15(a), and a free-carrier component α_{ca}:

$$\alpha = \alpha_{tr} + \alpha_{ca} \tag{1.146}$$

The free-carrier component α_{ca} stems from the plasma effect of the injected carriers in the laser structures. Henry et al. (1981) showed that the free-carrier component α_{ca} is much smaller than the interband-transition component α_{tr} in bulk structure lasers. Yamada and Haraguchi (1991) showed theoretically that the carrier fluctuation in the OCL region can lead to a significantly large free-carrier component of α in some SCH QW laser structures. The calculated α_{ca} at gain peak is plotted as a function of the modal gain in Fig. 1.15(b) for the assumed typical GaAs/AlGaAs QW and bulk DH lasers (Sect. 1.4.4). From Fig. 1.15 we find that α_{tr} is the dominant component in α for the assumed typical bulk DH and QW lasers. Due to the state-filling effect, the free-carrier component α_{ca} is larger in the SQW structure and is smaller in the MQW structures compared with bulk structures.

Equation (1.142) shows that in addition to the α parameter, the spontaneous emission factor n_{sp} also plays an important role in determining the spectral linewidth of semiconductor lasers. n_{sp} is the transition rate ratio of the spontaneous emission, coupled into the lasing mode, to the stimulated emission and is given by

$$n_{sp}(E) = \frac{\sum_{j=l,h}\int \delta_j \rho_{jr} f_e f_h \mathscr{V}(\mathscr{E}-E)\, d\mathscr{E}}{\sum_{j=l,h}\int \delta_j \rho_{jr}(f_e + f_h - 1)\mathscr{V}(\mathscr{E}-E)\, d\mathscr{E}} \quad (1.147)$$

The calculated spontaneous emission factor (TE mode) at the gain peak as a function of the modal gain for the assumed GaAs QW and bulk DH lasers is plotted in Fig. 1.16. It shows that QW lasers possess lower n_{sp} than bulk DH lasers. The SQW lasers have the smallest n_{sp}. This is due to the flat 2D density of states in the QW structures, which makes it necessary to have a larger inversion in the QW structures compared with bulk DH structures in order to achieve the same value of modal gain.

As shown in Eq. (1.142), $n_{sp}(1 + \alpha^2)$ can be used as the figure of merit for comparing the spectral linewidth between bulk and QW lasers. Since α decreases and n_{sp} increases in QW lasers as the number of quantum wells increases, there might exist an optimal quantum well number for achieving the narrowest spectral linewidth. To compare the linewidth enhancement in different laser structures, the plot of $n_{sp}(1 + \alpha^2)$ at the gain peak as a function of the modal gain is shown in Fig. 1.17, where both the interband-transition component and the free-carrier component of α are included. Figure 1.17 shows that the MQW lasers possess narrower linewidth compared with bulk lasers, and this is attributed mainly to the smaller n_{sp} in the MQW lasers. The linewidth enhancement of SQW lasers can be either smaller or larger than that of bulk lasers depending on the modal gain.

1.5 Reduction of state filling in QW lasers

In the preceding analysis, the influence of the upper energy bands on the the differential gain and carrier density has been discussed for QW lasers. The state-filling effects directly affect the threshold current and modulation dynamics due to their influence on the quasi-Fermi energies of injected carriers. The differential gain and the transparency current both depend on the increase rate of the quasi-Fermi energies with increasing injected carrier density. The presence of upper subbands with large density of states tends to clamp the Fermi energies, thus leading to low differential gain and high transparency current. The carrier population of the upper

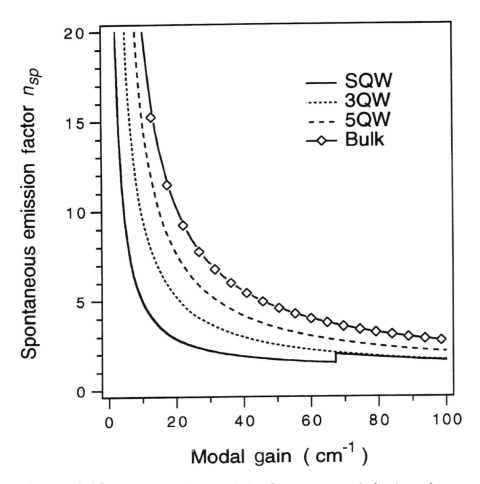

Figure 1.16: The spontaneous emission factor n_{sp} at optical gain peak as a function of the modal gain in bulk DH and QW lasers with different numbers of quantum wells.

energy subbands also opens various channels for carrier loss either radiatively or nonradiatively. In order to achieve a high value of differential gain (and thus high-speed modulation bandwidth) and a low value of injected carrier density to reach the necessary threshold gain (and thus low threshold current) in QW lasers, a major aim of QW laser design is to effectively increase the separation between the energy states for the laser transitions and other upper energy states (the states in either the

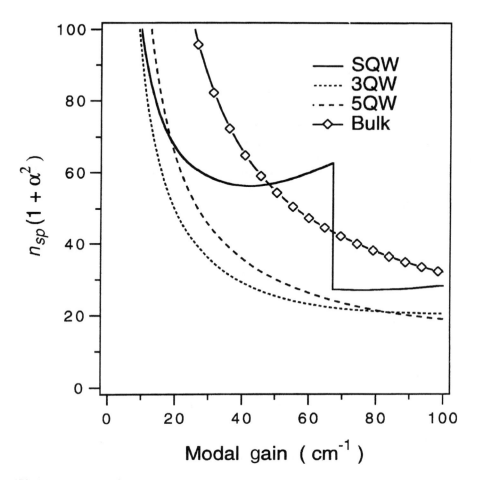

Figure 1.17: The figure of merit for the spectral linewidth enhancement in semiconductor lasers $n_{sp}(1 + \alpha^2)$ at optical gain peak as a function of the modal gain in bulk DH and QW lasers with different numbers of quantum wells.

SCH OCL region or the other subbands of the QW). In the following we discuss the methods that can reduce the state filling effects in QW lasers.

1.5.1 Multiple quantum well structures

As shown in Sects. 1.4.4 and 1.4.5, the use of an MQW structure as laser active region leads to a reduction of the state-filling effects on the modulation and spectral dynamics. In MQW structures, there is a larger fraction of injected carriers populated in the quantum wells and a smaller fraction of injected carriers "wasted" in the optical confining region in comparison with SQW structures. However, the penalty is that the threshold current density increases in MQW lasers at low value of threshold gain. For low value of threshold gain, the transparency current density is the major part of the threshold current density. The transparency current density in MQW structures is larger than that in SQW structures because of the multiple active region elements, i.e., the quantum wells, in MQW structures. However, it is necessary to point out that the transparency current density (J_{tr}) in MQW structures does not scale as $J_{tr}^{MQW} = N_{qw} J_{tr}^{SQW}$. The J_{tr} increase in MQW lasers is smaller than this scaling. This is due to the fact that *at transparency* the carrier population in the OCL region is almost the same in SQW and MQW lasers of the same OCL structures because the quasi-Fermi energies are about the same for SQW and MQW lasers *at transparency*. Thus the J_{tr} part caused by the radiative and nonradiative carrier loss in the OCL region does not scale by the number of quantum wells N_{qw}. Neither does the total J_{tr}.

To achieve high modulation bandwidth, the operation conditions (bias current and output power) to reach the maximum bandwidth must be considered as well. It has been shown that MQW structure is superior to SQW structure because the operating injection current density and optical power needed to reach the maximum modulation bandwidth are smaller in MQW lasers (Zhao et al., 1992a). Generally, the thermal degradation is very severe if the injection current density is beyond 10 kA/cm^2 in a semiconductor laser. The thermal effect will reduce the differential gain and thus the modulation bandwidth. The smaller operating optical power will reduce the damage to laser structures under high optical intensity. Essentially, the use of MQW as laser active region makes it much easier to approach the high-speed bandwidth limits by reduction of thermal effects and optical damage.

1.5.2 Quantum well barrier height

Equations (1.138) and (1.137) indicate that the increase in the energy separation (the parameter E_1) between the QW's ground lasing states and the large number of states in the OCL region will result in an increase in the differential gain and a reduction in the carrier population necessary to reach a certain threshold gain. Experimental investigations have been

1.5 Reduction of state filling in QW lasers

carried out to study the influence of QW barrier height or QW depth on the modulation dynamics in QW lasers by using different compositions for the optical confining layers (Zhao *et al.*, 1992d; Nagarajan *et al.*, 1992b). In QW lasers, the differential gain is not constant due to the sublinearity in the gain and carrier density relation. Caution must be taken when comparing the differential gain in different QW structures. There are a number of factors that can lead to quite different values of differential gain in QW lasers. For example, given the same threshold modal gain (e.g., the same cavity length, facet reflectivity, and internal loss) for QW lasers of different optical confinement, the differential gain could be different just because the carrier population in the QW is different. Note that the peak gain is mainly contributed by the injected carriers within the QW, and therefore, the different optical confinement leads to different material gain and thus a different carrier population in the QW for the same threshold modal gain. In such a case, even if other effects are not important to influence the differential gain, the differential gain is different because of the sublinearity in the gain and carrier density relation.

Two QW laser structures have been designed and fabricated to differentiate the state-filling effect on differential gain from other factors (Zhao *et al.*, 1992d). As shown schematically in Fig. 1.18, the two GaAs/AlGaAs GRIN-SCH SQW laser structures, referred to as sample A and B, were grown by MBE on (100) n^+-GaAs substrates. The GaAs QW thickness is 100 Å in both structures. The profiles and dimensions of the two GRIN-SCH structures are identical, as are the doping levels and profiles. The only difference between the two structures is the different Al composition in the GRIN-SCH OCL region. The identical profiles and dimensions of the GRIN-SCH OCL structures ensure a nearly identical optical confinement and similar carrier transport effect in these two structures. The slightly higher hole mobility in AlGaAs of lower Al concentration (Masu *et al.*, 1983) makes the carrier transport slightly better in structure B. The carrier transport effect in QW lasers is dominated by hole transport because the holes possess much lower mobility compared with the electrons. The main objective of the laser structure design was that the different Al composition in the GRIN-SCH OCL structures results in a quite different energy separation between the quantized states of the QW and the large number of states in the GRIN-SCH OCL region, which, in turn, results in very different state-filling effects in these two structures. For example, the energy separation between the first quantized state of the QW and the lowest energy state of the GRIN-SCH OCL region was esti-

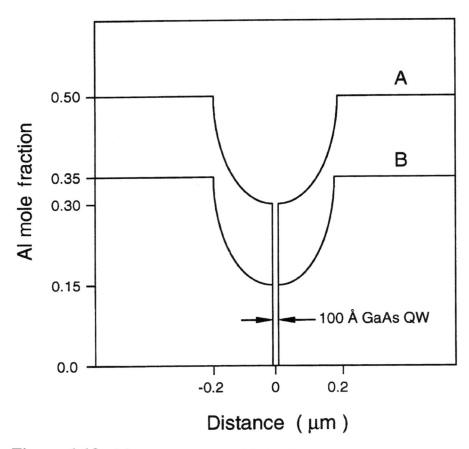

Figure 1.18: Schematic structures of GaAs/AlGaAs GRIN SCH SQW lasers used to investigate state-filling effects.

mated to be 225 meV in structure A and 102 meV in structure B for the electrons in the conduction band.

High-frequency modulation measurements were carried out for buried heterostructure (BH) lasers of structure A and structure B, which were fabricated by a liquid-phase epitaxy (LPE) regrowth technique. The internal loss constants were obtained as 8 cm^{-1} (structure A) and 12 cm^{-1} (structure B) from the measured differential quantum efficiency in these lasers. Figure 1.19 shows the square of the measured response peak frequency f_p as a function of the optical output power P for two typical

1.5 Reduction of state filling in QW lasers

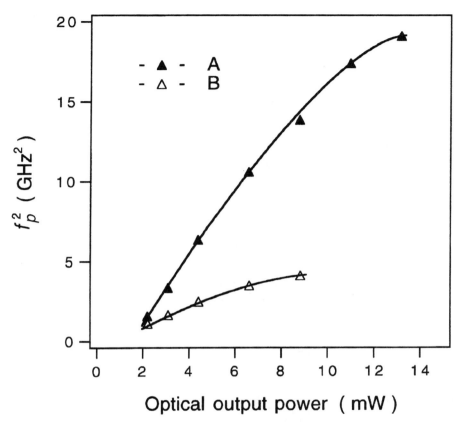

Figure 1.19: The square of the measured modulation-response peak frequency as a function of the optical output power for QW lasers of different GRIN SCH structure. The cavity lengths are 300 μm, and facets are uncoated.

lasers of the same cavity length 300 μm. The $f_p^2 - P$ curves do not go through the zero point due to the presence of the constant damping term in the modulation response. The mirror loss in these two lasers of cavity length 300 μm is about 40 cm^{-1}. From the measured internal loss constants, these two lasers have almost the same value of threshold modal gain. Thus the nearly identical optical confinement Γ ensured that the two lasers were biased at such a point that the carrier population gave almost the same peak material gain values G_0. Since the peak gain value G_0 is contributed mainly by the carriers in the QW, there exist roughly the same carrier populations in the QW for the two lasers. If other effects are not important to influence the differential gain, these two lasers should have nearly the same value of differential gain. However, the measured differential gain values, obtained from the $f_p^2 - P$ curves, were 2.8×10^{-16} cm^2 for the laser of structure A and 0.9×10^{-16} cm^2 for the laser of structure B.

The observed differential gain difference is due to the state-filling effect. In the laser of structure B, there is an appreciable carrier population in the SCH OCL region due to the smaller energy separation between the quantized states of the QW and the states in the SCH OCL region. The significant difference in the carrier population of the SCH OCL region leads to the very different differential gain values in the two lasers. In Fig. 1.20 the theoretically estimated ratio of the differential gain in these two structures is given as a function of the modal gain, where the state-filling effect has been taken into account. It is shown that the differential gains can differ by as much as a factor of 3.5 at the modal gain of 50 cm^{-1}, which is quite consistent with the experimental results. In addition, the difference in OCL carrier population is also responsible for the observed difference in threshold current of 10 mA (A) and 14 mA (B) in these two lasers.

The difference in differential gain leads to different modulation bandwidths in these two structures. The modulation bandwidth in the lasers of structure B is severely limited by the state-filling effect. The typical 3-dB modulation bandwidth were 3 to 4 GHz. For structure A, a 3-dB modulation bandwidth in excess of 9 GHz was obtained, which is the largest modulation bandwidth in unstrained GaAs SQW lasers with a uniform current pumping (Chen *et al.*, 1992c). These results bear directly on the design consideration of QW lasers for high-speed performance.

1.5.3 Separate confinement structures

In addition to the QW barrier height, the shape and dimensions of the SCH optical confining structure also influence the state-filling effect. First, the density of states in the SCH OCL structure is directly influenced by the shape and dimensions of the SCH OCL structure. Narrow SCH OCL structure reduces the number of states in the OCL region, thus leading to a reduction of carrier state filling in the OCL region. Second, the shape and the dimensions of the SCH OCL structure affect the optical confinement or the coupling factor Γ. For the same value of threshold

1.5 Reduction of state filling in QW lasers

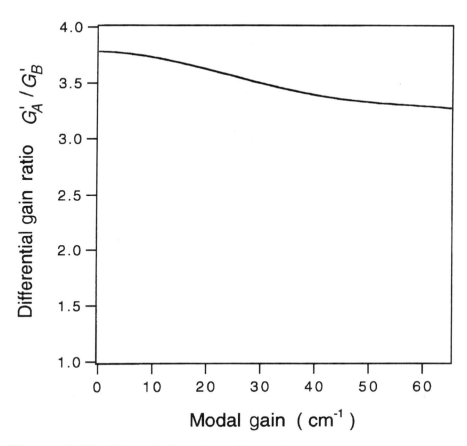

Figure 1.20: Theoretically estimated ratio of the differential gains between the two different GRIN SCH QW laser structures (shown in Fig. 1.18) as a function of modal gain.

modal gain, tighter optical confinement (a larger Γ) makes the threshold modal gain be reached at a lower injection carrier density within the QW. The lower Fermi energy levels lead to a smaller carrier population in the OCL region. Thus a larger value of differential gain can be obtained due to the higher differential gain at lower carrier density within the QW and the smaller carrier state filling in the OCL region. It was shown experimentally that the tight optical confinement and smaller density of states in narrow OCL structures resulted in threshold current reduction and modulation bandwidth enhancement (York et al., 1990; Bour et al.,

1990; Nagarajan *et al.*, 1992c). Third, a narrow SCH OCL region reduces the carrier transport effect, as evident in some laser structures (Nagarajan *et al.*, 1992c).

A new type of QW laser, called a *tunneling injection (TI) laser*, has been developed recently (Sun *et al.*, 1993; Lutz *et al.*, 1995). The band diagram of the TI-QW laser structure is shown schematically in Fig. 1.21. A resonant tunneling structure, such as the thin barrier layer shown in Fig. 1.21, is grown next to the QW on the side of n-type doping. The electrons are injected into the QW active region by optical phonon-assisted resonant tunneling through the barrier. The injected holes do not tunnel easily through this barrier layer because of the high scattering rate and larger effective mass of the holes. Since the barrier layer is very thin (typically <100 Å), its influence on optical confinement in the laser structure is negligible. Because the electrons are injected into the QW active region at a resonant tunneling energy just slightly above the lasing energy states of the QW, the hot carrier effects are reduced in TI-QW lasers. Recent experimental and theoretical work has shown that carrier heating could be more severe in conventional QW lasers than that in bulk lasers (Girardin *et al.*, 1995; Tsai *et al.*, 1996; Wang and Schweizer, 1996; Grupen and Hess, 1997). Due to the reduced hot carrier effects in TI-QW lasers, improved temperature dependence of threshold current, high differential gain, and low-frequency chirping have been demonstrated (Davis *et al.*, 1994; Yoon *et al.*, 1994, 1995; Bhattacharya *et al.*, 1996). A 3-dB modulation bandwidth of 43 GHz has been measured in both a strained InGaAs/AlGaAs TI-MQW laser and a strain-compensated InGaAs/GaAsP (QW/barrier) TI-MQW laser (Zhang *et al.*, 1996).

Looking at the structure of the TI-QW lasers, one can speculate on another mechanism leading to the performance improvement, i.e., the reduction of state filling. In most III-V semiconductor QW laser structures, the QW depth in the valence band is smaller than that in the conduction band because of the smaller bandgap offset at QW heterojunctions in the valence band. The holes have much larger effective mass, leading to a much larger number of states in the OCL region in comparison with the electrons. These two facts mean that the state-filling effects are dominated by the holes in the valence band. In TI-QW structures, the tunneling structure next to the QW on the n-type doping side reduces the number of energy states for the injected holes to occupy in the OCL region because the holes are very difficult to tunnel through the barrier layer. The reduced

1.5 Reduction of state filling in QW lasers

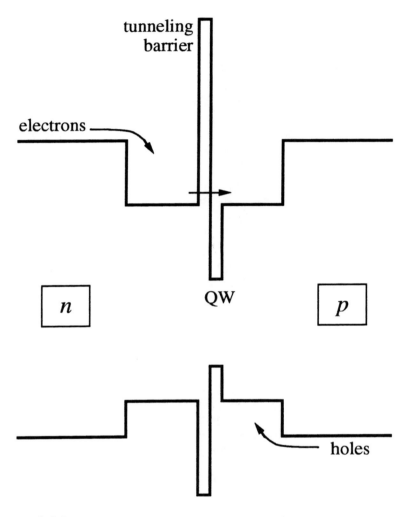

Figure 1.21: Schematic band diagram of a tunneling injection (TI) QW laser structure.

number of states for the holes in the OCL region will lead to a reduction in state-filling effects dominated by the holes, thus leading to performance improvement in temperature characteristics of threshold current, high-speed modulation, and frequency chirping in TI-QW lasers.

1.5.4 Strained quantum well structures

The improvement in crystal growth technology has made it possible to grow a semiconductor laser active layer with relatively large built-in strain. This is realized by growing a very thin QW layer with its thickness smaller than the critical thickness of the strained material. The built-in strain in the QW layer significantly modifies the electronic band structures, especially the valence band structures. Since the strained QW layer thickness is smaller than the critical thickness, there are no crystal defects that will cause injected carrier loss and optical property degradation in the strained QW layer.

The initial work on strained QW lasers showed reduced threshold current and improved modulation bandwidth compared with their unstrained counterparts (Chen et al., 1987, 1990; Eng et al., 1989; Choi and Wang, 1990; Williams et al. 1991; Chand et al., 1991; Ralston et al., 1991; Thijs et al., 1991; Zah et al., 1991; Weisser et al., 1992). The improvement can be attributed first to reduction of the effective mass of holes in the compressively strained QW lasers (Yablonovitch and Kane, 1986; Adams, 1986; Suemune et al., 1988; Ghiti et al., 1989; Lau et al., 1989). The smaller effective mass of holes makes the Fermi level of the holes increase more quickly with respect to the increase in injected carriers. This leads to a larger value of differential gain and a smaller value of injection carrier density to achieve the necessary threshold gain. In addition, the computed larger energy separation between the ground state and the other upper states of the strained QW in comparison with the unstrained QW (Ahn and Chuang, 1988; Corzine et al., 1990; Loehr and Singh, 1991; Suemune, 1991) indicates another reason for the improvement: reduction of state filling.

Shown in Fig. 1.22 are the calculated valence band structures for a 75-Å $In_{0.25}Ga_{0.75}As$ QW and a 75-Å GaAs QW, respectively, with $Al_{0.2}Ga_{0.8}As$ as the barrier layers. The valence band structures are obtained by solving the 4×4 Luttinger-Kohn Hamiltonian via the axial approximation, as we discussed in Sect. 1.3.1. It is evident that the subbands of the strained QW not only have smaller effective hole masses but also possess larger separation between the subbands of the QW. Most important, the energy separation between the ground state and the energy band edge of QW barrier, where there exists a large number of states from the OCL region, is larger in the strained QW because of the smaller energy band gap of InGaAs compared with GaAs. The dashed lines in the figures

1.5 Reduction of state filling in QW lasers

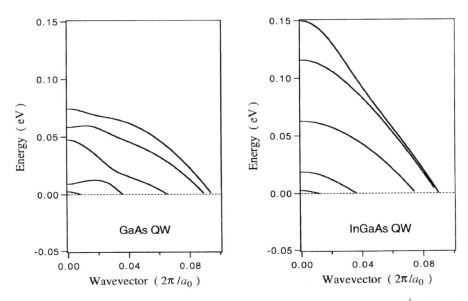

Figure 1.22: Calculated valence band structures for a 75-Å GaAs/$Al_{0.2}Ga_{0.8}As$ QW (left) and a 75-Å $In_{0.25}Ga_{0.75}As/Al_{0.2}Ga_{0.8}As$ QW (right), respectively. The dashed lines in the figures correspond to the energy band edges of the QW barriers in the valence band. a_0 is the crystal lattice constant.

correspond to the energy band edges of the QW barriers in the valence band. All these result in significant reduction of state filling in the compressively strained InGaAs QW structures.

In the quaternary InGaAsP/InP material system, both compressive and tensile strained QW are available. In the tensile strained QW, the gain for TM mode is larger than that for TE mode (Chong and Fonstad, 1989) because the first QW valence subband is a light-hole-like subband. For transitions between conduction band and light-hole-like subband, the transition matrix element (or polarization modification factor) for the TM mode is larger than that for the TE mode, thus resulting in larger gain in the TM mode. This forces lasing in the TM mode in the tensile-strained QW lasers. In unstrained or compressively strained QW lasers, the lasing usually happens in the TE mode because the ground valence subband of the QW is heavy-hole-like. Similar to the compressive-strained QW lasers, improved performance was observed in the tensile-strained QW lasers (Thijs *et al.*, 1991, 1994; Zah *et al.*, 1991; Kano *et al.*, 1994; Kikuchi *et* al., 1994; Wang et al., 1994). The reduction of hole effective mass and the larger subband separation induced by the tensile strain (Seki et al., 1993, 1994) lead to state-filling reduction in the ground lasing and other upper energy subbands, and they are responsible for the performance improvement in tensile-strained QW lasers. The strain-induced reduction in various nonradiative recombinations is also responsible for the improved performance, especially for InGaAsP long-wavelength lasers (O'Reilly and Adams, 1994; Lui et al., 1994).

The strain level in the QW and the number of QWs are limited by the critical thickness for misfit dislocation in the lattice mismatched semiconductors. The technique of strain compensation, by which QW barriers of an opposite strain with respect to the strain of the QW are grown in the active region, can increase the strain level in the QW and allow for growth of a large number of strained QWs with high crystal quality (Miller et al., 1991; Seltzer et al., 1991; Zhang and Ovtchinnikov, 1993; Takiguchi et al., 1994). The larger strain in the QW can lead to a larger modification of the valence band for the holes in the QW, i.e., smaller hole effective mass and larger energy separation between the subbands. The QW depth can be increased by the opposite strain in the QW and the barriers. In addition, the strain in the QW barriers can reduce the number of states in the OCL region. All these lead to reduction of state filling in strain-compensated QW lasers. Improved threshold current characteristics (Ougazzaden et al., 1995; Fukunaga et al., 1996), improved reliability (Toyonaka et al., 1995; Sagawa et al., 1995; Fukunaga et al., 1996), reduced linewidth enhancement factor α (Dutta et al., 1996), and improved modulation bandwidth (Kazmierski et al., 1993; Dutta et al., 1996; Han et al., 1996) were observed in the strain-compensated QW lasers. Recently, a 30-GHz modulation bandwidth has been achieved in a long-wavelength (1.55-μm) InGaAsP/InGaAlAs (QW/barrier) MQW laser by using the strain-compensation technique to the active region of 20 quantum wells (Matsui et al., 1997). Another 30-GHz modulation bandwidth was demonstrated in a 1.55-μm-wavelength strain-compensated InGaAsP MQW laser of 12 quantum wells (Kjebon et al., 1997). This achievement was made by using detuned operation of the laser from a monolithic distributed Bragg reflector (DBR). The detuned operation, i.e., detuning the lasing wavelength from the gain spectrum peak, can enhance the differential gain, reduce the damping, and improve other dynamic properties of a semiconductor laser (Vahala and Yariv, 1984). A more detailed discussion

1.5 Reduction of state filling in QW lasers

of strained QW lasers can be found in Chap. 2 (by Adams and O'Reilly) of this book.

1.5.5 Substrate orientation

In addition to the quantum effect, the energy subbands of a QW structure are dependent on the orientation of the QW. The energy-band structure is strongly dependent on the symmetry of a semiconductor crystal. The incorporation of a QW structure will modify the symmetry, and a different orientation of the QW will have a different impact on the symmetry modification. Thus the energy-band structures are dependent on the QW orientation.

Figure 1.23 shows the calculated valence subbands of GaAs/ $Al_{0.3}Ga_{0.7}As$ QW structures on (001) and (111) substrates, respectively (Batty *et al.*, 1989; Ghiti *et al.*, 1990). It is shown that the energy separation between the ground state and first excited state is larger in the (111) GaAs/AlGaAs QW structure than that in the (001) GaAs/AlGaAs QW structure if the GaAs QW width is less than 100 Å. At the same time, the ground state is further away from the band edge of QW barrier (near -100 meV) in the (111) GaAs/AlGaAs QW structure in comparison with the (001) GaAs/AlGaAs QW structure. Both effects will lead to a reduction of the state filling in the (111) GaAs/AlGaAs QW structure. Indeed, it has been observed in experiments that (111) GaAs/AlGaAs QW lasers possess consistently lower threshold current densities than (001) GaAs/AlGaAs QW lasers for QW width of less than 100 Å (Hayakawa *et al.*, 1987, 1988). A theoretical analysis on the influence of QW orientation was given by Niwa *et al.* (1995) for long-wavelength (InGaAsP system) strained QW lasers, and very similar results were shown.

1.5.6 Bandgap offset at QW heterojunctions

Although the state-filling effect can be reduced by employing MQW for the laser active region, the attendant penalty is an increase in threshold current due to the need to maintain the transparency density of carriers in a larger active volume. The state-filling effect can be reduced by increasing the bandgap offset at the QW heterojunction, i.e., increasing the QW barrier height. But this is limited by the materials available for the QW

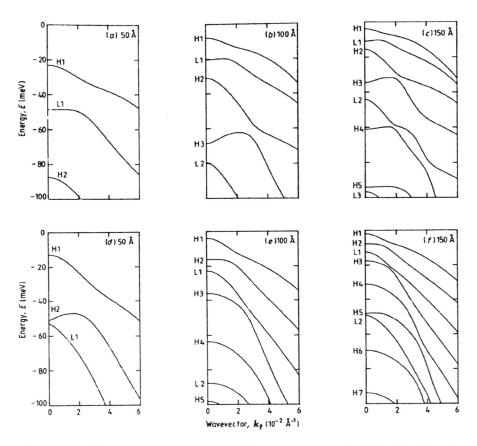

Figure 1.23: Calculated valence subband structures for GaAs/Al$_{0.3}$Ga$_{0.7}$As on (001) [(a)–(c)] and (111) [(d)–(f)] substrates. The quantum well thicknesses are 50, 100, and 150 Å, respectively. [From Batty et al. (1989).]

and the optical confining region. On the other hand, a very large barrier height might degrade the carrier injection and relaxation into the QW.

In most III-V semiconductors, there exists a very large asymmetry between the conduction-band and the valence-band structures. In an SCH QW laser, the larger effective mass of holes (compared with electrons) is responsible for a much larger number of energy states in the OCL region for the holes. Usually, the bandgap offset at a heterojunction between two III-V semiconductors is smaller in the valence band than that in the conduction band, resulting in smaller QW barrier height for the holes.

1.5 Reduction of state filling in QW lasers

These two facts lead to the fact that the state-filling effect is dominated by the holes in most III–V QW lasers.

The holes-dominated state-filling effect can be reduced by optimizing the bandgap offset ratio at the QW heterojunctions because the asymmetry between the conduction band and the valence band structures in the OCL region can be compensated by the corresponding optimal bandgap offset ratio at the QW heterojunctions (Zhao et al., 1993c). Assume that R_d is the conduction bandgap offset ratio, i.e., $R_d = \Delta E_c/\Delta E_g$. When R_d is large (i.e., smaller bandgap offset for the valence band), the state-filling effect is dominated by the hole population in the OCL region. As R_d decreases, the energy separation between the ground state and the states in the OCL region in the valence band increases. The state filling of holes is reduced, which leads to a smaller transparency carrier density and higher differential gain. However, if R_d is too small, the state filling will be dominated by the electrons in the conduction band, resulting in a larger transparency carrier density and a lower differential gain again.

Figure 1.24 shows the calculated carrier density and corresponding differential gain as functions of the bandgap offset ratio R_d for different values of modal gain g. The QW laser structure is assumed to have a 75-Å GaAs QW and a pair of 2000-Å $Al_{0.2}Ga_{0.8}As/Al_{0.5}Ga_{0.5}As$ GRIN-SCH optical confining layers. It is shown that the differential gain can be increased if R_d is reduced from its nominal value of ~0.7. For instance, the differential gain is doubled at $R_d = 0.5$ for the modal gain of 30 cm^{-1}, while the carrier density is reduced. These results indicate that the bandgap offset engineering will lead to enhanced performance in the QW lasers.

This analysis also applies to other material systems, such as strained InGaAs/AlGaAs and the quaternary InGaAsP/InP or InGaAlAs/InP QW structures. In general, bandgap offset engineering includes (1) the choice of the semiconductor materials for the QW and the optical confining layers to form the optimized heterojunctions and (2) artificially modifying the bandgap offset at the QW heterojunctions. Two methods have been proposed to artificially control the bandgap offset at a semiconductor heterointerface. The first employs a sheet of doping dipoles at the heterojunction that is formed by ultrathin ion layers. Changes in the bandgap offset at the heterojunction by as much as 140 meV were demonstrated at a GaAs/$Al_{0.26}Ga_{0.74}As$ heterojunction (Capasso et al., 1985). The second method is based on incorporation of an ultrathin atomic interlayer at the heterojunction, which modifies the charge distribution at the interface and cre-

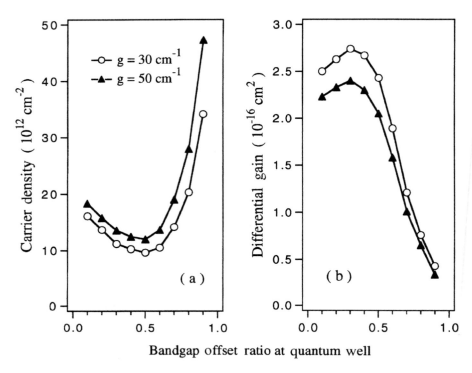

Figure 1.24: Calculated carrier density (a) and differential gain (b) as a function of bandgap offset ratio R_d for a typical GaAs/AlGaAs SCH SQW laser. g is the modal gain.

ates an interface dipole sheet. Bandgap offset change from -380 to 380 meV was reported by incorporation of an ultrathin Si interface layer at a GaAs/AlAs heterojunction (Sorba et al., 1991).

1.6 Some performance characteristics of QW lasers

1.6.1 Submilliampere threshold current

Low-threshold-current semiconductor lasers are key components in many high-density and high-performance applications, such as the optical interconnects in computer circuits and systems. High packing density and

1.6 Some performance characteristics of QW lasers

low power consumption are the two critical requirements for this kind of application. Therefore, threshold current in the submilliampere or microampere region is of importance. Low-threshold lasers often exhibit improved reliability and enhanced modulation bandwidth. They also can be used in direct digital modulation at rates of multigigabits per second under zero bias current conditions for lower electrical power consumption (Lau et al., 1987; Cutrer and Lau, 1995; Chen and Lau, 1996).

As discussed in Sect. 1.1, the threshold current density of a semiconductor laser will be reduced if the thickness of the active layer d is reduced. However, once d is reduced to the quantum regime of the injected carriers, any further reduction in d will not lead to significant reduction in the threshold current density because the carriers become 2D and are quantum-confined in the direction perpendicular to the active layer. The carrier state filling in the OCL region leads to an optimal range in the QW thickness for low threshold current density. If the QW thickness is too large, there exist many very closely packed (in the energy domain) subbands in the QW, resulting in a significant carrier population in these upper subbands at the threshold. If the QW thickness is too small, the first quantized states are raised up and close to the top of the QW, resulting in a significant carrier population in the OCL region at the threshold. In both cases, larger threshold current density results.

For bulk DH semiconductor lasers, the modal gain and injection current density approximately obey a linear relation (Thompson, 1980; Yariv, 1989):

$$g = \Gamma_{dh} B_{dh} (J - J_{tr}) \tag{1.148}$$

where Γ_{dh} is the optical confinement factor, B_{dh} is a constant, and the transparency current density J_{tr} includes both radiative carrier loss such as spontaneous emission and nonradiative carrier loss such as carrier leakage in a particular laser device structure used to confine the injection current. At the threshold, the modal gain is given by

$$g_{th} = \alpha_i + \frac{1}{L} \ln \frac{1}{R} \tag{1.149}$$

where α_i is the internal loss constant, L is the laser cavity length, $R = \sqrt{R_1 \cdot R_2}$, and R_1 and R_2 are the reflectivity of the two mirror facets. From Eqs. (1.148) and (1.149), the threshold current can be obtained as

$$I_{th} = wL J_{th} = w \left[L \left(\frac{\alpha_i}{\Gamma_{dh} B_{dh}} + J_{tr} \right) + \frac{1}{\Gamma_{dh} B_{dh}} \ln \frac{1}{R} \right] \tag{1.150}$$

From Eq. (1.150), the methods to achieve low threshold current in bulk DH lasers are to reduce the internal loss α_i, to reduce the transparency current density J_{tr}, to reduce the device dimensions w and L, to increase the optical mode confinement Γ_{dh}, to increase the differential gain ($\propto B_{dh}$), and to use high reflectivity coating ($R \to 1$).

The lowest threshold current density demonstrated in any semiconductor laser is in strained InGaAs QW lasers. Threshold current density as low as 65 A/cm^2 (Choi and Wang, 1990), 45 A/cm^2 (Williams et al., 1991), and 56 A/cm^2 (Chand et al., 1991) was achieved in long-cavity strained InGaAs SQW lasers. To achieve ultralow threshold currents, the laser device dimensions w and L need to be scaled down. However, as shown in the following, there exists an optimal cavity length to achieve the lowest threshold current in QW lasers. As the lateral dimension w is reduced to the quantum regime, the lasers become quantum wire lasers.

For QW lasers, the modal gain and injection current density relation is nonlinear due to the flat feature of the steplike density of states (Arakawa and Yariv, 1985; McIlroy et al., 1985; Kurobe et al., 1988; Cheng et al., 1988). Assume that the number of quantum wells N_{qw} is not very large, the quantum wells are decoupled, and the carrier injection is uniform with the QWs. Then the modal gain g will be roughly scaled by N_{qw} because, as shown schematically in Fig. 1.25, each quantum well contributes almost equally to the laser field. The nonlinear relation of the modal gain and injection current density can be approximated as

$$g = N_{qw} \Gamma_{qw} B_{qw} \ln \frac{J}{J_{tr}} \qquad (1.151)$$

where Γ_{qw} is the coupling factor for one QW and B_{qw} is a constant. From Eqs. (1.151) and (1.149), the threshold current in the QW lasers is

$$I_{th} = wL J_{tr} \exp\left[\frac{1}{N_{qw} \Gamma_{qw} B_{qw}} \left(\alpha_i + \frac{1}{L} \ln \frac{1}{R}\right)\right] \qquad (1.152)$$

From Eq. (1.152), small w and α_i, large Γ_{qw} and R, and large differential gain (i.e., large B_{qw}) will lead to low threshold current in QW lasers. The cavity-length dependence of I_{th} in QW lasers is very different from that in bulk lasers. From Eq. (1.152), it is easy to find that there exists a minimum threshold current

$$I_{th}(\min) = w \frac{J_{tr}}{N_{qw} \Gamma_{qw} B_{qw}} \exp\left(\frac{\alpha_i}{N_{qw} \Gamma_{qw} B_{qw}} + 1\right) \ln \frac{1}{R} \qquad (1.153)$$

1.6 Some performance characteristics of QW lasers

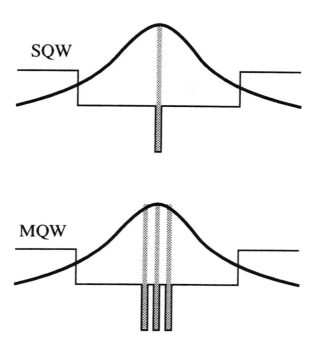

Figure 1.25: Schematic structures of SQW and MQW indicating that the modal gain roughly scales with the number of quantum wells N_{qw} in the case that N_{qw} is not too large.

that is achieved at a cavity length

$$L = \frac{1}{N_{qw}\Gamma_{qw}B_{qw}} \ln \frac{1}{R} \qquad (1.154)$$

From Eqs. (1.152) to (1.154) we can find how the number of QWs influences the threshold current. Both J_{tr} and α_i consist of two parts: One part scales with N_{qw}, and the other is independent of N_{qw}. For J_{tr}, the radiative loss of carriers within the QW scales with N_{qw}, but the radiative carrier loss of the carriers in the OCL region and the nonradiative carrier leakage are basically independent of N_{qw}. α_i consists of free carrier absorption due to the carriers within the QW, free carrier absorption due to the carriers in the OCL region, waveguide scattering, and impurity absorption. The first one scales with N_{qw}. The rest are independent of N_{qw}. The

N_{qw} dependence of I_{th} is also strongly influenced by the threshold modal gain. At very low threshold modal gain,

$$\exp\left[\frac{1}{N_{qw}\Gamma_{qw}B_{qw}}\left(\alpha_i + \frac{1}{L}\ln\frac{1}{R}\right)\right] \to 1$$

the SQW lasers have lower threshold current due to the smaller J_{tr}. At relatively high threshold modal gain, the term

$$\exp\left[\frac{1}{N_{qw}\Gamma_{qw}B_{qw}}\left(\alpha_i + \frac{1}{L}\ln\frac{1}{R}\right)\right]$$

dominates the threshold current. Use of MQW ($N_{qw} \neq 1$) will significantly reduce this term. This feature plus the extra differential gain enhancement in the MQW structure (i.e., larger B_{qw}) makes MQW lasers obtaining low threshold current in certain circumstances even though the transparency current density is larger in comparison with SQW lasers. Equations (1.153) and (1.154) indicate that compared with SQW lasers, lower minimum threshold current can be obtained at shorter cavity length in MQW lasers.

Figure 1.26 shows the threshold current distribution versus the cavity length for SQW, double quantum well (DQW), and triple quantum well (TQW) InGaAs/AlGaAs BH lasers (Chen et al., 1992a). The mirror facets of these lasers are not coated. The lasers were fabricated by a two-step hybrid crystal growth technique. The GRIN-SCH InGaAs strained SQW, DQW, and TQW materials were grown by MBE or MOCVD. After the MBE or MOCVD growth, mesas with an active stripe width of ~2 μm were wet-chemically etched in order to fabricate buried heterostructure (BH) lasers. A p-$Al_{0.4}Ga_{0.6}As$ layer and an n-$Al_{0.4}Ga_{0.6}As$ layer were grown successively in a liquid-phase epitaxy (LPE) system to form a blocking junction. The schematic structure for the BH lasers is shown in Fig. 1.27. From Fig.1.26 it is seen that the optimal cavity length for low-threshold-current operation in BH QW lasers decreases with an increasing number of wells. Approximately, the optimal lengths are 300 to 400 μm for SQW, 200 to 300 μm for DQW, and 100 to 200 μm for TQW lasers, respectively. The lowest threshold current obtained is 1 mA for an uncoated DQW laser at 200-μm cavity length. Although slightly lower threshold currents in BH strained InGaAs TQW lasers were reported by Xiao et al. (1992), the threshold currents of TQW lasers were higher than what had been expected. This could be caused by the degradation of crystal quality as

1.6 Some performance characteristics of QW lasers

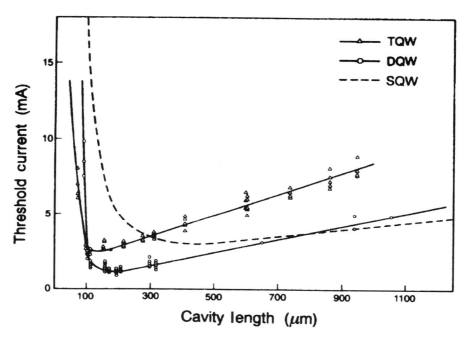

Figure 1.26: Threshold current versus cavity length for the SQW, DQW, and TQW strained InGaAs/AlGaAs BH lasers. The width of the active stripe is ~2 μm. [From Chen et al. (1992a).]

the number of strained layers increases or by the nonuniform carrier injection into the QWs as the number of QWs increases. To further reduce the threshold currents to the submilliampere range, high reflectivity (HR) dielectric coatings were applied to the facet mirrors of the SQW lasers (Chen et al., 1993a, 1993b, 1995). Threshold current behavior is shown in Fig. 1.28 for an SQW laser of 125-μm cavity length. The threshold current was 2.8 mA for the laser with cleaved facets. When an HR coating of 0.99 was applied to the rear facet, the threshold current dropped to 0.7 mA. A further application of an HR coating of 0.99 to the front facet led to a threshold current of 0.165 mA.

Another technique to fabricate low-threshold-current QW lasers is the patterned substrate (PS) approach, i.e., to grow QW laser structures on patterned substrate. In the PS approach, QW laser structures with good lateral optical and current injection confinement are obtained by

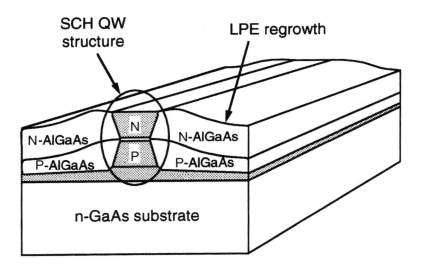

Figure 1.27: Schematic structure of a buried heterostructure (BH) QW laser.

using the very different crystal growth rate on the mesa top or at the groove bottom in comparison with that on the mesa sidewall, i.e., crystal facet selective epitaxy. In comparison with BH lasers, the PS approach can easily fabricate lasers of narrower active strip width, which thus have the potential for lower threshold current. In one PS approach to grow a patterned QW structure at the groove bottom, Kapon et al. (1990) demonstrated submilliampere threshold currents of 0.35 mA (pulsed) and 0.5 mA (CW) in a GaAs SQW laser with HR facet coatings. In another PS approach, a triangular prism-shaped laser active region was grown on the ridge top. By controlling the ridge height, the QW region and the n-type lower part of the active region could be buried in a pair of n-AlGaAs/p-AlGaAs layers that were grown in the area on each side of the ridge. The n-AlGaAs/p-AlGaAs layers provided a reversed p-n junction for current confinement. Without the HR facet coatings, a low threshold current of 0.88 mA was demonstrated in a three-quantum-well GaAs laser of this structure (Narui et al., 1992). Recently, low threshold currents of 0.6 mA

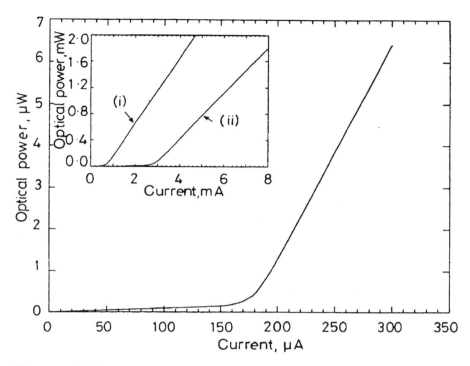

Figure 1.28: Optical power versus injection current for a strained InGaAs SQW BH laser of 125 μm cavity length and ~0.99 HR coatings to both facets. Inset: (i) only back facet with HR coating of 0.99; (ii) no HR coating to both facets. [From Chen et al (1995).]

(without HR facet coating) and 0.145 mA (with HR facet coating) were achieved in strained InGaAs SQW lasers of submicron active region width (Zhao et al., 1995). The submicron active region width was achieved by growing the QW active region at the top tip of the mesa structure on patterned substrate.

The threshold current in edge-emitting QW lasers can be further reduced at low temperatures (Eng et al., 1991; Zhao et al., 1994). This was motivated by the prospect of integration of semiconductor lasers with low-temperature electronics for high performance. Many electronic devices possess improved performance characteristics at cryogenic temperatures. For example, the high-temperature superconductor devices have advantages in high speed, high sensitivity, and low power consumption in comparison with many conventional room-temperature electronic devices as a result of the unique nature of superconductivity. Semiconductor field-effect devices and bipolar devices also have shown improved performance and reliability at low operating temperatures due to the improvement in carrier mobility, reduction in parasitic capacitance and interconnect resistance, and reduced electromigration and power consumption. Optical interconnects using semiconductor lasers in these low-temperature electronics appear very attractive because of the demonstrated extremely low threshold currents and the enhanced external quantum efficiency.

Figures 1.29 and 1.30 show the measured threshold current and external quantum efficiency as functions of the temperature for three 225-μm-long strained InGaAs SQW BH lasers with different facet HR dielectric coatings. It is shown that the threshold current can be reduced by a factor of 10 and the external quantum efficiency can be increased by a factor of 2 as the ambient temperature decreases from the room temperature ($\sim$300 K) to near liquid helium temperature ($\sim$4 K). The dramatic decrease in threshold current is due to the reduction in state-filling effects, modification of band structures, and suppression of various carrier loss mechanism at low temperatures in QW lasers. The increase in external quantum efficiency stems from both the increase in internal quantum efficiency and the decrease in internal loss constant in these lasers as the ambient temperature decreases. For the laser of mirror reflectivity of 0.75 and 0.99 (laser 3 shown in Figs. 1.29 and 1.30), threshold currents of 38 μA at 6 K and 56 μA at 77 K were demonstrated, and the external quantum efficiency was about 1 mW/mA at both 6 and 77 K.

Vertical cavity surface emitting lasers (VCSEL) are especially suitable for 2D high-density applications (Iga, 1992). The extremely small active volume in VCSEL QW lasers offers the advantage of obtaining extremely low threshold currents in submilliampere range (Geels and Coldren, 1990; Michalzik and Ebeling, 1993; Numai *et al.*, 1993). Huffaker *et al.* (1994a) first applied a selective oxidation technique in fabrication of VCSEL QW lasers to effectively reduce the lateral dimension of the active region for ultralow threshold current. In the selective oxidation fabrication process, layers of AlAs or AlGaAs were first grown in the laser structures and were partially oxidized in a subsequent process step to provide confinement for injection current and optical mode (Dallesasse *et al.*, 1990; Dallesasse and Holonyak, 1991). The oxidized AlAs and AlGaAs

1.6 Some performance characteristics of QW lasers 93

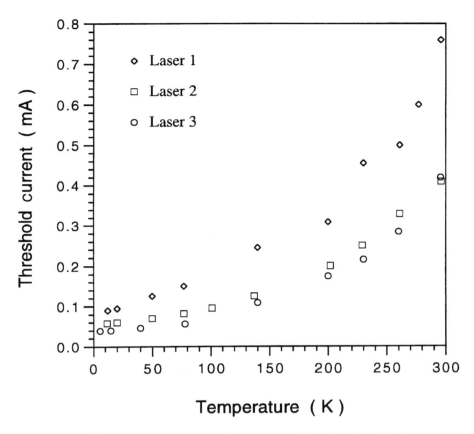

Figure 1.29: Measured threshold current as a function of ambient temperature for 225-μm-long strained InGaAs SQW BH lasers of different mirror facet HR coatings.

layers possess a lower refractive index in comparison with the nonoxidized ones. In the VCSEL structures, the selective oxidation layer introduces an intracavity dielectric aperture that laterally confines the optical mode and the current injection. In comparison with other VCSEL fabrication techniques, small lateral active dimension, high current injection efficiency, and low intracavity loss can be obtained easily by the selective oxidation method. Threshold currents below 0.1 mA were achieved in strained InGaAs QW VCSELs fabricated by the selective oxidation method (Huffaker *et al.*, 1994b, 1996; Hayashi *et al.*, 1995). To date, the lowest

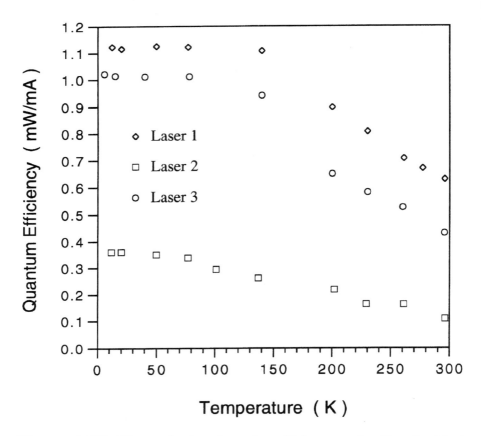

Figure 1.30: Measured external quantum efficiency as a function of ambient temperature for 225-μm-long strained InGaAs SQW BH lasers of different mirror facet HR coatings.

threshold current in a semiconductor laser is 8.7 μA, which was achieved in a strained InGaAs QW VCSEL fabricated by the selective oxidation method (Yang *et al.*, 1995). Although improvements in output power and external quantum efficiency were observed, the threshold current reduction in VCSEL QW lasers from room temperature to low temperatures was not as significant as that observed in edge-emitting lasers (Goobar *et al.*, 1995; Hornak *et al.*, 1995; Goncher *et al.*, 1996; Akulova *et al.*, 1997). This is attributed to the difficulty in alignment of the cavity mode to the QW gain spectrum peak from room temperature to low temperatures.

1.6 Some performance characteristics of QW lasers 95

In InGaAsP/InP long-wavelength lasers, the application of strain, either compressive or tensile, reduces the density of states and nonradiative carrier loss, leading to reduction in threshold current density and threshold current. Threshold current density in the vicinity of 90 to 100 A/cm^2 was observed in both compressive and tensile strained SQW lasers of 1.55 μm wavelength (Thijs *et al.*, 1992b; Yamamoto *et al.*, 1994a; Mathur and Dapkus, 1996). For 1.3-μm wavelength compressive strained SQW lasers, the threshold current density is also in the vicinity of 90 to 100 A/cm^2 (Zah *et al.*, 1992; Yamamoto, *et al.*, 1994b). However, the threshold current density in 1.3-μm wavelength tensile strained SQW lasers was about 200 A/cm^2 (Zah *et al.*, 1992; Yokouchi *et al.*, 1996). Submilliampere threshold currents in the range of 0.4 to 0.98 mA have been demonstrated by using dielectric HR facet coatings to the 1.3- and 1.55-μm wavelength strained QW lasers (Zah *et al.*, 1990; Temkin *et al.*, 1990; Osinski *et al.*, 1992; Thijs *et al.*, 1992a, 1992b; Uomi *et al.*, 1994; Terakado *et al.*, 1995). For low-temperature operation, significant threshold current reduction (from 27 mA at 300 K to 0.9 mA at 10 K) and modulation bandwidth increase (from 10 GHz at 290 K to 27 GHz at 77 K) were observed in the 1.55-μm wavelength InGaAsP strained MQW lasers (Yu *et al.*, 1994). By using GaAs/Al(Ga)As Bragg mirrors with strained InGaAsP active layers bonded by wafer fusion, submilliampere threshold currents (~0.8 mA) have been demonstrated for both 1.3- and 1.55-μm long-wavelength VCSEL strained QW lasers under pulsed operation (Margalit *et al.*, 1996; Qian *et al.*, 1997).

1.6.2 High-speed modulation at low operation current

High-speed modulation of semiconductor lasers is very important because of their applications in optical telecommunications and optical interconnects. A number of research and development efforts have been devoted to exploring high-speed performance. Chapter 3 (by Nagarajan and Bowers) in this book deals with the subject of high-speed modulation of semiconductor lasers. Here we limit our discussion to high-speed modulation at low operation current.

It is appreciated that ultralow threshold lasers are going to play an important role in optical interconnects for computer circuits due to their low power consumption. It must be borne in mind, however, that these lasers need be modulated at rates up to a few gigabits per second, and the low power advantage will be lost if the bias current needed to reach the required modulation bandwidth is excessive. At steady state, the total number of photons emitted from a laser cavity is

$$\frac{P_0 V_{\text{opt}}}{\tau_p} = \eta_i \frac{I - I_{th}}{e} \tag{1.155}$$

where P_0 is the photon density at the active region, V_{opt} is the effective optical mode volume ($V_{\text{opt}} = wtL$), τ_p is the photon lifetime, and η_i is the internal quantum efficiency. From Eqs. (1.129) and (1.155), the 3-dB modulation bandwidth at low optical power levels can be obtained as

$$f_{3\,\text{dB}} \approx 1.55 f_r = \frac{1.55}{2\pi} \sqrt{\frac{v_g \eta_i}{e V_{\text{opt}}} (I - I_{th}) G_0'} \tag{1.156}$$

The *modulation current efficiency factor* (MCEF) was introduced to better characterize the high-speed modulation at low operation current in semiconductor lasers (Chen et al., 1993b), which is defined as

$$\text{MCEF} = \frac{f_{3\,\text{dB}}}{\sqrt{I - I_{th}}} \approx \frac{1.55}{2\pi} \sqrt{\frac{G_0' v_g \eta_i}{e V_{\text{opt}}}} \tag{1.157}$$

For the same effective injection current level ($I - I_{th}$), a higher MCEF value will enable a higher modulation bandwidth in a semiconductor laser.

From Eq. (1.156), it follows that for a given injection current I, the achievement of high $f_{3\,\text{dB}}$ requires a low threshold I_{th}, a large differential gain G_0', a small optical mode volume V_{opt}, and a high internal quantum efficiency η_i. QW lasers become the main contenders for this kind of application because of the demonstrated submilliampere threshold currents. Equations (1.153) and (1.154) indicate that a lower threshold current can be achieved by increasing the QW number, using a shorter cavity length, and using a narrower active strip width. At the same time, a shorter cavity length and a narrower active strip width lead to smaller optical mode volume, which, in turn, leads to a larger modulation bandwidth. However, shorter cavity length entails operation at higher current (and carrier) densities because of the large threshold modal gain required. This inevitably reduces the differential gain G_0' in QW lasers due to the nonlinear relationship of gain versus carrier density. High-reflectivity coatings on the facets of such lasers with short cavity lengths can reduce the threshold modal gain dramatically (and also the threshold current), thus retaining a high value of differential gain. In addition, the use of MQW as the active region will lead to a higher differential gain at low

1.6 Some performance characteristics of QW lasers

threshold modal gain compared with SQW structures because of the reduced state-filling effects. The use of strained QW leads to further improvement because strained QW lasers have demonstrated lower threshold current density and higher differential gain due to the strain-induced valence band modification.

Using design criteria inspired by the preceding arguments, a 5-GHz, 3-dB modulation bandwidth was demonstrated at an operating current of 2.1 mA in a strained InGaAs double quantum well BH laser (Zhao et al., 1992e). The modulation response is shown in Fig. 1.31. The high-speed bandwidth at low operation current is attributed to the lower threshold current (0.5 mA), smaller optical mode volume [short cavity (150 μm) and narrow active strip width (2 μm)], and high differential gain at low modal gain in this laser. The 0.95 HR coatings on both facets of this laser led to very small threshold modal gain. The MCEF value in this laser was about 4 GHz/$\sqrt{mA}$. By using a narrower laser strip width and a shorter cavity, a 5 GHz/$\sqrt{mA}$ MCEF was demonstrated in a strained InGaAs SQW laser (Chen et al., 1993b).

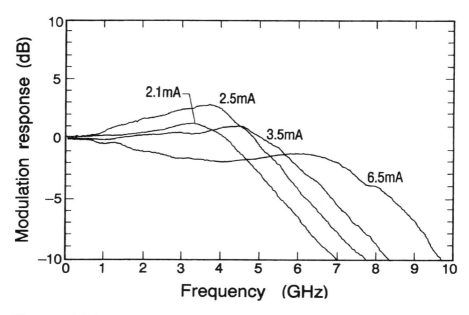

Figure 1.31: Measured modulation response of a strained InGaAs double quantum well BH laser. A 5-GHz, 3-dB modulation bandwidth was achieved at an operating current 2.1 mA.

The VCSELs are superior to edge-emitting lasers for high modulation speed at low operation current. This is so primarily because the optical mode volume in the VCSELs is much smaller than that in edge-emitting lasers. From Eqs. (1.156) and (1.157), the significantly smaller optical mode volume V_{opt} would enable a much larger MCEF and thus a much larger modulation bandwidth at low injection current. A 3-dB modulation bandwidth of 12.1 GHz at an operation current of 2.8 mA was demonstrated in a GaAs VCSEL of 9.5 GHz/$\sqrt{mA}$ MCEF (Lehman et al.; 1995). Recently, 3-dB modulation bandwidth of 15 GHz has been achieved at a bias current of 2.1 mA in a strained InGaAs 3QW VCSEL of 3.1 μm diameter (Thibeault et al., 1997). The threshold current was 0.3 mA, and the MCEF was 14 GHz/$\sqrt{mA}$. In addition, Lear et al. (1996) reported a 3-dB modulation bandwidth in excess of 16 GHz at an operation current of 4.5 mA in another strained InGaAs 3QW VCSEL of 0.37 mA threshold current and 16.8 GHz/$\sqrt{mA}$ MCEF.

These results are attractive for the optical interconnects in computer circuits and short-distance local-area communication systems. Due to the damping effect on modulation response, at high operation current levels, the operation current increase does not efficiently increase the modulation bandwidth. Let us examine two extreme cases to construct a short-distance communication system. First, we assume that one uses a laser in which a high-speed modulation bandwidth of 40 GHz can be obtained at an operation current of 155 mA (Weisser et al., 1996). To obtain the 40-GHz communication bandwidth, one can use a system consisting of three lasers, and each of them has a 15-GHz modulation bandwidth at an operating current of 2 mA (Thibeault et al., 1997). The typical series resistance R_{laser} of a packaged semiconductor laser is about a few ohms. From $P_{el} = I^2 R_{laser}$, the electrical power consumption P_{el} in the second configuration is about three orders of magnitude less than that in the first configuration of using one single 40-GHz laser. On the other hand, the requirements for high-speed (40-GHz) photodetectors and other supporting electronics can be relaxed to 15 GHz.

1.6.3 Amplitude-phase coupling and spectral linewidth

The amplitude-phase coupling factor α (or linewidth enhancement factor) of a semiconductor laser determines its spectral linewidth at steady state and the wavelength chirping under high-speed modulation. A reduction

1.6 Some performance characteristics of QW lasers

in α will lead to a reduction in spectral linewidth and wavelength chirping, thus leading to an increase in optical telecommunications capacity.

As we discussed earlier, the state-filling effect strongly influences the α parameter of QW lasers. By using a large QW barrier height, reductions in the α parameter and spectral linewidth have been observed in QW lasers (Zhao et al., 1993a, 1993b). Figure 1.32 shows the measured linewidth ($\Delta \nu$) of two typical QW lasers of different QW barrier heights (see structures A and B in Fig. 1.18) as a function of the inverse optical power from one facet of the lasers ($1/P$). The lasers were both 300 μm long, and their facets were uncoated. It is seen that the slope of the curve for

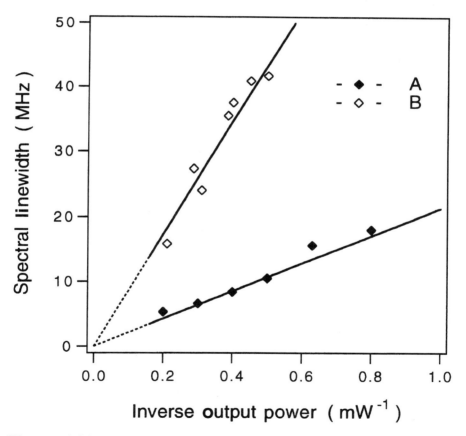

Figure 1.32: Measured spectral linewidth behavior in QW lasers with different QW barrier heights. Laser A has a larger QW barrier height.

structure B is about 4 times as large as that for structure A. Equation (1.142) indicates that the slope of the $\Delta\nu - 1/P$ curve depends on the threshold modal gain g, the mirror loss α_m, the spontaneous emission factor n_{sp}, and the α parameter in the form of $(1 + \alpha^2)$. The effective distributed mirror loss of these two 300-μm-long, uncoated lasers was the same α_m of $\sim$40 cm^{-1}. From the measured internal loss constant in the two laser structures, it was found that these two lasers had almost the same threshold gain value g of $\sim$50 cm^{-1}. The estimated n_{sp} values in structure A and structure B at modal gain of 50 cm^{-1} were very close to each other (Zhao et al., 1993a). Since the geometries of these lasers were essentially identical, it is concluded that the observed difference in the dependence of linewidth on 1/P is caused mainly by the different amplitude-phase coupling (i.e., different α) in the two laser structures.

Indeed, these arguments have been confirmed by direct measurement of the α parameter in these lasers. The measured α values versus wavelength are shown in Fig. 1.33 for the 300-μm-long, uncoated lasers of structures A and B, respectively. The α values were extracted from the spontaneous emission spectra and Fabry-Perot mode wavelength shifts under CW operation of the lasers (Zhao et al., 1993b). In comparison with the α parameter measurement under pulsed operation (Henning and Collins, 1983), the measurement under CW operation can significantly improve the signal-to-noise ratios and enable a much higher spectral resolution for accurate measurement. The lasing wavelength for the laser of A structure was about 8500 Å, and the lasing wavelength for the laser of B structure was about 8570 Å. From Fig. 1.33, the difference in α at the lasing wavelength between the two lasers is evident.

Notice that the optical confinement in these two structures is almost identical; at lasing, the two lasers were biased at such a point that the carrier population within the QW gives almost the same peak material gain. However, the separation between the quasi-Fermi energies and the energy states in the OCL region is quite different at such a bias point due to the different QW barrier heights in the two structures. This difference has two consequences for amplitude-phase coupling. First, in a structure with smaller barrier height, such as structure B, the smaller separation between the quasi-Fermi energies and the large number of states in the OCL region will cause a larger carrier population in the OCL region. As discussed in Sect. 1.4.5, a larger interband-transition component of α will result due to the larger OCL carrier state-filling effect on the laser transitions. Second, the larger carrier population in the OCL

1.6 Some performance characteristics of QW lasers

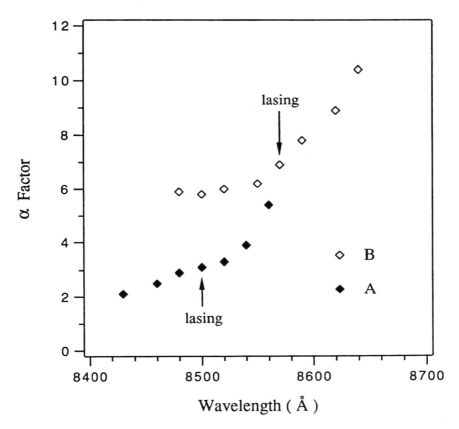

Figure 1.33: Directly measured α parameters in QW lasers of different QW barrier heights. Laser A has a larger QW barrier height.

region will result in a larger free-carrier component of α through the plasma effect.

As discussed earlier, use of the MQW structure as the active region and the addition of strain to QW layers lead to reduction of state filling. In addition to the improvement in modulation dynamics, these also lead to a reduction in the α parameter. The strain effects on α have been studied extensively in various strained QW lasers (Rideout *et al.*, 1990; Dutta *et al.*, 1990a, 1990b; Banerjee *et al.*, 1991; Kikuchi *et al.*, 1991; Kano *et al.*, 1991; Tiemeijer *et al.*, 1991; Schonfelder *et al.*, 1993). The experimental results in these works show that the α parameter in strained (either compressive or tensile) QW lasers is consistently smaller than that in unstrained QW lasers. Figure 1.34 shows the α parameter as a function of the detuning from the lasing photon energy for bulk InGaAsP, lattice-matched InGaAs/InGaAsP MQW, compressive strained InGaAs/InGaAsP MQW, and tensile strained InGaAs/InGaAsP MQW lasers (Thijs et al., 1994). At the lasing wavelength, the α factors for bulk and unstrained MQW lasers are 5 and 3, respectively. In comparison, the α factors for compressive strained MQW and tensile strained MQW lasers are 1.7 and 1.5, respectively, at the lasing wavelength. The slightly smaller α factor in tensile strained QW lasers is attributed to the quantum-squeezed free-carrier component of α (Tiemeijer et al., 1992). In the tensile strained QW structures, the lasing mode is TM mode with its polarization direction perpendicular to the QW plane. Due to the quantum confinement, the carriers in the QW can only move within the QW plane. Thus these carriers in the QW do not have the plasma effect for the lasing TM mode of polarization direction perpendicular to the QW plane.

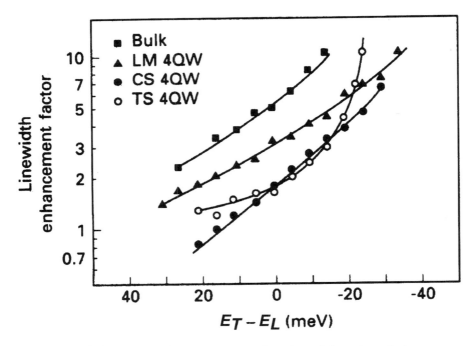

Figure 1.34: Measured α factors as a function of detuning photon energy from the gain peak for different InGaAsP long-wavelength lasers. LM, lattice-matched; CS, compressive-strained; TS, tensile-strained. [From Thijs et al. (1994).]

It has been found experimentally that the p-type doping in the compressive strained InGaAs MQW led to a reduction in the α parameter (Schonfelder et al., 1994). The reduction was attributed to the p-doping-induced reduction of differential index and enhancement of differential gain. It is shown in Figs. 1.33 and 1.34 that the α parameters are strongly wavelength or photon-energy dependent. It was shown theoretically that in p-type doped compressive strained QW lasers, an α parameter of zero value might be obtained at lasing wavelength detuned from the gain peak wavelength to the shorter wavelength (Yamanaka et al., 1993). The detuned operation can be achieved by employing a distributed Bragg reflector (DBR) or distributed feedback (DFB) structure in the lasers.

1.6.4 Wavelength tunability and switching

A QW laser can provide wide and flat gain spectra, as shown in Fig. 1.10. The onset of the second quantized state results in more than one local gain maximum in the gain spectrum. A wide and flat gain spectrum of a QW laser structure will result in a large lasing wavelength tuning range with proper device configuration. Tunable semiconductor lasers are desired for spectroscopy, heterodyne detection, wavelength division multiplexing (WDM) communications, and optical memory.

Lasing wavelength tuning ranges of 105 nm in GaAs SQW lasers (Mehuys et al., 1989) and 170 nm in strained InGaAs SQW lasers (Eng et al., 1990) were demonstrated, respectively, with a grating-coupled external cavity configuration (shown in Fig. 1.35). Figure 1.36 shows the thresh-

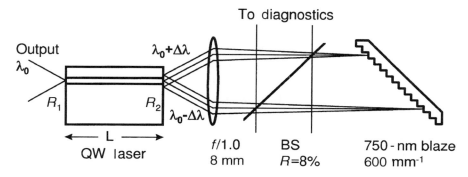

Figure 1.35: Schematic diagram of external cavity configuration for a wide tuning range of lasing wavelength.

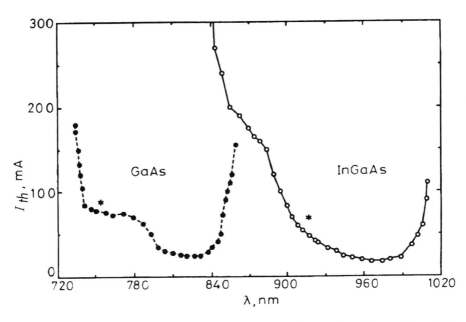

Figure 1.36: Lasing threshold current at grating selected wavelength in GaAs and InGaAs SQW lasers. [From Eng et al. (1990).]

old current as a function of tuned lasing wavelength in GaAs and InGaAs lasers. If we can monolithically integrate the GaAs and InGaAs QW laser structures into one optical cavity, very wide wavelength tuning from 740 to 1010 nm would result (a tuning range of 270 nm). Using similar external cavity configurations, tuning ranges of more than 200 nm were observed in InGaAsP/InP MQW long-wavelength lasers (Lidgard et al., 1990; Bagley et al., 1990; Tabuchi and Ishikawa, 1990). The single-mode operation, reduced spectral linewidth due to the external feedback, and wide tuning range of lasing wavelength in the external cavity configuration are very attractive for many applications. The compact and monolithic versions of this configuration have been realized in various multielectrode DFB or DBR laser structures. But the lasing wavelength tuning range is smaller than that demonstrated in the external cavity configuration. By monolithic integration of various DBR and other mode-coupling structures, continuous or quasi-continuous wavelength tuning ranges of 74 nm (Kim et al., 1994) in a tensile strained MQW laser, 62 nm (Ishii et al., 1996) in

1.6 Some performance characteristics of QW lasers

a compressive strained MQW laser, and 58 nm (Lealman et al., 1996) in a strain compensated MQW laser were demonstrated. The material system was InGaAsP/InP, and the lasing wavelength was near 1.55 μm. Detailed discussion on tunable semiconductor lasers can be found in Chapter 4 (by Amann) of this book.

Another interesting phenomenon in lasing wavelength behavior of QW lasers is the observed lasing wavelength switching controlled by injection current or temperature. It was found that in Fabry-Perot (FP) QW lasers of a certain cavity length, the lasing wavelength suddenly jumped to a shorter wavelength by about a few tens of nanometers as the injection current increased beyond a certain value. This phenomenon was first observed in GaAs SQW lasers (Tokuda et al., 1986) and then in InGaAs MQW lasers (Chen et al., 1992b). In order to have the lasing wavelength switching controlled by the injection current, one needs to control the cavity loss (i.e., the threshold modal gain) to such a point that the local maximum gain from the first quantized state is slightly larger than that from the second quantized state in the gain spectrum. Initially, the lasing is on the first quantized state. As the current increases, the temperature of the laser increases. Higher temperature leads to smaller local gain maximum from the first quantized state, and more carriers need to populate the QW to retain the threshold modal gain. Most of the added carriers will populate in the second quantized state due to the high Fermi energy levels and larger number of states available for the carriers to populate. This carrier state filling into the second quantized state will increase the corresponding local maximum gain from the second quantized state. Finally, as the injection current (and the temperature) increases beyond a certain value, the local maximum gain from the second quantized state is larger than that from the first quantized state. The lasing wavelength thus switches from the first quantized state (long wavelength) to the second quantized state (short wavelength). Figure 1.37 shows the calculated gain spectra of a GaAs SQW laser at different temperatures for a given value of threshold modal gain (peak gain). It is clearly shown how the lasing wavelength switches between the first quantized state and the second quantized state as the temperature (or injection current) changes.

The realization of monolithic lasing wavelength control in FP QW lasers should be much more economical than that in DFB lasers. However, to equalize the two local gain maxima in QW lasers, one needs to operate the lasers under very high injection conditions. Sometimes it is difficult

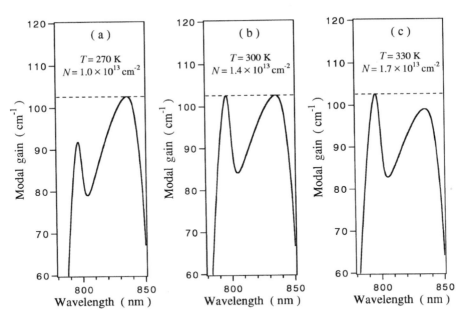

Figure 1.37: Gain spectra of a GaAs SQW laser showing the mechanism of wavelength switching. The dotted line marks the value of threshold modal gain. The temperature and the corresponding two-dimensional threshold carrier density are indicated. (a) Lasing at the first quantized state; (b) lasing simultaneously at the first and the second quantized states; (c) lasing at the second quantized state alone.

to achieve CW operation under such high injection levels, and operation at very high injection levels can lead to reliability problems. By using asymmetric double quantum well structures, lasing wavelength switching was observed under lower injection conditions and CW operation (Ikeda et al., 1989; Ikeda and Shimizu, 1991). In the laser structures, two non-identical quantum wells were separated by a high and/or thick barrier. This could result in non-quasi-thermal equilibration between the carriers in the two wells; i.e., the quasi-Fermi energies for the carriers in the two wells are different. The injected carrier density in each well was controlled by changing the applied voltage to the lasers. The carrier distribution change in each well resulted in a total gain profile change and gain peak wavelength change, leading to the lasing wavelength switching.

1.6 Some performance characteristics of QW lasers

The wide gain spectrum and light radiation spectrum from a QW laser are also useful in generating short mode-locked optical pulses. Ideally, the pulse-width limit of a 100-nm-wide spectrum should be on the order of 10 femtoseconds, as deduced from the transform limit. Since the subject of the mode locking of semiconductor lasers is covered in more detail in Chapter 3 (by Nagarajan and Bowers) of this book, here we only discuss briefly the wavelength tunability of mode-locked QW lasers.

Very short optical pulses (in the picosecond to femtosecond range) have found many applications in studying ultrafast phenomena in physics, chemistry, and biology. In some of these applications, one requirement for the short optical pulses is that the pulse spectrum is not too wide and the center wavelength of the pulse spectrum can be tuned over a certain wavelength range. The typical spectral width of passively mode-locked QW lasers is on the order of a few nanometers. Optical pulses as short as 1.6 ps and tuning ranges of 8.8 nm were demonstrated in a temperature-tuned monolithic colliding pulse mode-locked (CPM) QW laser (Wu *et al.*, 1991). Using a two-section passively mode-locked QW laser with an external cavity configuration, optical pulses of 4.5 ps and tuning ranges of 26 nm were achieved (Schrans *et al.*, 1992). A wider wavelength tuning range of 115 nm was achieved for relatively long pulses (~30 ps) in a mode-locked MQW laser with external cavity (Hofmann *et al.*, 1994). Figure 1.38 shows a setup for generation of tunable subpicosecond optical

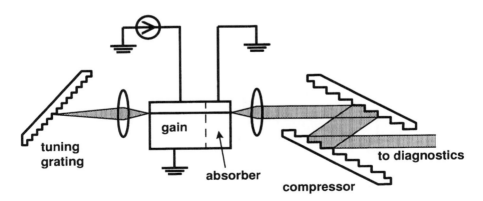

Figure 1.38: Configuration for wavelength-tunable two-section mode-locked QW lasers with external tuning grating and grating pair compressor.

pulses (Salvatore et al., 1993). The laser was a two-section MQW laser under passively mode-locked operation. The external grating is used for centering the pulse spectrum to the desired wavelength. The grating pair compressor is used to compress the linearly chirped optical pulses from the laser by about one order of magnitude to subpicosecond range. Figure 1.39 shows the measured pulse width as a function of spectrum center wavelength. The center wavelength was controlled by the tuning grating. Subpicosecond (<0.6 ps) optical pulses with a tuning range of 16 nm have been obtained.

1.7 Conclusion and outlook

In this chapter we have discussed a number of interesting theoretical and experimental results in QW lasers with emphasis on the basic physical phenomena involved in gain, threshold current, modulation dynamics,

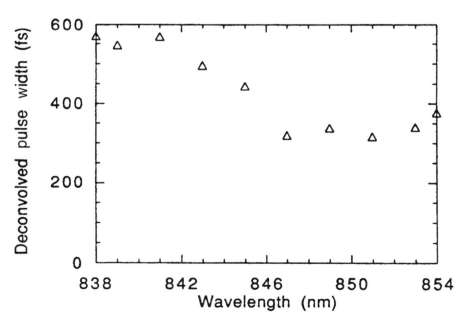

Figure 1.39: Subpicosecond optical pulse width as a function of its spectrum central wavelength. [From Salvatore et al. (1993).]

spectral dynamics, and lasing wavelength tunability. The main challenges in the development of QW lasers at the moment include the ever-larger direct-current modulation bandwidth, the extremely small dynamic spectral linewidth under direct-current modulation, the very-high-speed operation with monolithic lasing wavelength tunability, the monolithic integration of ultralow threshold and high-speed QW lasers on IC chips for interconnect applications, the monolithic controllable ultrashort optical pulse generation, the monolithic wide lasing wavelength tuning, the mode and coherence control of high-power wide-area lasers including laser arrays, the single-channel lasers with mode-controlled multiwatt output, and the ever-shorter lasing wavelengths. The current level of research on QW lasers should guarantee continued progress at an increased pace in the future.

References

Adams, A. R. (1986). *Electron. Lett.,* **22**, 249.

Agrawal, G. P. (1987). *IEEE J. Quantum Electron.,* **QE-23**, 860.

Agrawal, G. P. (1988). *J. Appl. Phys.,* **63**, 1232.

Agrawal, G. P., and Dutta, N. K. (1993). *Long-Wavelength Semiconductor Lasers,* Van Nostrand Reinhold, New York.

Ahn, D., and Chuang, S. L. (1988). *IEEE J. Quantum Electron.,* **QE-24**, 2400.

Ahn, D., Chuang, S. L., and Chang, Y.-C. (1988). *J. Appl. Phys.,* **64**, 4056.

Akhtar, A. I., and Xu, J. M. (1995). *J. Appl. Phys.,* **78**, 2962.

Akulova, Y. A., Thibeault, B. J., Ko, J., and Coldren, L. A. (1997). *IEEE Photon. Tech. Lett.,* **9**, 277.

Altarelli, M. (1985). In *Heterojunctions and Semiconductor Superlattices* (G. Allan, G. Bastard, N. Boccara, M. Lannoo, and M. Voos, eds.), p. 12, Springer-Verlag, New York.

Altarelli, M., Ekenberg, U., and Fasolino, A. (1985). *Phys. Rev. B,* **32**, 5138.

Arakawa, Y., and Yariv, A. (1985). *IEEE J. Quantum Electron.,* **QE-21**, 1666.

Arakawa, Y., and Yariv, A. (1986). *IEEE J. Quantum Electron.,* **QE-22**, 1887.

Arakawa, Y., Vahala, K., and Yariv, A. (1984). *Appl. Phys. Lett,* **45**, 950.

Asada, M., and Suematsu, Y. (1985). *IEEE J. Quantum Electron.,* **QE-21**, 434.

Asada, M., Kameyama, A., and Suematsu, Y. (1984). *IEEE J. Quantum Electron.,* **QE-20**, 745.

Ashcroft, N. W., and Mermin, N. D. (1976). *Solid State Physics,* Saunders College Publishing, Philadelphia.

Bagley, M., Wyatt, R., Elton, D. J., Wickes, H. J., Spurdens, P. C., Seltzer, C. P., Cooper, D. M., and Devlin, W. J. (1990). *Electron. Lett.,* **26**, 267.

Baldereschi, A., and Lipari, N. O. (1973). *Phys. Rev. B,* **8**, 2697.

Banerjee, S., Srivastava, A. K., and Chand, N. (1991). *Appl. Phys. Lett.,* **58**, 2198.

Basov, N. G., Krokhin, O. N., and Popov, Y. M. (1961). *Soviet Physics, JETP,* **13**, 1320.

Batty, W., Ekenberg, U., Ghiti, A., and O'Reilly, E. P. (1989). *Semicond. Sci. Technol.,* **4**, 904.

Bernard, M. G., and Duraffourg, G. (1961). *Phys. Status Solidi,* **1**, 699.

Bhattacharya, P., Singh, J., Yoon, H., Zhang, X. K., Gutierrezaitken, A., and Lam, Y. L. (1996). *IEEE J. Quantum Electron.,* **32**, 1620.

Bour, D. P., Martinelli, R. U., Hawrylo, F. Z., Evans, G. A., Carlson, N. W., and Gilbert, D. B. (1990). *Appl. Phys. Lett.,* **56**, 318.

Bowers, J. E., Hemenway, B. R., Gnauck, A. H., and Wilt, D. P. (1986). *IEEE J. Quantum Electron.,* **QE-22**, 833.

Broido, D. A., and Sham, L. J. (1985). *Phys. Rev. B,* **31**, 888.

Burt, M. G. (1983). *Electron. Lett.,* **19**, 210.

Burt, M. G. (1984). *Electron. Lett.,* **20**, 27.

Capasso, F., Cho, A. Y., Mohammed, K., and Foy, P. W. (1985). *Appl. Phys. Lett.,* **46**, 664.

Chand, N., Becker, E. E., van der Ziel, J. P., Chu, S. N. G., and Dutta, N. K. (1991). *Appl. Phys. Lett.,* **58**, 1704.

Channin, D. P. (1979). *J. Appl. Phys.,* **50**, 3858.

Chen, L. P., and Lau, K. Y. (1996). *IEEE Photon. Technol. Lett.,* **8**, 185.

Chen, H. Z., Ghaffari, A., Morkoç, H., and Yariv, A. (1987). *Electron. Lett.,* **23**, 1334.

Chen, T. R., Eng, L. E., Zhao, B., Zhuang, Y. H., Sanders, S., Morkoç, H. and Yariv, A. (1990). *IEEE J. Quantum Electron.,* **QE-26**, 1183.

Chen, T. R., Zhao, B., Zhuang, Y. H., Yariv, A., Ungar, J. E., and Oh, S. (1992a). *Appl. Phys. Lett.,* **60**, 1782.

Chen, T. R., Zhuang, Y. H., Xu, Y. J., Zhao, B., Yariv, A., Ungar, J. E., and Oh, S. (1992b). *Appl. Phys. Lett.,* **60**, 2954.

Chen, T. R., Zhao, B., Yamada, Y., Zhuang, Y. H., and Yariv, A. (1992c). *Electron. Lett.* **28**, 1989.

References

Chen, T. R., Eng, L. E., Zhao, B., Zhuang, Y. H., and Yariv, A. (1993a). *Appl. Phys. Lett.,* **63**, 2621.

Chen, T. R., Zhao, B., Eng, L. E., Zhuang, Y. H., O'Brien, J., and Yariv, A. (1993b). *Appl. Phys. Lett.,* **63**, 2621.

Chen, T. R., Zhao, B., Eng, L. E., Feng, J., Zhuang, Y. H., and Yariv, A. (1995). *Electron. Lett.,* **31**, 285.

Cheng, S. P., Brillouet, F., and Correc, P. (1988). *IEEE J. Quantum Electron.,* **QE-24**, 2433.

Chinn, S. R., Zory, P. S., and Reisinger, A. R. (1988). *IEEE J. Quantum Electron.,* **QE-24**, 2191.

Cho, A. Y. (1971). *J. Vac. Sci.,* **8**, S31.

Cho, A. Y., Dixon, R. W., Casey, H. C., Jr., and Hartman, R. L. (1976). *Appl. Phys. Lett.,* **28**, 501.

Choi, H. K., and Wang, C. A. (1990). *Appl. Phys. Lett.,* **57**, 321.

Chong, T. C., and Fonstad, C. G. (1989). *IEEE J. Quantum Electron.,* **QE-25**, 171.

Chow, W. W., Dente, G. C., and Depatie, D. (1987). *IEEE J. Quantum Electron.,* **QE-23**, 1314.

Chow, W. W., Koch, S. W., and Sargent, M. (1994). *Semiconductor-Laser Physics,* Springer-Verlag, New York.

Coldren, L. A., and Corzine, S. W. (1995). *Diode Lasers and Photonic Integrated Circuits,* John Wiley & Sons, New York.

Coleman, J. J. (1995). In *Semiconductor Lasers: Past, Present, and Future* (G. P. Agrawal, ed.), Chap. 1, AIP Press, Woodbury, New York.

Corzine, S. W., Yan, R. H., and Coldren, L. A. (1990). *Appl. Phys. Lett.,* **57**, 2835.

Cutrer, D. M., and Lau, K. Y. (1995). *IEEE Photon. Technol. Lett.,* **7**, 4.

Dallesasse, J. M., and Holonyak, N. Jr. (1991). *Appl. Phys. Lett.,* **58**, 394.

Dallesasse, J. M., Holonyak, N. Jr., Sugg, A. R., Richard, T. A., and Elzein, N. (1990). *Appl. Phys. Lett.,* **57**, 2844.

Davis, L., Sun, H. C., Yoon, H., and Bhattacharya, P. K. (1994). *Appl. Phys. Lett.,* **64**, 3222.

Derry, P. L., Chen, T. R., Zhuang, Y. H., Paslaski, J., Mittelstein, M., Vahala, K., Yariv, A., Lau, K. Y., and Bar-Chaim, N. (1988). *Optoelectronics: Dev. and Technol.,* **3**, 117.

Dong, H., Williamson, F., and Gopinath, A. (1996). *IEEE Photon. Technol. Lett.,* **8**, 46.

Dresselhaus, G. (1955). *Phys. Rev.,* **100**, 580.

Dupuis, R. D., and Dapkus, P. D. (1977). *Appl. Phys. Lett.,* **31**, 466.

Dupuis, R. D., Dapkus, P. D., Holonyak, N., Jr., Rezek, E. A., and Chin, R. (1978). *Appl. Phys. Lett.*, **32**, 295.

Dupuis, R. D., Dapkus, P. D., Chin, R., Holonyak, N., Jr., and Kirchoefer, S. (1979a). *Appl. Phys. Lett.*, **34**, 265.

Dupuis, R. D., Dapkus, P. D., Holonyak, N., Jr., and Kolbas, R. M. (1979b). *Appl. Phys. Lett.*, **35**, 487.

Dutta, N. K. (1982). *J. Appl. Phys.*, **53**, 7211.

Dutta, N. K., Wynn, J., Sivo, D. L., and Cho, A. Y. (1990a). *Appl. Phys. Lett.*, **56**, 2293.

Dutta, N. K., Temkin, H., Tanbun-Ek, T., and Logan, R. (1990b). *Appl. Phys. Lett.*, **57**, 1390.

Dutta, N. K., Hobson, W. S., Vakhshoori, D., Han, H., Freeman, P. N., Dejong, J. F., and Lopata, J. (1996). *IEEE Photon. Technol. Lett.*, **8**, 852.

Eng, L. E., Chen, T. R., Sanders, S., Zhuang, Y. H., Zhao, B., Yariv, A., and Morkoç, H. (1989). *Appl. Phys. Lett.*, **55**, 1378.

Eng, L. E., Mehuys, D. G., Mittelstein, M., and Yariv, A. (1990). *Electron. Lett.*, **26**, 1675.

Eng, L. E., Sa'ar, A., Chen, T. R., Grave, I., Kuze, N., and Yariv, A. (1991). *Appl. Phys. Lett.*, **58**, 2752.

Fukunaga, T., Wada, M., and Hayakawa, T. (1996). *Appl. Phys. Lett.*, **69**, 248.

Geels, R. S., and Coldren, L. A. (1990). *Appl. Phys. Lett.*, **57**, 1605.

Ghiti, A., O'Reilly, E. P., Adams, A. R. (1989). *Electron. Lett.*, **25**, 821.

Ghiti, A., Batty, W., and O'Reilly, E. P. (1990). *Superlatt. Microstruct.*, **7**, 353.

Girardin, F., Duan, G. H., Gallion, P., Talneau, A., and Ougazzaden, A. (1995). *Appl. Phys. Lett.*, **67**, 771.

Goncher, G., Lu, B., Luo, W. L., Cheng, J., Hersee, S., Sun, S. Z., Schneider, R. P., and Zolper, J. C. (1996). *IEEE Photon. Technol. Lett.*, **8**, 316.

Goobar, E., Peters, M. G., Fish, G., and Coldren, L. A. (1995). *IEEE Photon. Tech. Lett.*, **7**, 851.

Grupen, M., and Hess, K. (1997). *Appl. Phys. Lett.*, **70**, 808.

Hall, R. N., Fenner, G. E., Kingsley, J. D., Soltys, T. J., and Carlson, R. O. (1962). *Phys. Rev. Lett.*, **9**, 366.

Han, H., Freeman, P. N., Hobson, W. S., Dutta, N. K., Lopata, J., Wynn, J. D., and Chu, S. N. G. (1996). *IEEE Photon. Technol. Lett.*, **8**, 1133.

Harder, C., Vahala, K., and Yariv, A. (1983). *Appl. Phys. Lett.*, **42**, 328.

Hasuo, S., and Ohmi, T. (1974). *Jpn. J. Appl. Phys.*, **13**, 1429.

Hayakawa, T., Kondo, M., Suyama, T., Takahashi, K., Yamamoto, S., and Hijikata, T. (1987). *Jpn. J. Appl. Phys.,* **26**, L302.

Hayakawa, T., Suyama, T., Takahashi, K., Kondo, M., Yamamoto, S., and Hijikata, T. (1988). *Appl. Phys. Lett.,* **52**, 339.

Hayashi, Y., Mukaihara, T., Hatori, N., Ohnoki, N., Matsutani, A., Koyama, F., and Iga, K. (1995). *Electron. Lett.,* **31**, 560.

Henning, I. D., and Collins, J. V. (1983). *Electron. Lett.,* **19**, 927.

Henry, C. H., Logan, R. A., and Bertness, K. A. (1981). *J. Appl. Phys.,* **52**, 4457.

Henry, C. H. (1982). *IEEE J. Quantum Electron.,* **QE-18**, 259.

Hirayama, H., and Asada, M. (1994). *Optical Quantum Electron.,* **26**, S719.

Hofmann, M., Weise, W., Elsasser, W., and Gobel, E. O. (1994). *IEEE Photon. Tech. Lett.,* **6**, 1306.

Holonyak, N., Jr., Kolbas, R. M., Dupuis, R. D., and Dapkus, P. D. (1980). *IEEE J. Quantum Electron.,* **QE-16**, 170.

Hornak, L. A., Barr, J. C., Cox, W. D., Brown, K. S., Morgan, R. A., and Hibbs-Brenner, M. K. (1995). *IEEE Photon. Technol. Lett.,* **7**, 1110.

Huffaker, D. L., Deppe, D. G., Kumar, K., and Rogers, T. J. (1994a). *Appl. Phys. Lett.,* **65**, 97.

Huffaker, D. L., Shin, J., and Deppe, D. G. (1994b). *Electron. Lett.,* **30**, 1946.

Huffaker, D. L., Graham, L. A., Deng, H., and Deppe, D. G. (1996). *IEEE Photon. Technol. Lett.,* **8**, 974.

Iga, K. (1992). *IEICE Trans.,* **E75-A**, 12.

Ikeda, S., and Shimizu, A. (1991). *Appl. Phys. Lett.,* **59**, 504.

Ikeda, S., Shimizu, A., and Hara, T. (1989). *Appl. Phys. Lett.,* **55**, 1155.

Ishii, H., Tanobe, H., Kano, F., Tohmori, Y., Kondo, Y., and Yoshikuni, Y. (1996). *Electron. Lett.,* **32**, 454.

Kan, S. C., Vassilovski, D., Wu, T. C., and Lau, K. Y. (1992a). *IEEE Photon. Technol. Lett.,* **4**, 428.

Kan, S. C., Vassilovski, D., Wu, T. C., and Lau, K. Y. (1992b). *Appl. Phys. Lett.,* **61**, 752.

Kane, E. O. (1957). *J. Phys. Chem. Solids,* **100**, 580.

Kane, E. O. (1966). In *Semiconductors and Semimetals* (R. K. Willardson and A. C. Beer, eds.), Chap. 3, Academic Press, New York.

Kano, F., Yoshikuni, Y., Fukuda, M., and Yoshida, J. (1991). *IEEE Photon. Technol. Lett.,* **3**, 877.

Kano, F., Yamanaka, T., Yamamoto, N., Mawatari, H., Tohmori, Y., and Yoshikuni, Y. (1994). *IEEE J. Quantum Electron.,* **30**, 533.

Kapon, E., Simhony, S., Harbison, J. P., and Florez, L. T. (1990). *Appl. Phys. Lett.,* **56**, 1825.

Kazmierski, C., Ougazzaden, A., Robein, D., Mathoorasing, D., Blez, M., and Mircea, A. (1993). *Electron. Lett.,* **29**, 1290.

Kikuchi, K., Kakui, M., Zah, C. E., and Lee, T. P. (1991). *IEEE Photon. Technol. Lett.,* **3**, 314.

Kikuchi, K., Amano, M., Zah, C. E., Lee, T. P. (1994). *IEEE J. Quantum Electron.,* **30**, 571.

Kim, I., Alferness, R. C., Koren, U., Buhl, L. L., Miller, B. I., Young, M. G., Chien, M. D., Koch, T. L., Presby, H. M., Raybon, G., and Burrus, C. A. (1994). *Appl. Phys. Lett.,* **64**, 2764.

Kittel, C. (1987). *Quantum Theory of Solids,* John Wiley & Sons, New York.

Kjebon, O., Schatz, R., Lourdudoss, S., Nilsson, S., Stalnacke, B., and Backbom, L. (1997). *Electron. Lett.,* **33**, 488.

Koch, T. L., and Bowers, J. E. (1984). *Electron. Lett.,* **20**, 1038.

Koch, T. L., and Linke, R. A. (1986). *Appl. Phys. Lett.,* **48**, 613.

Kurobe, A., Furuyama, H., Naritsuka, S., Sugiyama, N., Kokubun, Y., and Nakamura, M. (1988). *IEEE J. Quantum Electron.,* **QE- 24**, 635.

Lau, K. Y. (1993). In *Quantum Well Lasers* (P. S. Zory, ed.), Chap. 5, Academic Press, San Diego.

Lau, K. Y., and Yariv, A. (1985a). *IEEE J. Quantum Electron.,* **QE-21**, 121.

Lau, K. Y., and Yariv, A. (1985b). In *Semiconductors and Semimetals,* Vol. 22B (W. T. Tsang, ed.), p. 69, Academic Press, Orlando.

Lau, K. Y., Bar-Chaim, N., Ury, I., and Yariv, A. (1984). *Appl. Phys. Lett.,* **45**, 316.

Lau, K. Y., Bar-Chaim, N., Derry, P. L., and Yariv, A. (1987). *Appl. Phys. Lett.,* **51**, 69.

Lau, K. Y., Derry, P. L., and Yariv, A. (1988). *Appl. Phys. Lett.,* **52**, 88.

Lau, K. Y., Xin, S., Wang, W. I., Bar-Chaim, N., and Mittelstein, M. (1989). *Appl. Phys. Lett.,* **55**, 1173.

Lealman, I. F., Bagley, M., Cooper, D. M., Fletcher, N., Harlow, M., Perrin, S. D., Walling, R. H., and Westbrook, L. D. (1991). *Electron. Lett.,* **27**, 1191.

Lealman, I. F., Okai, M., Robertson, M. J., Rivers, L. J., Perrin, S. D., and Marshall, P. (1996). *Electron. Lett.,* **32**, 339.

Lear, K. L., Mar, A., Choquette, K. D., Kilcoyne, S. P., Schneider, R. P., and Geib, K. M. (1996). *Electron. Lett.,* **32**, 457.

Lee, T. P., Burrus, C. A., Copeland, J. A., Dentai, A. G., and Marcuse, D. (1982). *IEEE J. Quantum Electron.,* **QE-18**, 1101.

Lehman, J. A., Morgan, R. A., Hibbs-Brenner, M. K., and Carlson, D. (1995). *Electron. Lett.,* **31**, 1251.

Lester, L. F., and Ridley, B. K. (1992). *J. Appl. Phys.,* **72**, 2579.

Lester, L. F., O'Keefe, S. S., Schaff, W. J., and Eastman, L. F. (1992). *Electron. Lett.,* **28**, 383.

Lidgard, A., Tanbun-Ek, T., Logan, R. A., Temkin, H., Wecht, K. W., and Olsson, N. A. (1990). *Appl. Phys. Lett.,* **56**, 816.

Lipsanen, H., Coblentz, D. L., Logan, R. A., Yadvish, R. D., Morton, P. A., and Temkin, H. (1992). *IEEE Photon. Technol. Lett.,* **4**, 673.

Loehr, J. P., and Singh, J. (1991). *IEEE J. Quantum Electron.,* **QE-27**, 708.

Lui, W. W., Yamanaka, T., Yoshikuni, Y., Seki, S., and Yokoyama, K. (1994). *IEEE J. Quantum Electron.,* **30**, 392.

Luttinger, J. M., and Kohn, W. (1955). *Phys. Rev.,* **97**, 869.

Luttinger, J. M. (1956). *Phys. Rev.,* **102**, 1030.

Lutz, C. R., Jr., Agahi, F., and Lau, K. M. (1995). *IEEE Photon. Technol. Lett.,* **7**, 596.

McIlroy, P. W. A., Kurobe, A., and Uematsu, Y. (1985). *IEEE J. Quantum Electron.,* **QE-21**, 1958.

Margalit, N. M., Babic, D. I., Streubel, K., Mirin, R. P., Naone, R. L., Bowers, J. E., and Hu, E. L. (1996). *Electron. Lett.,* **32**, 1675.

Masu, K., Tokumitsu, E., Konagai, M., and Takahashi, K. (1983). *J. Appl. Phys.,* **54**, 5785.

Mathur, A., and Dapkus, P. D. (1996). *IEEE J. Quantum Electron.,* **32**, 222.

Matsui, Y., Murai, H., Arahira, S., Kutsuzawa, S., and Ogawa, Y. (1997). *IEEE Photon. Technol. Lett.,* **9**, 25.

Mehuys, D. G., Mittelstein, M., Yariv, A., Sarfaty, R., and Ungar, J. E. (1989). *Electron. Lett.,* **25**, 143.

Meland, E., Holmstrom, R., Schlafer, J., Lauer, R. B., and Powazinik, W. (1990). *Electron. Lett.,* **26**, 1827.

Michalzik, R., and Ebeling, K. J. (1993). *IEEE J. Quantum Electron.,* **QE-29**, 1963.

Miller, B. I., Koren, U., Young, M. G., and Chien, M. D. (1991). *Appl. Phys. Lett.,* **58**, 1952.

Mittelstein, M., Arakawa, Y., Larsson, A., and Yariv, A. (1986). *Appl. Phys. Lett.,* **49**, 1689.

Mittelstein, M., Mehuys, D., Yariv, A., Ungar, J. E., and Sarfaty, R. (1989). *Appl. Phys. Lett.,* **54**, 1092.

Moreno, J. B. (1977). *J. Appl. Phys.*, **48**, 4152.

Morin, S., Deveaud, B., Clerot, F., Fujiwara, K., Mitsunaga, K. (1991). *IEEE J. Quantum Electron.*, **27**, 1669.

Morton, P. A., Logan, R. A., Tanbun-Ek, T., Sciortino, P. F., Jr., Sergent, A. M., Montgomery, R. K., and Lee, B. T. (1992). *Electron. Lett.*, **28**, 2156.

Nagarajan, R., Kamiya, T., and Kurobe, A. (1989). *IEEE J. Quantum Electron.*, **QE-25**, 1161.

Nagarajan, R., Fukushima, T., Bowers, J. E., Geels, R. S., and Coldren, L. A. (1991a). *Electron. Lett.*, **27**, 1058.

Nagarajan, R., Fukushima, T., Corzine, S. W., and Bowers, J. E. (1991b). *Appl. Phys. Lett.*, **59**, 1835.

Nagarajan, R., Ishikawa, M., Fukushima, T., Geels, R. S., and Bowers, J. E. (1992a). *IEEE J. Quantum Electron.*, **QE-28**, 1990.

Nagarajan, R., Mirin, R. P., Reynolds, T. E., and Bowers, J. E. (1992b). *IEEE Photon. Technol. Lett.*, **4**, 832.

Nagarajan, R., Fukushima, T., Ishikawa, M., Bowers, J. E., Geels, R. S., and Coldren, L. A. (1992c). *IEEE Photon. Technol. Lett.*, **4**, 121.

Nagle, J., Hersee, S., Krakowski, M., and Weisbuch, C. (1986). *Appl. Phys. Lett.*, **49**, 1325.

Narui, H., Hirata, S., and Mori, Y. (1992). *IEEE J. Quantum Electron.*, **28**, 4.

Nathan, M. I., Dumke, W. P., Burns, G., Dills, F. H., and Lasher, G. (1962). *Appl. Phys. Lett.*, **1**, 62.

Niwa, A., Ohtoshi, T., Kuroda, T. (1995). *IEEE J. Selected Topics Quantum Electron.*, **1**, 211.

Numai, T., Kawakami, T., Yoshikawa, T., Sugimoto, M., Sugimoto, Y., Yokoyama, H., Kasahara, K., and Asakawa, K. (1993). *Jpn. J. Appl. Phys.*, **32**, L1533.

O'Reilly, E. P., and Adams, A. R. (1994). *IEEE J. Quantum Electron.*, **30**, 366.

Osinski, M., and Buus, J. (1987). *IEEE J. Quantum Electron.*, **QE-23**, 9.

Osinski, J. S., Grodzinski, P., Zou, Y., Dapkus, P. D., Karim, Z., and Tanguary, A. R., Jr. (1992). *IEEE Photon. Technol. Lett.*, **4**, 1313.

Ougazzaden, A., Mircea, A., and Kazmierski, C. (1995). *Electron. Lett.*, **31**, 803.

Qian, Y., Zhu, Z. H., Lo, Y. H., Huffaker, D. L., Deppe, D. G., Hou, H. Q., Hammons, B. E., Lin, W., and Yu, Y. K. (1997). *IEEE LEOS Newsletter*, **11**(3), 16.

Ralston, J. D., Gallagher, D. F. G., Tasker, P. J., Zappe, H. P., Esquivias, I., and Fleissner, J. (1991). *Electron. Lett.*, **27**, 1720.

Ralston, J. D., Esquivias, I., Weisser, S., Gallagher, D. F. G., Tasker, P. J., Larkins, E. C., Rosenzweig, J., Zappe, H. P., Fleissner, J., and As, D. J. (1992). *High-Speed Electronics and Optoelectronics,* Proc. SPIE, Vol 1680, p. 127.

Rideout, W., Yu, B., LaCourse, J., York, P. K., Beernink, K. J., and Coleman, J. J. (1990). *Appl. Phys. Lett.,* **56**, 706.

Rideout, W., Sharfin, W. F., Koteles, E. S., Vassell, M. O., and Elman, B. (1991). *IEEE Photon. Technol. Lett.,* **3**, 784.

Sagawa, M., Toyonaka, T., Hiramoto, K., Shinoda, K., and Uomi, K. (1995). *IEEE J. Selected Topics Quantum Electron.,* **1**, 189.

Salvatore, R. A., Schrans, T., and Yariv, A. (1993). *IEEE Photon. Technol. Lett.,* **5**, 756.

Sargent, M., III, Scully, M. O., and Lamb, W. E., Jr. (1974). *Laser Physics,* Addison-Wesley, Reading, Mass.

Schonfelder, A., Weisser, S., Ralston, J. D., and Rosenzweig, J. (1993). *Electron. Lett.,* **29**, 1685.

Schonfelder, A., Weisser, S., Ralston, J. D., and Rosenzweig, J. (1994). *IEEE Photon. Technol. Lett.,* **6**, 891.

Schrans, T., Sanders, S., and Yariv, A. (1992). *IEEE Photon. Technol. Lett.,* **4**, 323.

Seki, S., Yamanaka, T., Liu, W., Yoshikuni, Y., and Yokoyama, K. (1993). *IEEE Photon. Technol. Lett.,* **5**, 500.

Seki, S., Yamanaka, T., Lui, W., Yoshikuni, Y., and Yokoyama, K. (1994). *IEEE J. Quantum Electron.,* **30**, 500.

Seltzer, C. P., Perrin, S. D., Tatham, M. C., and Cooper, D. M. (1991). *Electron. Lett.,* **27**, 1268.

Sorba, L., Bratina, G., Ceccone, G., Antonini, A., Walker, J. F., Micovic, M., and Franciosi, A. (1991). *Phys. Rev. B,* **43**, 2450.

Su, C. B., and Lanzisera, V. A. (1986). *IEEE J. Quantum Electron.,* **QE-22**, 1568.

Suemune, I. (1991). *IEEE J. Quantum Electron.,* **QE-27**, 1149.

Suemune, I., Coldren, L. A., Yamanishi, M., and Kan, Y. (1988). *Appl. Phys. Lett.,* **53**, 1378.

Sun, H. C., Davis, L., Sethi, S., Singh, J., and Bhattacharya, P. (1993). *IEEE Photon. Technol. Lett.,* **5**, 870.

Tabuchi, H., and Ishikawa, H. (1990). *Electron. Lett.,* **26**, 742.

Takahashi, T., Nishioka, M., and Arakawa, Y. (1991). *Appl. Phys. Lett.,* **58**, 4.

Takiguchi, T., Goto, K., Takemi, M., Takemoto, A., Aoyagi, T., Watanabe, H., Mihashi, Y., Takamiya, S., and Mitsui, S. (1994). *J. Crystal Growth,* **145**, 892.

Tauber, D. A., Spickermann, R., Nagarajan, R., Reynolds, T., Holmes, A. L., and Bowers, J. E. (1994). *Appl. Phys. Lett.*, **64**, 1610.

Temkin, H., Dutta, N. K., Tanbun-Ek, T., Logan, R. A., and Sergent, A. M. (1990). *Appl. Phys. Lett.*, **57**, 1610.

Terakado, T., Tsuruoka, K., Ishida, T., Nakamura, T., Fukushima, K., Ae, S., and Uda, A. (1995). *Electron. Lett.*, **31**, 2182.

Tessler, N., and Eisenstein, G. (1993). *Appl. Phys. Lett.*, **62**, 10.

Tessler, N., Nagar, R., Abraham, D., Eisenstein, G., Koren, U., and Raybon, G. (1992a). *Appl. Phys. Lett.*, **60**, 665.

Tessler, N., Nagar, R., and Eisenstein, G. (1992b). *IEEE J. Quantum Electron.*, **QE-28**, 2242.

Thibeault, B. J., Bertilsson, K. Hegblom, E. R., Strzelecka, E., Floyd, P. D., Naone, R., and Coldren, L. A. (1997). *IEEE Photon. Technol. Lett.*, **9**, 11.

Thijs, P. J. A., Tiemeijer, L. F., Kuindersma, P. I., Binsma, J. J. M., and van Dongen, T. (1991). *IEEE J. Quantum Electron.*, **QE-27**, 1426.

Thijs, P. J. A., Binsma, J. J. M., Tiemeijer, L. F., Slootweg, R. W. M., van Roijen, R., and van Dongen, T. (1992a). *Appl. Phys. Lett.*, **60**, 3217.

Thijs, P. J. A., Binsma, J. J. M., Tiemeijer, L. F., and van Dongen, T. (1992b). *Electron. Lett.*, **28**, 829.

Thijs, P. J. A., Tiemeijer, L. F., Binsma, J. J. M., and van Dongen, T. (1994). *IEEE J. Quantum Electron.*, **QE-30**, 477.

Thompson, G. H. B. (1980). *Physics of Semiconductor Laser Devices*, John Wiley & Sons, New York.

Tiemeijer, L. F., Thijs, P. J. A., de Waard, P. J., Binsma, J. J. M., and van Dongen, T. (1991). *Appl. Phys. Lett.*, **58**, 2738.

Tiemeijer, L. F., Thijs, P. J. A., Binsma, J. J. M., and van Dongen, T. (1992). *Appl. Phys. Lett.*, **60**, 2466.

Tiwari, S. (1994). *IEE Proc. Optoelectronics,* **141**, 163.

Tokuda, Y., Tsukada, N., Fujiwara, K., Hamanaka, K., and Nakayama, T. (1986). *Appl. Phys. Lett.*, **49**, 1629.

Toyonaka, T., Sagawa, M., Hiramoto, K., Shinoda, K., Uomi, K., and Ohishi, A. (1995). *Electron. Lett.*, **31**, 198.

Tsai, C.-Y., Tsai, C.-Y., Lo, Y.-H., and Eastman, L. F. (1994). *IEEE Photon. Technol. Lett.*, **6**, 1088.

Tsai, C. Y., Tsai, C. Y., Spencer, R. M., Lo, Y. H., and Eastman, L. F. (1996). *IEEE J. Quantum Electron.*, **32**, 201.

Tsang, W. T. (1978). *Appl. Phys. Lett.*, **34**, 473.

Tsang, W. T. (1982). *Appl. Phys. Lett.,* **40**, 217.
Tsang, W. T., Weisbuch, C., Miller, R. C., and Dingle, R. (1979). *Appl. Phys. Lett.,* **35**, 673.
Tucker, R. S. (1985). *IEEE J. Lightwave Technol.,* **LT-3**, 1180.
Twardowski, A., and Hermann, C. (1987). *Phys. Rev. B,* **35**, 8144.
Ulbrich, R., and Pilkuhn, M. H. (1970). *Appl. Phys. Lett.,* **16**, 516.
Uomi, K., Chinone, N., Ohtoshi, T., and Kajimura, T. (1985). *Jpn. J. Appl. Phys.,* **24**, L539.
Uomi, K., Mishima, T., and Chinone, N. (1987). *Appl. Phys. Lett.,* **51**, 78.
Uomi, K., Tsuchiya, T., Komori, M., Oka, A., Shinoda, K., and Oishi, A. (1994). *Electron. Lett.,* **30**, 2037.
Vahala, K., and Newkirk, M. (1990). *Appl. Phys. Lett.,* **57**, 974.
Vahala, K., and Yariv, A. (1983a). *IEEE J. Quantum Electron.,* **QE-19**, 1096.
Vahala, K., and Yariv, A. (1983b). *IEEE J. Quantum Electron.,* **QE-19**, 1102.
Vahala, K., and Yariv, A. (1984). *Appl. Phys. Lett.,* **45**, 501.
Vahala, K., Chiu, L. C., Margalit, S., and Yariv, A. (1984). *Appl. Phys. Lett.,* **42**, 631.
van der Ziel, J. P., Dingle, R., Miller, R., Wiegmann, W., and Nordland, W. A., Jr. (1975). *Appl. Phys. Lett.,* **26**, 463.
Vassell, M. O., Sharfin, W. F., Rideout, W., and Lee, J. (1992). *Appl. Phys. Lett.,* **61**, 1145.
Wang, J. A., and Schweizer, H. (1996). *IEEE Photon. Technol. Lett.,* **8**, 1441.
Wang, Z., Darby, D. B., Whitney, P., Panock, R., and Flanders, D. (1994). *Electron. Lett.,* **30**, 1413.
Weisbuch, C., and Nagle, J. (1987). *Physica Scripta,* **T19**, 209.
Weisbuch, C., and Vinter, B. (1991). *Quantum Semiconductor Structures: Fundamentals and Applications,* Academic Press, San Diego.
Weisser, S., Ralston, J. D., Larkins, E. C., Esquivias, I., Tasker, P. J., Fleissner, J., and Rosenzweig, J. (1992). *Electron. Lett.,* **28**, 2141.
Weisser, S., Larkins, E. C., Czotscher, K., Benz, W., Daleiden, J., Esquivias, I., Fleissner, J, Ralston, J. D., Romero, B., Sah, R. E., Schonfelder, A., and Rosenzweig, J. (1996). *IEEE Photon. Technol. Lett.,* **8**, 608.
Willatzen, M., Takahashi, T., and Arakawa, Y. (1992). *IEEE Photon. Technol. Lett.,* **4**, 682.
Williams, R. L., Dion, M., Chatenoud, F., and Dzurko, K. (1991). *Appl. Phys. Lett.,* **58**, 1816.

Wolf, H. D., Lang, H., and Korte, L. (1989). *Electron. Lett.*, **25**, 1249.

Wu, B., Georges, J. B., Cutrer, D. M., and Lau, K. Y. (1995). *Appl. Phys. Lett.*, **67**, 467.

Wu, M. C., Chen, Y. K., Tanbun-Ek, T., Logan, R. A., and Chin, M. A. (1991). *IEEE Photon. Technol. Lett.*, **3**, 874.

Xiao, J. W., Xu, J. Y., Yang, G. W., Zhang, J. M., Xu, Z. T., and Chen, L. H. (1992). *Electron. Lett.*, **28**, 154.

Yablonovitch, E., and Kane, E. O. (1986). *IEEE J. Lightwave Technol.*, **LT-4**, 504.

Yamada, M. (1983). *IEEE J. Quantum Electron.*, **QE-19**, 1365.

Yamada, M., and Haraguchi, Y. (1991). *IEEE J. Quantum Electron.*, **QE-27**, 1676.

Yamada, M., and Suematsu, Y. (1981). *J. Appl. Phys.*, **52**, 2653.

Yamada, M., Ogita, S., Yamagishi, M., Tabata, K., Nakaya, N., Asada, M., and Suematue, Y. (1984). *Appl. Phys. Lett.*, **45**, 324.

Yamada, M., Ogita, S., Yamagishi, M., and Tabata, K. (1985). *IEEE J. Quantum Electron.*, **QE-21**, 640.

Yamamoto, N., Yokoyama, K., Yamanaka, T., and Yamamoto, M. (1994a). *Electron. Lett.*, **30**, 243.

Yamamoto, M., Yamamoto, N., and Nakano, J. (1994b). *IEEE J. Quantum Electron.*, **30**, 554.

Yamanaka, T., Yoshikuni, Y., Lui, W., Yokoyama, K., and Seki, S. (1993). *Appl. Phys. Lett.*, **62**, 1191.

Yamanishi, M., and Suemune, I. (1984). *Jpn. J. Appl. Phys.*, **23**, L35.

Yang, G. M., MacDougal, M. H., and Dapkus, P. D. (1995). *Electron. Lett.*, **31**, 886.

Yariv, A. (1988). *Appl. Phys. Lett.*, **53**, 1033.

Yariv, A. (1989). *Quantum Electronics,* 3rd ed., John Wiley & Sons, New York.

Yokouchi, N., Yamanaka, N., Iwai, N., Nakahira, Y., and Kasukawa, A. (1996). *IEEE J. Quantum Electron.*, **QE-32**, 2148.

Yoon, H., Sun, H. C., and Bhattacharya, P. K. (1994). *Electron. Lett.*, **30**, 1675.

Yoon, H., Gutierrezaitken, A. L., Jambunathan, R., Singh, J., and Bhattacharya, P. K. (1995). *IEEE Photon. Technol. Lett.*, **7**, 974.

York, P. K., Lansjoen, S. M., Miller, L. M., Beernink, K. J., Alwan, J. J., and Coleman, J. J. (1990). *Appl. Phys. Lett.*, **57**, 843.

Yu, R. C., Nagarajan, R., Reynolds, T., Holmes, A., Bowers, J. E., DenBaars, S. P., and Zah, C. E. (1994). *Appl. Phys. Lett.*, **65**, 528.

Zah, C. E., Favire, F. J., Bhat, R., Menocal, S. G., Andreadakis, N. C., Hwang, D. M., Koza, M., and Lee, T. P. (1990). *IEEE Photon. Technol. Lett.*, **2**, 852.

Zah, C. E., Bhat, R., Pathak, B., Caneau, C., Favire, F. J., Andreadakis, N. C., Hwang, D. M., Koza, M. A., Chen, C. Y., and Lee, T. P. (1991). *Electron. Lett.,* **27**, 1414.

Zah, C. E., Bhat, R., Favire, F. J., Koza, M., Lee, T. P., Darby, D., Flanders, D. C., and Hsieh, J. J. (1992). *Electron. Lett.,* **28**, 2323.

Zhang, G., and Ovtchinnikov, A. (1993). *Appl. Phys. Lett.,* **62**, 1644.

Zhang, X., Gutierrezaitken, A. L., Klotzkin, D., Bhattacharya, P., Caneau, C., and Bhat, R. (1996). *Electron. Lett.,* **32**, 1715.

Zhao, B., Chen, T. R., and Yariv, A. (1991). *Electron. Lett.,* **27**, 2343.

Zhao, B., Chen, T. R., and Yariv, A. (1992a). *Appl. Phys. Lett.,* **60**, 313.

Zhao, B., Chen, T. R., and Yariv, A. (1992b). *IEEE Photon. Technol. Lett.,* **4**, 124.

Zhao, B., Chen, T. R., and Yariv, A. (1992c). *Appl. Phys. Lett.,* **60**, 1930.

Zhao, B., Chen, T. R., Yamada, Y., Zhuang, Y. H., Kuze, N., and Yariv, A. (1992d). *Appl. Phys. Lett.,* **61**, 1907.

Zhao, B., Chen, T. R., Zhuang, Y. H., Yariv, A., Ungar, J. E., and Oh, S. (1992e). *Appl. Phys. Lett.,* **60**, 1295.

Zhao, B., Chen, T. R., and Yariv, A. (1992f). *IEEE J. Quantum Electron.,* **QE-28**, 1479.

Zhao, B., Chen, T. R., Iannelli, J., Zhuang, Y. H., Yamada, Y., and Yariv, A. (1993a). *Appl. Phys. Lett.,* **62**, 1200.

Zhao, B., Chen, T. R., Wu, S., Zhuang, Y. H., Yamada, Y., and Yariv, A. (1993b). *Appl. Phys. Lett.,* **62**, 1591.

Zhao, B., Chen, T. R., Shakouri, A., and Yariv, A. (1993c). *Appl. Phys. Lett.,* **63**, 432.

Zhao, B., Chen, T. R., and Yariv, A. (1993d). *IEEE J. Quantum Electron.,* **QE-29**, 1027.

Zhao, B., Chen, T. R., Eng, L. E., Zhuang, Y. H., Shakouri, A., and Yariv, A. (1994). *Appl. Phys. Lett.,* **65**, 1805, 1994.

Zhao, H. M., Cheng, Y., Macdougal, M. H., Yang, G. M., and Dapkus, P. D. (1995). *IEEE Photon. Technol. Lett.,* **7**, 593.

Zory, P. S., ed. (1993). *Quantum Well Lasers,* Academic Press, San Diego.

Chapter 2

Strained Layer Quantum Well Lasers

Alfred R. Adams
Eoin P. O'Reilly
Mark Silver

Department of Physics, University of Surrey, Guildford, United Kingdom

2.1 Introduction

When a forward bias voltage is applied to a semiconductor diode laser, the injected current of electrons and holes flows directly into states in the active region, where they can recombine with the emission of light. This direct conversion from electrical to optical power leads to high efficiency and the possibility of fast modulation. However, the broad energy bands in semiconductors that, on the one hand, make the injection and carrier transport possible, on the other hand, allow the electrons and holes to occupy a wide range of energy and momentum values. Thus, because the useful laser output occurs as a beam of photons with a sharply defined energy, direction, and polarization, only a very small fraction of the injected carriers has the correct energy and momentum to contribute directly to the laser gain. Nevertheless, most of the carriers can take part in nonradiative loss processes and in spontaneous emission, both of which reduce the efficiency of the device. One approach toward solving the first problem of wide energy distributions is to progressively confine the carriers to lower dimensions, forming quantum wire and quantum dot structures, which concentrate the density of states and hence the carriers to the band edge, as described in Chap. 4. The second problem of unwanted spontaneous emission could be controlled through the formation of microcavity lasers [see e.g. Sale (1995)]. Here we consider how diode laser characteristics can be greatly improved by introducing strain within the active region. We describe below how strain can decrease the density of states at the valence band maximum and so reduce the carrier density required to reach threshold. It also reduces the three-dimensional symmetry of the lattice and helps match the wavefunctions of the holes to the polarization of the laser beam, thus reducing the spontaneous emission. These modifications were predicted by Adams (1986) and by Yablonovitch and Kane (1986) to have significant benefits for semiconductor laser performance, including reduced threshold current density, improved efficiency and temperature sensitivity, and enhanced dynamic response and high-speed performance (Suemune et al., 1988; Ghiti et al., 1989; Lau et al., 1989). Many of these proposed advantages have now been demonstrated, and strained lasers are available commercially at a wide range of wavelengths.

To examine more closely the first problem, concerning the energy spread of the electrons, it is useful to consider the situation at transparency in a semiconductor laser. The Bernard-Duraffourg (1961) condition tells us that transparency occurs when the quasi-Fermi levels for electrons E_{fc} and for holes E_{fv} are separated by more than the band-gap energy E_g:

$$E_{fc} - E_{fv} > E_g \quad (2.1)$$

However, in a bulk direct-gap III–V semiconductor such as GaAs, the conduction band has a low effective mass m_c, whereas the highest valence band is always heavy-hole-like, with a large effective mass m_h. This causes the quasi-Fermi level for holes E_{fv} to move downward much more slowly than E_{fc} moves upward as the injected carrier density is increased. Thus E_{fc} is well into the conduction band before E_{fv} has reached the valence band edge, as illustrated for a bulk laser in Fig. 2.1(a) and an idealized unstrained quantum well laser in Fig. 2.1(b). If it were possible to make the densities of states of the conduction and valence bands equal, transparency would be achieved as E_{fc} and E_{fv} simultaneously cross the band edges, and the carriers would only be spread in energy by thermal effects.

2.1 Introduction

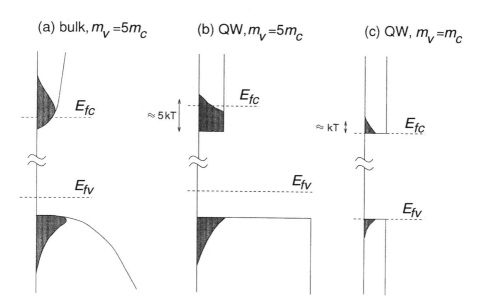

Figure 2.1: Illustration of the electron and hole distributions at transparency in the case of (a) an unstrained bulk semiconductor, where the density of states (plotted horizontally) varies as $E^{1/2}$ and the conduction band density of states (DOS) is typically at least a factor of 5 smaller than the valence band DOS, (b) a model unstrained quantum well, where we again assume the valence band DOS is 5 times larger than the conduction band DOS, and (c) an ideal quantum well laser structure, where the valence band DOS equals the conduction band DOS. The shaded regions indicate the electron and hole distributions in the conduction and valence bands, respectively.

This situation is illustrated in Fig. 2.1(c). When a layer is grown in a state of strain, both the cubic symmetry of the lattice and the degeneracy of the so-called heavy- and light-hole bands are removed. Large strains, either compressive or tensile, can lead to a considerable reduction in the hole mass (Osbourn, 1986; Schirber *et al.*, 1985; O'Reilly, 1989; Lancefield *et al.*, 1990; Krijn *et al.*, 1992) and hence in the density of states at the top of the valence band (Krijn *et al.*, 1992; Silver and O'Reilly, 1994) so that the laser characteristics may then approach the more ideal situation illustrated in Fig. 2.1(c).

We now consider the second problem of matching the symmetry of the carrier distibution to the uniaxial symmetry of the laser beam. The states composing the top of the valence band have almost pure *p*-like symmetry (Harrison, 1980), and therefore, in a bulk semiconductor the holes are equally distributed between p_x, p_y, and p_z states, and only one-third, e.g., those in the p_z states, can recombine to emit light along the laser cavity with the required polarization, e.g., in the TM mode. However, since the uniaxial strain removes the cubic symmetry of the lattice, it changes the relative energies of the p_x, p_y, and p_z states. This can be used to selectively engineer the states at the valence band maximum so that almost all the injected holes are in states of the correct symmetry to take part in the lasing action (O'Reilly et al., 1991; Sugawara, 1992; Jones and O'Reilly, 1993).

Both the reduced density of states and the improved symmetry of the states at the valence band maximum lead to a considerable reduction in the radiative threshold current and threshold carrier density in all types of semiconductor lasers. In addition, because all nonradiative loss mechanisms depend on the threshold carrier density, these are also reduced and the efficiency of the lasers increased. This is particularly marked in lasers operating at 1.5 μm, where Auger recombination is reduced (Thijs et al., 1994) and intervalence band absorption is effectively eliminated by the introduction of strain (Ring et al., 1992).

Since the strain helps assemble the injected carriers into those states contributing to the laser gain, it greatly increases the rate of change of gain G with carrier concentration n (Suemune et al., 1988; Ghiti et al., 1989). Thus modulation of the laser output requires a smaller change in n than in unstrained devices and therefore results in a smaller change in refractive index and hence in chirp. Similarly, the linewidth enhancement factor is reduced.

Because strain can be accommodated successfully in thin epitaxial layers, it adds another degree of freedom to the combination of materials that can be grown. Thus it is possible to extend the operating wavelength of quantum well lasers grown on GaAs out to about 1.1 μm by using compressively strained InGaAs, as reviewed by Coleman (1993). This brings them into a useful wavelength range for pumping erbium- and praseodymium-doped optical-fiber amplifiers, and they also enjoy the improved band structure described above. Since compressive strain leads to enhanced gain of the TE mode, while tensile strain leads to enhanced gain of the TM mode, it is possible by an appropriate combination of both to produce a semiconductor optical amplifier that is practically insensitive to the polarization of the light (Tiemeijer et al., 1993).

We shall see below how the gain and threshold properties of lattice-matched structures vary more rapidly with the introduction of a small amount of strain than is the case for structures that are already highly strained. Since device processing and mounting can introduce additional and variable strains into the active region, the properties of highly strained devices then show less variation with processing than is the case for lattice-matched structures. Highly strained devices therefore are more suited for incorporation into optoelectronic integrated devices.

We begin the next section by describing the theoretical and experimental limits on strained growth. We then review in Sect. 2.3 the electronic structure of strained layers, concentrating in particular on the way in which strain improves the valence band structure. We give a brief overview of visible lasers in Sect. 2.4 and then focus in Sect. 2.5 on what has been learned and achieved through the incorporation of strain in lasers operating in the region of 1.5 μm. We give a short review in Sect. 2.6 of the improvements that strain has brought for high-speed operation of lasers and then consider optical amplifiers in Sect. 2.7. Finally, we summarize our conclusions in Sect. 2.8. The properties of visible lasers are described in more detail by Valster et al. in Volume 2, chapter 1, while the dynamic characteristics of semiconductor lasers are also reviewed in more detail by Nagarajan and Bowers in Chap. 3.

2.2 Strained layer structures

2.2.1 Elastic properties

It is now possible to grow high-quality strained epitaxial layer structures in which, for example, a single layer is composed of a semiconductor that normally would have a significantly different lattice constant a_e to the substrate lattice constant a_s. For a sufficiently thin epilayer (i.e., below the critical thickness), almost all the strain will be incorporated in the layer [Fig. 2.2(a)]. The epilayer is under a biaxial stress such that its in-plane lattice constant $a_\parallel$ equals the substrate lattice constant a_s. We consider growth along the [001] direction, which is the one normally used. The net strain in the layer plane $\varepsilon_\parallel$ is given by

$$\varepsilon_\parallel = \varepsilon_{xx} = \varepsilon_{yy} = \frac{a_s - a_e}{a_e} \qquad (2.2)$$

In response to the biaxial stress, the layer relaxes along the growth direction, the strain $\varepsilon_\perp$ ($= \varepsilon_{zz}$) being of opposite sign to $\varepsilon_\parallel$ and given by

$$\varepsilon_\perp = -\frac{2\sigma}{1-\sigma}\varepsilon_\parallel \qquad (2.3)$$

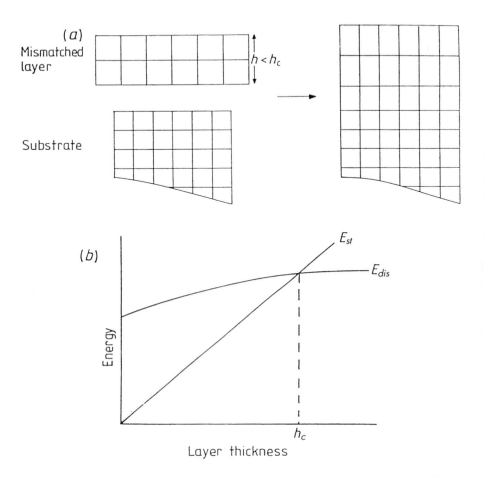

Figure 2.2: (a) Biaxial compression occurs when a mismatched layer has a larger lattice constant than the substrate; the layer is compressed in the x and y directions (the growth plane) and relaxes by expanding along the z axis (the growth direction). Biaxial tension occurs when the mismatched layer has a smaller lattice constant than the substrate. (b) Energy stored per unit area versus layer thickness h in a strained layer (E_{st}) and in a dislocation network relieving the strain (E_{dis}). The strained layer is thermodynamically stable for $h < h_c$.

where σ is Poisson's ratio. For tetrahedral semiconductors stressed along one of the principal axes, σ is approximately $\frac{1}{3}$, so $\varepsilon_\perp \approx -\varepsilon_\parallel$. The total strain can be resolved into a purely axial component ε_{ax}:

$$\varepsilon_{ax} = \varepsilon_\perp - \varepsilon_\parallel \approx -2\varepsilon_\parallel \tag{2.4}$$

2.2 Strained layer structures

and a hydrostatic component ε_{vol} ($= \Delta V/V$)

$$\varepsilon_{\text{vol}} = \varepsilon_{xx} + \varepsilon_{yy} + \varepsilon_{zz} \approx \varepsilon_\| \tag{2.5}$$

This resolution of the strain into two components is very useful because we can treat independently the effects of each on the band structure, as discussed below.

2.2.2 Critical layer thickness

The strain energy stored in a mismatched layer is linearly dependent on the thickness of the layer, whereas a certain minimum energy is associated with the formation of a dislocation and plastic relaxation [Fig. 2.2(b)]. Therefore, below a certain critical thickness h_c, the elastically strained layers are predicted to be thermodynamically stable, and high-quality "pseudomorphic" growth can be achieved. Studies of the critical thickness for the onset of dislocations in various materials systems have turned out to be quite difficult and sometimes even controversial. Part of this controversy arises as a result of the sensitivities of the different characterization techniques used to study the coherent growth of strained-layer structures (e.g., PL, TEM, x-ray diffraction). In practice, the experimentally observed critical thickness to introduce a small number of dislocations is comparable with the predicted value, and practical strained-layer structures are used widely to produce reliable laser devices.

The energy stored per unit area E_{st} increases linearly with layer thickness h as

$$E_{st} = 2G \frac{1 + \sigma}{1 - \sigma} \varepsilon_\|^2 h \tag{2.6}$$

where G is the shear modulus of the layer material. Expressions vary as to the energy required to introduce a network of dislocations (Frank and van der Merwe, 1949; Matthews and Blakeslee, 1974, 1975, 1976; People and Bean, 1985; Dunstan et al., 1991), with the most widely applied approach being that of Matthews and Blakeslee. The energy required to introduce a dislocation is large, since it involves breaking a line of bonds. Once present, the dislocation energy increases slowly with layer thickness, as $\ln(h)$, due to the strain field associated with it [Fig. 2.2(b)]. There will thus be a critical layer thickness h_c below which a dislocation-free strained layer is thermodynamically stable. This was calculated by Voisin (1988), for instance, to be

$$h_c = \frac{1 - \sigma/4}{4\pi(1 + \sigma)} b \varepsilon_\|^{-1}[\ln(h_c/b) + \theta] \tag{2.7}$$

where b is the dislocation Burgers vector (≈ 4 Å), and θ is a constant (~ 1) reflecting the dislocation core energy. This predicts $h_c \approx 90$ Å for $\varepsilon_\| = 1\%$, so good-quality stable layers may be expected with significant built-in strain and reasonable layer thicknesses. Comparison of theory and experiment is complicated by the expectation of a thickness-dependent energy barrier to the introduction of dislocations (Voisin, 1988), so the measured critical layer thickness will then depend on growth conditions (People and Bean, 1985; Kasper, 1986) and often can be larger than the predicted value. As an example, Andersson et al. (1987) demonstrated that high-quality growth of $In_xGa_{1-x}As$ on GaAs can be achieved if the product of lattice mismatch times layer thickness is less than about 200 Å%.

Two approaches are possible to the growth of strained multi-quantum well structures. A strained-layer superlattice with zero net stress can be grown if alternating layers are under appropriate biaxial compression and tension. Such a structure can, in principle, be grown *ad infinitum* as long as the individual layer thicknesses are each less than the relevant critical layer thickness. The theoretical critical layer thickness h_c^{SL} for an individual layer in a superlattice where the lattice mismatch is shared equally between the two materials is four times that of a single layer (Matthews and Blakeslee, 1976; Voisin, 1988). For a net strain superlattice (e.g., an $In_xGa_{1-x}As$-GaAs superlattice on GaAs), the situation is more complicated, but if the individual layers are sufficiently thin, then the theoretical critical thickness H_c for the complete superlattice structure is that expected for an alloy of the equivalent average composition on the given substrate.

It was feared initially that the excess energy that is dissipated during operation of a laser structure would encourage dislocation formation and migration even in high-quality, moderately strained layers and lead to rapid degradation of the material quality and laser characteristics. This is not the case. Bour et al. (1990), for instance, reported 10,000 hours of operation of InGaAs/GaAs lasers with a degradation rate of under 1% per kh. This degradation rate is significantly better than that of comparable lattice-matched GaAs/AlGaAs quantum well lasers (Harnagel et al., 1985; Waters and Bertaska, 1988). Coleman (1993) grew a series of 1.75%

strained InGaAs quantum well lasers on GaAs with different well widths L_z. Life-test measurements showed that the devices were at least as reliable as GaAs/AlGaAs lasers for $L_z = 100$ Å but that the threshold current increased rapidly with testing for $L_z = 125$ Å, giving a critical strain–well width product for these SQW devices of between 175 and 220 Å%. Impressive results also have been obtained at longer wavelengths, including the demonstration of 10,000 hours of reliable operation of 1480-nm strained-layer lasers employing four 1.8% compressively strained 25-Å quantum wells (Thijs et al., 1992). These lasers have a projected median lifetime at 70°C operation of 3.25×10^5 hours (37 years). This reflects the improved efficiency of strained-layer lasers and the mechanical stability of layers grown below the critical thickness.

2.3 Electronic structure and gain

2.3.1 Requirements for efficient lasers

The growth of strained-layer lasers has mostly taken place on (001) substrates or, for visible lasers, on substrates misoriented slightly from (001). We choose a coordinate axis in this chapter in which the growth (and strain) direction is along the z axis, with the x axis along the axis of the laser cavity, as shown in Fig. 2.3. Thus TM stimulated emission involves photons polarized along the z axis and, as we shall see below, is due to recombination with z-like valence states, while TE stimulated emission is due to recombination with y-like valence states.

The band structure of a bulk unstrained direct-gap III-V semiconductor is illustrated in Fig. 2.4(a). The light- and heavy-hole bands are degenerate at the zone center Γ, and the spin-split-off band lies at an energy E_{so} below the two highest bands. The lowest conduction band is approximately parabolic, and the electron dispersion at small k can be described by

$$E(k) = E_c + \frac{\hbar^2 k^2}{2m^*} \tag{2.8}$$

where E_c is the conduction band edge energy, and m^* is the electron effective mass. The valence band structure of Fig. 2.4(a) has several disadvantages for laser operation. We have already discussed in the introduction how (1) the large density of states associated with the heavy-hole band results in a large carrier density and a large spread in electron

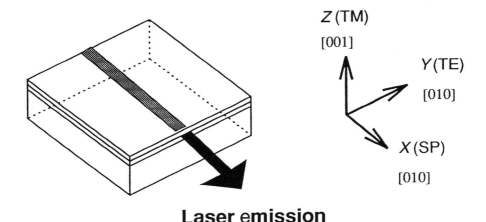

Figure 2.3: Laser structure (not to scale) showing coordinate geometry used in this chapter. The Z and Y directions indicate the directions of the electric field for stimulated emission in the TM and TE modes, respectively. Photons with their electric field in the X direction cannot contribute to gain along the cavity but only to spontaneous (SP) emission.

energies at population inversion and (2) how the isotropic polarization of the spontaneous emission increases the carrier and current density required for threshold, as compared with the case where only one polarization component contributes significantly to spontaneous emission and gain. In addition, (3) because the heavy and light bands are degenerate at the valence band maximum, both bands will experience a population inversion. While the carrier density in the light-hole band is much smaller than that in the heavy-hole band, the carrier recombination lifetime is shorter (Haug, 1987; Ghiti et al., 1992), and the two bands make approximately equal contributions to the radiative current density. Unfortunately, since the heavy-hole band predominantly provides TE gain and the light-hole band TM gain, only one of the two bands is thus contributing efficiently to the TE (or TM) polarized stimulated emission of the laser beam.

In summary, for an ideal semiconductor laser we require that (1) there is only one band at the valence band maximum, with (2) a density-of-states effective mass equal to that of the electrons, in order to minimize the carrier and current density for population inversion. Above transpar-

2.3 Electronic structure and gain

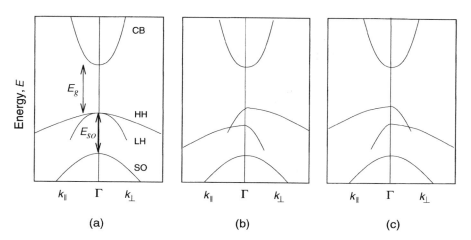

Figure 2.4: (a) A schematic representation of the band structure of an unstrained direct-gap tetrahedral semiconductor. The light-hole (LH) and heavy-hole (HH) bands are degenerate at the Brillouin zone center Γ, and the spin-split-off (SO) band lies E_{so} lower in energy. The lowest conduction band (CB) is separated by the band-gap energy from the valence bands. (b) Under biaxial compression, the hydrostatic component of the stress increases the mean band gap, while the axial component splits the degeneracy of the valence band maximum and introduces an anisotropic valence band structure. Note that this introduces an ambiguity in terminology, since the highest band is now *heavy* along $k_\perp$, the strain axis (= growth direction) but *light* along $k_\parallel$, in the growth plane. We label bands in this chapter by their mass along the growth direction, so a band that is referred to as heavy-hole may in fact have a low in-plane mass, as here. (c) Under biaxial tension, the mean band gap decreases and the valence band splitting is reversed, so the highest band is now *light* along $k_\perp$, the strain axis, and comparatively *heavy* along $k_\parallel$.

ency, we require (3) that the polarization of the gain is anisotropic, with spontaneous emission and gain being suppressed along all directions except along the axis contributing to the laser beam. The growth of strained quantum well structures can bring benefits in all three areas.

2.3.2 Strained-layer band structure

Axial strain breaks the cubic symmetry of the semiconductor lattice, introducing a tetragonal distortion that splits the degeneracy of the light- and heavy-hole states at the valence band maximum Γ, typically by about 60 to 80 meV for a 1% lattice mismatch. For strained epitaxial growth, the strain axis is parallel to the growth (z) axis. The resulting valence band structure is highly anisotropic [Fig. 2.4(b) and (c)], with the band that is heavy along the strain (z) axis, $k_\perp$ being comparatively light perpendicular to that direction, $k_\parallel$, and vice versa. The polarization of the spontaneous emission and gain spectra also becomes anisotropic (Jones and O'Reilly, 1993). The band structure anisotropy leads to a reduction in the density of states at the valence band maximum. For quantum well structures, the mass along the strain axis (= growth direction) determines the confinement energy, while the density of states is determined by the effective mass in the quantum well plane (i.e., perpendicular to the strain axis). Figure 2.4(b) thus predicts that there should be a marked reduction in the valence band density of states in compressively strained quantum wells, and this has indeed been demonstrated experimentally, with the hole effective mass m_h^*, for instance, being reduced from the bulk value of over 0.5 to a value of m_h^* of 0.155 in strained InGaAs quantum wells grown on GaAs substrates (Lancefield et al., 1990).

The combination of axial strain and quantum confinement allows a wide range of valence subband structures, with both tensile and compressive strain bringing benefits for laser applications. The variation of quantum well band structure with strain is illustrated in Fig. 2.5, where we show the calculated in-plane valence subband dispersion for lattice-matched and strained $In_{1-x}Ga_xAs$ quantum wells with lattice-matched InGaAsP barriers whose bandgap is 1.08 eV (Silver and O'Reilly, 1994). The band structures were calculated using the 6 × 6 Luttinger-Kohn Hamiltonian, which is described in more detail below, and the well widths were adjusted to give laser emission at a wavelength of about 1.5 μm. In the lattice-matched case [Fig. 2.5(a)], quantum confinement splits the degeneracy of the heavy- and light-hole bands and brings the heavy-hole band to the valence band maximum. This band then has a low effective mass and low in-plane density of states near the valence band maximum, but only over a very limited energy range. The incorporation of compressive strain [Fig. 2.5(b)] increases the splitting between the heavy- and light-hole zone center states, leading to a reduced density of states over a much wider energy range, of obvious benefit for lasers. The situation is more complicated in tensile-strained quantum wells. The incorporation of a moderate amount of tensile strain can bring the light- and heavy-hole bands into coincidence [Fig. 2.5(c)], and subband mixing effects cause

2.3 Electronic structure and gain

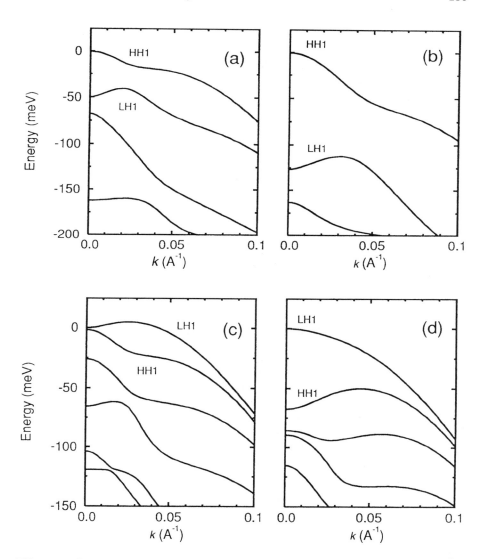

Figure 2.5: Valence subband structure calculated for (a) unstrained (50 Å), (b) 1.2% compressively strained (25Å), (c) 0.4% tensile strained (90 Å), and (d) 1.2% tensile strained (110 Å) InGaAs quantum wells designed to emit at 1.5 μm.

the highest valence band to have an electron-like dispersion near the zone center (negative effective mass) (Chang et al., 1986; O'Reilly, 1989; Viña et al., 1993). This band structure is particularly unsuited for conventional laser operation. For larger tensile strain, however [Figure 2.5(d)], the highest light-hole band shifts significantly above the heavy-hole states, and although the effective mass is larger than in the compressive case, the highest band is nevertheless approximately parabolic over a wide energy range and is well separated from the second valence subband, favoring laser action.

The band structure of bulk semiconductors and quantum wells can be calculated by many different methods, including the pseudo-potential (e.g., Jaros, 1985; Ninno et al., 1986) and tight-binding (Schulman and Chang, 1985) techniques. The technique, however, that is most widely used and has proved most fruitful is the envelope-function method both because of its adaptability and because of the many insights that it gives into valence band mixing and dispersion (O'Reilly, 1989; Corzine et al., 1993, Foreman, 1994). Because many of the benefits of strained lasers are due to the way in which strain modifies the valence band structure, we now consider the calculation of the strained valence band structure, focusing for simplicity mainly on the band structure of bulk strained layers but considering also the case of quantum wells.

2.3.3 Strained valence band Hamiltonian

For general treatment of the valence bands, we use the 6×6 Luttinger-Kohn (LK) Hamiltonian (Luttinger and Kohn, 1955; Luttinger, 1956). This includes the heavy-hole, light-hole, and spin-split-off bands and describes the dispersion of and interactions between these bands up to order k^2. Some authors include also the lowest conduction band (Krijn et al., 1992; Bastard and Brum, 1986; Eppenga et al., 1987) to give an 8×8 Hamiltonian. We omit the conduction band states from the Hamiltonian here because we are interested primarily in elucidating valence effects. We require six basis functions to describe the valence states at the zone center, which are formed from linear combinations of $\uparrow$ and $\downarrow$ spin functions and spatial functions X, Y, and Z, which have the same symmetry as atomic p states under the operation of the tetrahedral group. We consider here the 6×6 LK Hamiltonian, in an (X, Y, Z) basis, for a tetrahedral

2.3 Electronic structure and gain

semiconductor strained along the (001) direction, which (neglecting terms linear in k) is given by (Jones and O'Reilly, 1993)

$$\begin{array}{c} X\uparrow \\ Y\uparrow \\ Z\uparrow \\ X\downarrow \\ Y\downarrow \\ Z\downarrow \end{array} \begin{vmatrix} \alpha + a - b & c - i\Delta & e & 0 & 0 & \Delta \\ c + i\Delta & \alpha + a + b & d & 0 & 0 & -i\Delta \\ e & d & \alpha - 2a & -\Delta & i\Delta & 0 \\ 0 & 0 & -\Delta & \alpha + a - b & c + i\Delta & e \\ 0 & 0 & -i\Delta & c - i\Delta & \alpha + a + b & d \\ \Delta & i\Delta & 0 & e & d & \alpha - 2a \end{vmatrix}$$
(2.9)

where we have set $\hbar = m = 1$ and

$$\begin{aligned}
\alpha &= -\tfrac{1}{2}\gamma_1(k_x^2 + k_y^2 + k_z^2) - \Delta & c &= -3\gamma_3 k_x k_y \\
a &= \gamma_2(k_z^2 - \tfrac{1}{2}k_t^2) + S & d &= -3\gamma_3 k_y k_z \\
b &= 1.5\gamma_2(k_x^2 - k_y^2) & e &= -3\gamma_3 k_x k_z \\
\Delta &= E_{so}/3 & k_t^2 &= k_x^2 + k_y^2
\end{aligned}$$
(2.10)

with E_{so} equal to the spin-orbit splitting for zero strain and $S = -b\varepsilon_{ax}$ describing the strain-induced shift in the zone center states, where b is the axial deformation potential (O'Reilly, 1989) and ε_{ax} is the built-in axial strain. It can be seen how an (001) axial strain shifts the Z-like state in the opposite direction to the X- and Y-like valence band maximum states. This splitting can be understood from Fig. 2.2. In the unstrained cubic crystal, the x, y, and z directions are equivalent, and so the three valence states are at the same energy; strain differentiates the (001) (z) direction from the other two, so the Z-like state need no longer be at the same energy as the X- and Y-like valence states.

This Hamiltonian gives the same band dispersion as the more usual 6×6 Hamiltonian (Luttinger and Kohn, 1955; Eppenga et al., 1987; O'Reilly 1989) but gives a clearer insight into aspects of laser gain in bulk semiconductors because the x, y, or z character of each state directly reflects the polarization of photons associated with transitions to or from the given state (Jones and O'Reilly, 1993). TM gain in an (001) grown laser structure is, for example, associated with recombination between electrons in s-like conduction states and holes in z-like valence states. It is possible to transform from Eq. (2.9) to the Hamiltonian based on the heavy-hole $(|\tfrac{3}{2}, \pm\tfrac{3}{2}\rangle)$, light-hole $(|\tfrac{3}{2}, \pm\tfrac{1}{2}\rangle)$, and split-off $(|\tfrac{1}{2}, \pm\tfrac{1}{2}\rangle)$ zone center states quantized along the z direction. We recover, for instance, the Hamiltonian used by O'Reilly (1989) by using as basis states

$$|\tfrac{3}{2}, \tfrac{3}{2}\rangle = \tfrac{1}{\sqrt{2}} |X\uparrow + iY\uparrow\rangle$$

$$|\tfrac{3}{2}, -\tfrac{3}{2}\rangle = \tfrac{i}{\sqrt{2}} |X\downarrow - iY\downarrow\rangle$$

$$|\tfrac{3}{2}, \tfrac{1}{2}\rangle = \tfrac{i}{\sqrt{6}} |X\downarrow + iY\downarrow - 2Z\uparrow\rangle$$

$$|\tfrac{3}{2}, -\tfrac{1}{2}\rangle = \tfrac{1}{\sqrt{6}} |X\uparrow - iY\uparrow + 2Z\downarrow\rangle$$

$$|\tfrac{1}{2}, \tfrac{1}{2}\rangle = \tfrac{1}{\sqrt{3}} |X\downarrow + iY\downarrow + Z\uparrow\rangle$$

$$|\tfrac{1}{2}, -\tfrac{1}{2}\rangle = -\tfrac{i}{\sqrt{3}} |X\uparrow - iY\uparrow - Z\downarrow\rangle$$

(2.11)

The 6×6 LK Hamiltonian of Eq. (2.9) can be further simplified for specific applications. Laser gain calculations often use the axial approximation (Altarelli et al., 1985), which retains the exact dispersion along the (001) direction (when $k_x = k_y = 0$) but modifies the Hamiltonian so that it is axially symmetric in the xy plane about the z axis. This is achieved by replacing γ_2 and γ_3 by γ_{av} in b and c, the terms involving $(k_x^2 - k_y^2)$ and $k_x k_y$, where

$$\gamma_{av} = \tfrac{1}{2}(\gamma_2 + \gamma_3) \qquad (2.12)$$

The axial approximation works well for bulk layers and is also particularly suited to quantum well structures because it gives the correct confinement energies and still allows a straightforward expression for the average matrix elements as a function of k_t. The gain curves agree closely with those calculated by retaining the angular dependence (Colak et al., 1987), so the axial approximation is now widely used to study gain in semiconductor lasers, as described below. If we set $\mathbf{k} = (k_t \cos \phi, k_t \sin \phi, k_z)$, the eigenvalues for a given k_t and k_z are then independent of ϕ, and the 6×6 Hamiltonian of Eq. (2.9) can be decoupled into two independent 3×3 Hamiltonians (Broido and Sham, 1985; Ahn and Chuang, 1988; Jones and O'Reilly 1993), one of which takes the form

$$\begin{vmatrix} \alpha + A - B & -i\Delta & C + i\Delta \\ i\Delta & \alpha + A + B & \Delta \\ C - i\Delta & \Delta & \alpha - 2A \end{vmatrix} \qquad (2.13)$$

2.3 Electronic structure and gain

where

$$\alpha = -\tfrac{1}{2}\gamma_1 k^2 - \Delta \qquad B = 1.5\gamma_{\mathrm{av}}k_t^2$$
$$A = \gamma_2(k_z^2 - \tfrac{1}{2}k_t^2) + S \qquad C = -3\gamma_3 k_t k_z \qquad (2.14)$$
$$k^2 = k_t^2 + k_z^2$$

We used the Hamiltonian of Eq. (2.13) to calculate laser gain in strained bulklike layers (Jones and O'Reilly, 1993).

In an unstrained bulklike semiconductor [Fig. 2.4(a)], the heavy- and light-hole bands are degenerate at the valence band maximum, which has equal contributions from X-, Y-, and Z-like states so that spontaneous emission comes from two bands and has equal components polarized along the three principal axes. Hence, when carriers are injected into such a laser, equal proportions contribute to TM gain (polarized along z), to TE gain (polarized along y), and to spontaneous emission polarized along the direction of the laser cavity (taken here to be x). Therefore, only one-third of the holes at the correct energy are in the right polarization state to contribute to the lasing mode.

With biaxial compression [Fig. 2.4(b)], the heavy-hole state shifts upward in energy; since this state has no z character at the center of the Brillouin zone and equal x and y character, TE gain will be enhanced and TM gain suppressed. One in two carriers near the band edge now contributes directly to the dominant polarization component, so the threshold current density should decrease and the differential gain should increase with increasing biaxial compression. For biaxial tension, the light-hole state shifts upward in energy. This state has two-thirds z-like character at the zone center for small strains, and thus two of every three carriers are able to contribute to the dominant gain mechanism. However, under biaxial strain, the light-hole and spin-split-off states become coupled by the strain causing a change in the character of the light-hole band. It can be shown by calculating the eigenvectors of the Hamiltonian that the fractional z-like character of the light-hole band varies with strain-splitting energy S as (Jones and O'Reilly, 1993)

$$f_z = \frac{1}{2}\left(1 + \frac{\Delta - 3S}{\sqrt{9\Delta^2 - 6S\Delta + 9S^2}}\right) \qquad (2.15)$$

Figure 2.6 shows the calculated variation of f_z, the z-like character of the light-hole band, with tensile strain for 1.55-μm GaInAsP layers on InP

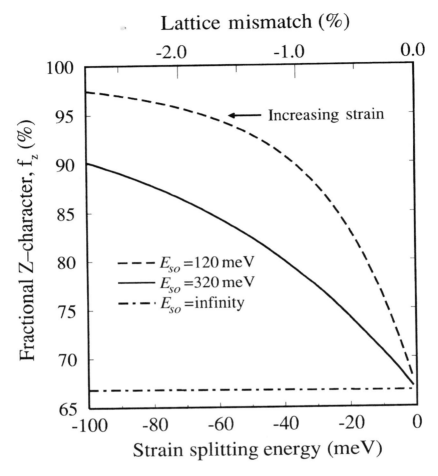

Figure 2.6: Plot of fractional z-like character f_z of the light-hole state as a function of tensile strain for GaInAsP on InP (solid line), GaInP on GaAs (dashed line), and assuming infinite spin-orbit splitting (dot-dashed line).

(solid line), for visible-wavelength GaInP layers on GaAs (dashed line), and assuming an infinite spin-orbit splitting (dot-dashed line).

We note that since f_z depends on the ratio of the strain-splitting energy S (related to the lattice mismatch) to the spin-orbit splitting energy $E_{so} = 3\Delta$, the enhancement of z-like character with strain will be most pronounced in phosphide-based tensile lasers, such as the tensile

$Ga_{1-x}In_xP$ on GaAs structures that have demonstrated significant improvements for short-wavelength (visible red) semiconductor lasers. The value of f_z relates directly to the transition matrix elements in the expression for the radiative gain (Jones and O'Reilly, 1993), and this has a pronounced effect on the performance of the TM gain in tensile-strained structures.

2.3.4 Laser gain

To illustrate the characteristics of the strained valence band structure outlined above and its effects on laser performance, we initially calculate the variation in peak gain with carrier density and radiative current density in strained bulk layers. This allows us to concentrate on strain effects alone, without the added complication of quantum confinement, which will be described later. Figure 2.7 shows the peak gain as a function of (a) carrier density and (b) radiative current density for bulk strained and unstrained layers. The calculations were based on the GaInAsP material system, assuming an optical gap of 800 meV, spin-orbit splitting of 320 meV, and the strain-splitting energy S set equal to ± 50 meV in the compressive and tensile layers, respectively. This is equivalent to about a 1.2% lattice mismatch between the strained layers and the substrate.

The TE and TM gains are equal in the unstrained layer (long dashed line). When the layer is under biaxial compression, with $S = 50$ meV, the heavy-hole states at the valence band maximum contribute equally to emission polarized along the x and y directions. The TE gain (dashed line) is then enhanced compared with the unstrained case, since one of every two carriers near the band maximum contributes to TE gain, compared with one in three in the unstrained case. By contrast, the TM gain (o) is strongly suppressed, since only states from the light-hole band contribute to it, and such states have been shifted away from the valence band maximum. The transparency carrier density for TE gain is also reduced because of the reduced density of states at the valence band maximum, and the transparency radiative current density is reduced because recombination is predominantly with one valence band.

For layers under biaxial tension, with $S = -50$ meV, the states at the valence band maximum are predominantly z-like, and from Eq. (2.15), 83% of the carriers there contribute to TM gain, with the remainder contributing equally to emissions polarized along the x and y directions. The TM gain (solid line) is then almost three times larger than in the

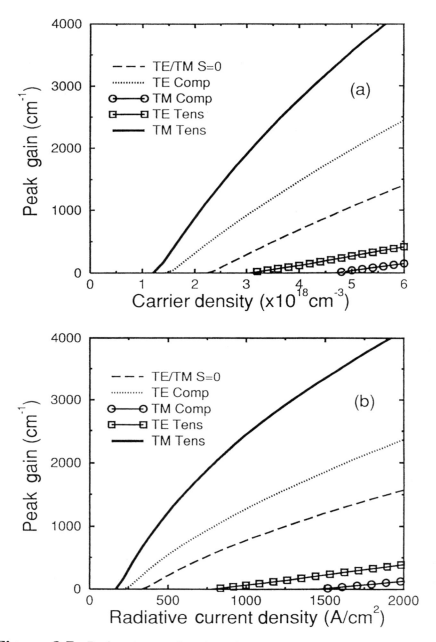

Figure 2.7: Peak gain as a function of (a) carrier density and (b) radiative current density in strained and unstrained laser structures (Jones and O'Reilly, 1993). The axial strain energy S is ± 50 meV for layers under biaxial compression and tension, respectively, corresponding to a lattice mismatch of about 1.2%.

2.3 Electronic structure and gain

unstrained case at the same carrier or current density, whereas the TE gain (□) is strongly suppressed for large biaxial tension. The transparency carrier and current densities are further reduced in the tensile example compared with the compressive one because the interaction between the light-hole and spin-split-off bands leads to a larger splitting of the valence band maximum in the tensile than compressive case.

The differential gain dG/dn shown by the slopes of the curves in Fig. 2.7(a) is significantly enhanced with axial strain in the bulklike layers considered here, being almost three times greater for TM gain in the tensile-strained structure than in the unstrained case. The TE differential gain is also enhanced in the compressively strained structure but is still only approximately half the value found for TM gain under biaxial tension, due to the different character of the valence band maximum in the two cases.

We see from Fig. 2.7 that significant improvements can be obtained in lasers with a built-in strain. In practice, the incorporation of any strain in a bulklike layer will reduce the carrier and radiative current densities compared with equivalent unstrained structures (Jones and O'Reilly, 1993). Similar results are obtained for strained quantum well structures. The threshold current of quantum well structures depends on many factors that may vary with strain in a series of practical devices, including the well width (to maintain a particular emission wavelength), the band offsets, subband splittings, and carrier spillover, all of which influence the laser characteristics. Nevertheless, it is found experimentally (Thijs et al., 1991, 1994; Zah et al., 1991; Valster et al., 1992; Smith et al., 1993; Jones et al., 1998) and theoretically (Krijn et al., 1992; Silver and O'Reilly, 1994) that strain is often the dominant factor influencing the threshold characteristics, with the maximum threshold current generally found in quantum wells with moderate tensile strain, where quantum confinement and strain effects are balanced so that the heavy- and light-hole bands are degenerate at the valence band maximum. This is illustrated in Fig. 2.8, which shows the calculated radiative current density at transparency (dashed line) and at threshold (solid line) as a function of strain for structures designed to emit at 1.5 μm, each of which has four $In_{1-x}Ga_xAs$ quantum wells grown on InP (Silver and O'Reilly, 1994). It can be seen that the largest threshold current density occurs in a quantum well structure for moderate tensile strain, when the highest heavy- and light-hole states are approximately degenerate. The lattice-matched characteristics are improved compared with this worst case owing to quantum confine-

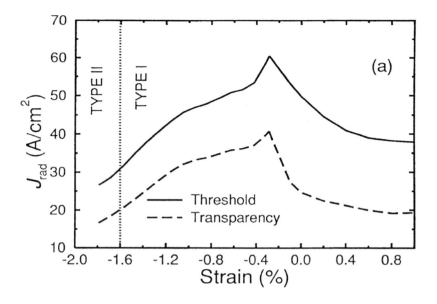

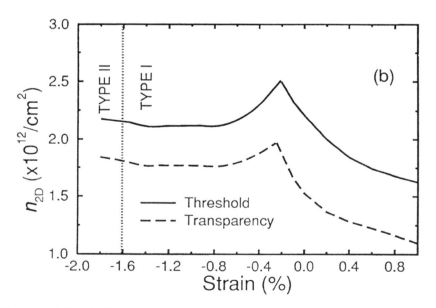

Figure 2.8: (a) Radiative component of the current density and (b) of the 2D carrier density as a function of strain at transparency (dashed lines) and threshold (solid lines) for $In_{1-x}Ga_xAs$ laser structures designed to emit at 1.5 μm (from Silver and O'Reilly, 1994, based on lasers reported in Thijs *et al.*, 1991).

2.3 Electronic structure and gain

ment effects in Fig. 2.8, and the further improvement in going to, say, 1.2% compressive strain is then less marked than was the case in Fig. 2.7, where the lattice-matched layer is the "worst case" bulk layer. Because the bandgap of strained $In_{1-x}Ga_xAs$ increases with decreasing x, it is necessary to change the quantum well width from 160 Å for 2.2% tension to 25 Å for 1.2% compression. Since the critical layer thickness is expected to be similar for layers under compressive or tensile strain, it is to be expected, and it is indeed found experimentally (Thijs et al., 1994), that the widest, highest strained tensile layers considered here are above the critical thickness. The conduction band line-up is also calculated to become type II at about −1.4% strain in the structures considered, with electrons at low carrier densities then being confined in the barrier layers. Despite these complications, we found that the peak radiative current density occurs at moderate tensile strain and that the radiative component of the threshold current density decreases both for compressive strain and at larger tensile strain. The threshold carrier density is more strongly influenced by the change in well width and also by the reduced electron confinement, resulting in little change in the two-dimensional carrier density with strain. We shall return to discuss these results further in Sect. 2.5.

2.3.5 Strained layers on non-[001] substrates

We have focused here on the advantages of strain for lasers grown on (001) substrates, since this is by far the most common substrate orientation used. It is interesting, however, to consider the benefits or otherwise of strained growth for other substrate orientations. Previous experimental (Hayakawa et al., 1988a, 1988b) and theoretical (Ghiti et al., 1990) work has shown that the threshold current density is reduced in [111]-oriented *lattice-matched* lasers compared with equivalent [001] devices. This arises primarily because the heavy-hole mass is anisotropic in a bulk semiconductor, being, for instance, approximately twice as large along [111] as along [001] in GaAs (Mollenkamp et al., 1988; Shanabrook et al., 1989). The increased heavy-hole mass then leads to an enhanced splitting between the highest heavy-hole state, HH1, and the highest light-hole state, LH1, in lattice-matched [111]-oriented structures compared with the [001] case. This enhanced splitting leads to a reduced density of states at and near the valence band maximum in the [111] lattice-matched case and consequently to improved lasing characteristics. By contrast, the threshold characteristics in compressively strained lasers are in large part limited by the splitting between the two highest heavy-hole states, HH1 and HH2 (Seki et al., 1994), and since this energy separation is maximized along [001], we do not expect a further reduction in threshold current due to changing growth direction in this case. The situation is slightly more complicated in the tensile-strained case, where the relevant splitting is between LH1 at the valence band maximum and HH1 or LH2 as the next confined state; since the light-hole mass varies more weakly with growth direction, we do not expect an enhanced splitting contributing to a reduced threshold current in this case.

Growth on non-[001] substrates does, however, allow at least two possible interesting effects: the introduction of piezoelectric fields and polarization anisotropy. It was originally proposed theoretically (Mailhiot and Smith, 1987) and has been well confirmed experimentally (e.g., Laurich et al., 1989) that strained-layer structures grown along [111] [and in fact most non-[001] directions] have a built-in piezoelectric field due to the relative displacement of the cation and anion sublattices. The induced fields can be on the order of 10^5 V/cm for a 1% lattice mismatch. Such large fields have a significant effect on the band edge energy and can lead to Stark shifts on the order of 50 meV and to spatial separation of the electron and hole wavefunctions. Screening causes the built-in field to decrease with increasing carrier density, but it is calculated that typically only half the field would be screened at threshold for room-temperature laser operation. This carrier-dependent field effect has potential applications both for nonlinear devices (Mailhiot and Smith, 1987) and for lasers (Pabla et al., 1996).

Turning to consider polarization anisotropy, we saw above how the polarization characteristics in a bulk semiconductor are isotropic, whereas the incorporation of strain or growth of a quantum well along the (001) direction differentiates the z-direction from the xy plane, leading to an enhancement or suppression of TM polarized emission from a laser structure. However, with (001) growth, the two orthogonal directions remain equivalent in the xy plane, and the gain characteristics for vertical emission (along the (001) direction) are in principle isotropic. This can lead to polarization instability and switching just above threshold in vertical cavity lasers. The same problem is also expected in [111]-grown vertical cavity structures, where the two orthogonal polarization directions are again equivalent. We discuss below how the growth of chemically ordered layers on a [001] substrate may overcome this problem. Another route to polarization selectivity is to grow on a lower-symmetry substrate, such as $01\bar{1}$, where it has been shown both theoretically (Los et al., 1996) and experimentally (Sun et al., 1995) that the two orthogonal directions (100) and $(01\bar{1})$ are no longer equivalent, leading to an enhancement of the $(01\bar{1})$ polarized gain and a consequent suppression of the (100) polarization direction. The introduction of such novel effects as piezoelectric fields and polarization anisotropy should ensure that growth on non-[001] substrates will remain a field of active interest in the coming years.

2.4 Visible lasers

Visible lasers fabricated using the materials system AlGaInP grown on GaAs have received considerable attention because of their enormous commercial potential. Bulk devices operating CW at room temperature were demonstrated as early as 1985 by Kobayashi, but the best results have generally been obtained using strained-layer QW structures (see, for example, the chapter by Valster et al. in Volume 2). These wide-gap devices can be grown under either compressive (In-rich) or tensile (Ga-rich) strain. They do not suffer from nonradiative losses due to Auger recombination and intervalence band absorption, which are present in longer-wavelength devices, and therefore, they provide a good system for comparison with the theory outlined above. In addition, it is possible to achieve an ordered or a disordered crystal structure in this material system, depending on the growth conditions (Mascarenhas et al., 1989). In the disordered structure, the gallium and indium atoms occupy the group III sublattice at random, whereas in the ordered case, gallium and indium preferentially occupy alternate (111)-type planes in the crystal structure. This reduced symmetry has interesting consequences for polarization engineering, as discussed below.

A systematic study of the effects of strain in disordered GaInP/AlGaInP/GaAs QW lasers operating at 633 nm has been undertaken by Valster et al. (1992). The structures contained two quantum wells of $Ga_xIn_{1-x}P$, with x varying between 0.65 (-1% tension) and 0.38 ($+1\%$ compression). In order to keep the operating wavelength the same for all devices, it was necessary to vary the well width from about 100 Å (-1% tension) to 26 Å ($+1\%$ compression), thus compensating for the change in alloy composition with a change in the quantum confinement energy. The observed variation at 300 K of the threshold current I_{th} for 300-μm-long lasers with stripe widths of 7 μm is shown in Fig. 2.9 (from Thijs, 1992).

As can be seen, there is a close similarity between the theoretical curve shown, for instance, in Fig. 2.8(a) for InGaAs and the observed decrease in threshold current at moderate values of both tensile and compressive strain. Also, the laser beam is found to be TE polarized in devices with compressive strain and TM polarized in those with tensile strain, as predicted. However, in both directions of strain, the threshold current goes through a minimum near 0.5% strain and beyond this begins to increase. The causes for this are believed to be different on the tensile and compressive sides.

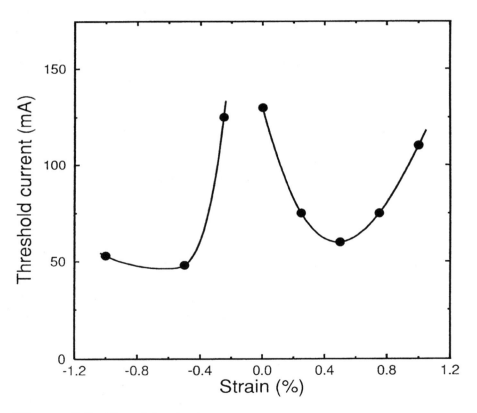

Figure 2.9: Plot of threshold current versus strain for $Ga_xIn_{1-x}P$ lasers designed to emit at 633 nm (from Thijs, 1992). (© 1992 IEEE)

2.4 Visible lasers

The factor limiting the increase in the bandgap of InGaP lasers is the presence of the indirect X and L conduction band minima, whose average energy alters little with alloy composition. However, these minima can be strongly influenced by the presence of strain, depending on its direction. The predicted variation of the Γ and X band edges for $Ga_xIn_{1-x}P$ grown on (001) GaAs is shown in Fig. 2.10 as a function of composition/

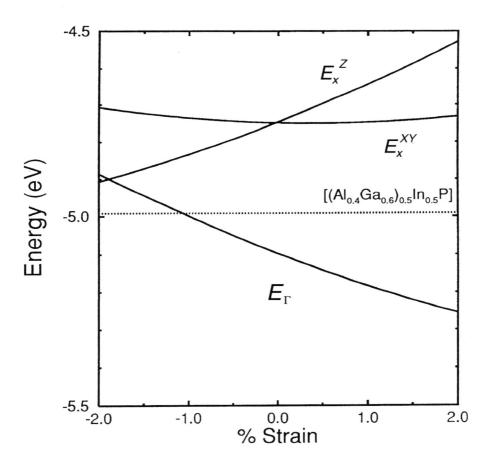

Figure 2.10: Calculated variation of the absolute energy of the Γ and X conduction band edge energies for $Ga_xIn_{1-x}P$ layers grown on GaAs. The horizontal dotted line indicates for comparison the calculated position of the conduction band minimum in $(Al_{0.4}Ga_{0.6})_{0.5}In_{0.5}P$, the typical quantum confining barrier layer used in this material system.

strain (from Hawley et al., 1993). There is a strong splitting of the X minima by the (001) strain so that in the tensile material the X minima along the z (growth) direction approach the Γ minimum. It is therefore believed that the increase in I_{th} observed at large tensile strain is due to the thermal excitation of electrons to the X_z minimum within the well. This is one of the few deleterious effects of strain on semiconductor lasers.

It is also clear from Fig. 2.10 that the direct bandgap decreases considerably as the compressive strain (In content) is increased. To keep the wavelength constant, this must be compensated by a decrease in the well width, to increase the quantum confinement energies. Unfortunately, this leads to a correspondingly large decrease in the optical confinement factor. The resulting increase in the threshold gain cannot be matched at large strain by improvements in the material gain, since the splitting between the heavy-hole and light-hole band edges saturates at about 60 meV for compressive strain in these materials due to the small spin-orbit splitting. We believe that it is this gain saturation that leads to the increased threshold current density in visible lasers with larger compressive strains. The observed behavior and improvements of disordered visible lasers due to strain therefore can be well accounted for by theory.

It is possible with appropriate growth conditions to grow GaInP with a highly ordered crystal structure. Two types of ordered domains occur, with gallium and indium atoms preferentially occupying alternate group III layers either in the ($\bar{1}11$) or ($1\bar{1}1$) direction. It is observed with highly ordered GaInP active layers that the threshold current density is strongly dependent on the laser cleavage direction (Fujii et al., 1992), with the threshold current density increasing by 50% for laser diodes with stripes in the (110) direction compared with diodes with stripes in the ($\bar{1}10$) direction, while Forstmann et al. (1994) also showed how the polarization of the emitted laser beam depended on the stripe direction. This is a direct consequence of and also a direct experimental verification of polarization engineering of the type discussed above in Sect. 2.3 (Ueno, 1993).

We investigated the effects of ordering by considering a structure ordered along the ($\bar{1}11$) direction, with the orthogonal coordinate axes then chosen along (110) and ($1\bar{1}2$). Ordering reduces the symmetry of the crystal and shifts the ($\bar{1}11$) valence state down in energy, enhancing the (110) and ($1\bar{1}2$) character of the valence band maximum. This is illustrated in Fig. 2.11, which shows the character of the state at the valence band maximum as a function of strain, assuming an ordering-induced splitting of 22 meV at the valence band maximum in the unstrained case (O'Reilly

2.4 Visible lasers

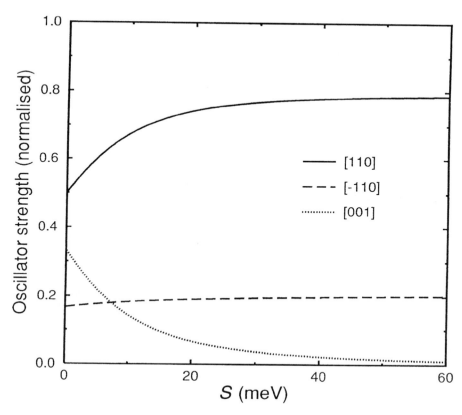

Figure 2.11: Normalized oscillator strengths calculated for (110), ($\bar{1}$10) and (001) polarized light at the valence band maximum in ordered GaInP, as a function of valence band strain-splitting energy S for compressive strain. The matrix element for (110)-polarized emission is enhanced at all strain values compared with that for ($\bar{1}$10)-polarized emission.

and Meney, 1995). TM-polarized emission is associated with (001) states, whereas TE emission is associated with (110) states for a ($\bar{1}$10) stripe and with ($\bar{1}$10) states for a (110) stripe, respectively. It can be seen from Fig. 2.11 that the band edge matrix element for TE emission is considerably enhanced in lasers with ($\bar{1}$10) stripes and suppressed in the (110) case.

This modification of the matrix elements is directly responsible for the poor performance of the lasers with (110) stripes. This result emphasizes the important role that polarization engineering can play in achieving reduced threshold current densities in semiconductor lasers. The incorporation of such chemically ordered GaInP layers into the active region of surface-emitting lasers can also bring benefits (Meney *et al.*, 1995, Chen *et al.*, 1997), enabling strong selectivity between the two orthogonal polarization directions, reducing noise, and enabling stable polarization control for applications such as magneto-optic disks and coherent detection systems, where the laser polarization state must be well defined.

2.5 Long-wavelength lasers

2.5.1 Introduction

Much effort has been devoted to developing semiconductor lasers for operation at 1.3 and at 1.55 μm, the respective wavelengths at which pulse dispersion and optical losses are minimized in silica-based optical fibers. Despite these efforts, both bulk and quantum well lasers operating at these wavelengths suffer from significant intrinsic losses, and as a consequence, both the threshold current and quantum efficiency are very temperature sensitive. It was predicted originally that the improved band structure due to strain should significantly reduce the losses in strained quantum well lasers compared with unstrained devices (Adams, 1986), leading to reduced threshold current and improved temperature sensitivity and differential efficiency. In practice, useful improvements have been obtained, but not on the scale that was first hoped. In this section we first review the dominant loss mechanisms in long-wavelength lasers. We then consider how strain influences these loss mechanisms. Finally, we address the influence of strain on temperature sensitivity.

2.5.2 The loss mechanisms of Auger recombination and intervalence band absorption

The most important mechanism contributing to the intrinsic loss in long-wavelength lasers is Auger recombination (Beattie and Landsberg, 1959), with intervalence band absorption (Adams *et al.*, 1980; Asada *et al.*, 1981) believed also to play an important role in bulk heterostructure devices. The intervalence band absorption (IVBA) process is illustrated by the

2.5 Long-wavelength lasers

solid lines in Fig. 2.12. When a photon is emitted in the lasing process, it may be reabsorbed by lifting an electron from the spin-split-off band into a hole that has been injected into the heavy-hole band. The absorption increases the gain required at threshold and hence increases the threshold carrier density n_{th}. This leads to an increase in the radiative current density and, more important in long-wavelength lasers, to an increase in the Auger recombination current, which is described below. Examples of the two major band-to-band Auger recombination (AR) processes are illustrated in Fig. 2.13. When an electron and hole recombine, they can

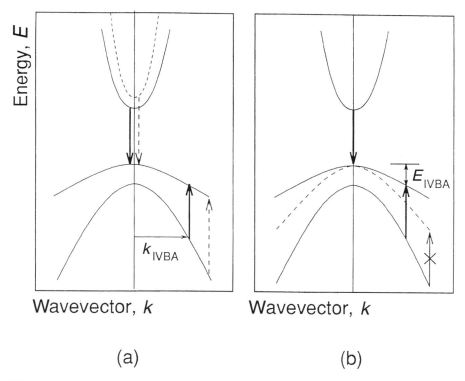

Figure 2.12: (a) In the intervalence band absorption (IVBA) process, a photon emitted in the lasing process is reabsorbed by exciting an electron from the spin-split-off band into a hole state in the heavy-hole band (solid vertical arrow). The dotted arrows indicate how IVBA moves to larger wavevector k_{IVBA} and is thereby reduced with increasing band gap. (b) The elimination of IVBA in strained structures, where holes are confined in a smaller region of k space.

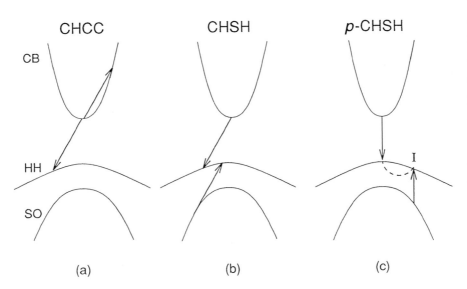

Figure 2.13: (a) In the direct CHCC Auger recombination process, the energy and momentum released when a conduction electron and heavy hole recombine across the band gap is used to excite a conduction electron to a higher conduction band state. (b) In the CHSH process, the energy and momentum released excite an electron from the spin-split-off band into a state in the heavy-hole band. (c) An example of a phonon-assisted CHSH Auger transition. The electron excited from the split-off band passes through a forbidden intermediate state (I) and is then scattered with the absorption or emission of a phonon to the final state. The phonon allows conservation of energy and momentum in the overall process.

transfer their energy and momentum to another electron by lifting it higher into the conduction band. This is called the *CHCC process* and is illustrated in Fig. 2.13(a). Alternatively, a hole can be excited deeper into the valence band. This is most likely to involve the spin-split-off band, as shown in Fig. 2.13(b), and is then called the *CHSH process*. Many other possible Auger processes may occur, including phonon-assisted Auger, for which the *CHSH process* is illustrated in Fig. 2.13(c). It is very difficult to obtain theoretically an accurate value of the total Auger recombination rate. However, we have estimated that in 1.5-μm devices AR is responsible for more than 80% of the total current at room temperature (Silver and O'Reilly, 1994, Silver et al., 1997). Time-resolved photoluminescence measurements suggest that band-to-band AR is the dominant loss mechanism in bulk devices (Hausser et al., 1990). This was originally expected to be the case in quantum well devices as well, but experimental evidence now indicates that phonon-assisted Auger may be the dominant loss process both in lattice-matched and strained QW structures (Fuchs et al., 1993; Wang et al., 1993). It also has been suggested that the temperature and carrier density dependence of the direct Auger recombination process is much weaker in QW structures than that predicted using parabolic band models and Boltzmann statistics (Lui et al., 1993).

2.5.3 Influence of strain on loss mechanisms

The magnitude of IVBA depends directly on the density of holes at k_{IVBA} (see Fig. 2.12). If we assume a parabolic band structure, the kinetic energy of the holes taking part in IVBA $E_{IVBA} = \hbar^2 k^2_{IVBA}/2m_h$ is determined by the wavevector k_{IVBA}, where the energy separation between the heavy-hole and spin-split-off bands $(E_{so} + \hbar^2 k^2/2m_s) - \hbar^2 k^2/2m_h$ equals the bandgap energy E_g. This occurs when

$$E_{IVBA} = \frac{m_s(E_g - E_{so})}{m_h - m_s} \qquad (2.16)$$

where m_h is the heavy-hole effective mass, and m_s is the spin-split-off effective mass. Equation (2.16) shows that when the heavy-hole mass is reduced toward that of the spin-split-off band, the corresponding hole energy for IVBA increases, as illustrated by the dashed "heavy hole" band in Fig. 2.12(b). Thus IVBA should decrease swiftly with decreasing heavy-hole mass in strained structures and become impossible if the two masses are equal.

On the assumption that the carrier concentration in the active region becomes pinned at threshold, Auger recombination should not affect the external differential efficiency above threshold η_{ex}, which is related to the internal differential efficiency η_{in} by (Yariv, 1989)

$$\eta_{ex}^{-1} = \eta_{in}^{-1} [\alpha L/\ln(1/R) + 1] \qquad (2.17)$$

where α is the effective internal losses, L the device length, and R is the mirror reflectivity. However, IVBA contributes directly to the internal losses, so η_{ex} depends strongly on the magnitude of the IVBA coefficient, and this has been studied in unstrained and strained QW lasers using hydrostatic pressure techniques. The results are shown in Fig. 2.14 (from Ring *et al.*, 1992 and Adams *et al.*, 1993). When hydrostatic pressure is applied to a III–V semiconductor, the direct energy gap increases, as shown by the dotted line in Fig. 2.12(a). In 1.5-μm devices, hydrostatic pressure causes the photon energy at lasing to increase by about 9 meV/kbar. From Eq. (2.16), this increases the hole energy at which IVBA occurs and hence decreases the magnitude of the IVBA losses. As shown in Fig.

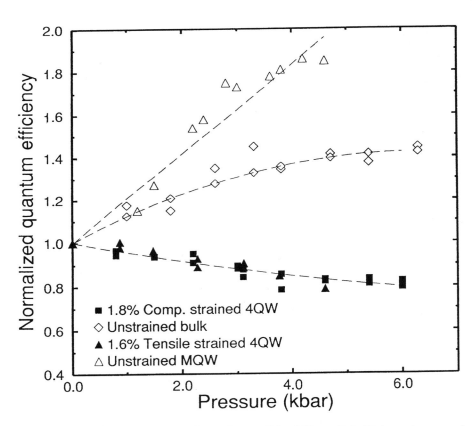

Figure 2.14: The pressure dependence of the differential efficiency in several lasers emitting at about 1.5 μm. The efficiency of the unstrained devices is seen to increase with pressure, in accordance with the removal of IVBA. The efficiency of both compressively and tensiley strained devices remains relatively unchanged with pressure, indicating that IVBA has already been removed by the built-in strain.

2.14, this causes a considerable increase in η_{ex} in lattice-matched QW lasers, as observed previously in bulk devices. However, when the same pressure is applied to compressively or tensiley strained QW devices, no such increase is seen, and indeed, η_{ex} even decreases somewhat. This may be interpreted by assuming that IVBA was negligible in these QW devices, at room temperature and so any further increase in bandgap due to external pressure has no additional effect. Fuchs et al. (1992) and Joindot and Beylat (1993) also have measured marked reductions in the IVBA coefficient in going from lattice-matched to compressively strained 1.5-μm QW lasers.

Auger recombination involves three carriers, so the Auger current J_{NR} in a device with an undoped active region varies approximately as

$$J_{NR}(T) = C(T)n_{th}^3 \qquad (2.18)$$

where $C(T)$ is the temperature-dependent Auger coefficient, and n_{th} is the threshold carrier density, which is also a function of temperature. The influence of strain on Auger recombination may be conveniently divided into two aspects. First, J_{NR} is very sensitive to any reduction in n_{th} brought about by strain through either a decrease in hole mass or an increase in the optical transition strength. Second, strain may change the magnitude of the Auger coefficient $C(T)$. We consider first the band-to-band Auger processes, for which the Auger coefficient increases exponentially with temperature (Taylor et al., 1985) and is given by

$$C(T) = C_o \exp(-E_a/kT) \qquad (2.19)$$

where the activation energy E_a is determined as being the sum of the minimum carrier kinetic energies required to conserve both energy and momentum in the transitions of Fig. 2.13(a) and (b) and is given by

$$E_a(\text{CHCC}) = \frac{m_c E_g}{m_c + m_h} \qquad (2.20a)$$

and

$$E_a(\text{CHSH}) = \frac{m_s(E_g - E_{so})}{2m_h + m_c - m_s} \qquad (2.20b)$$

assuming parabolic bands and Boltzmann statistics. This very simple model implies that both the CHCC and CHSH band-to-band Auger coefficients should be reduced by several orders of magnitude due to a reduction in the hole mass m_h due to strain (Adams, 1986). More detailed calculations show that the effect of strain is generally less marked (Silver et al., 1997).

The phonon-assisted Auger recombination rate $C_P(T)$ also depends on the band structure, varying as (Haug, 1990a, 1990b)

$$C_p(T) = \frac{B}{e^x - 1}\left[\frac{1}{(E_1 + \hbar\omega)^2} + \frac{e^x}{(E_1 - \hbar\omega)^2}\right] \quad (2.21)$$

where E_1 is the kinetic energy associated with the forbidden intermediate state, $x = \hbar\omega/kT$, $\hbar\omega$ is the LO phonon energy, and B is proportional to the transition matrix elements, with E_1 given by

$$E_1 = \frac{m_s}{m_h(E_g - E_{so})} \quad (2.22a)$$

for the CHSH-P process [see Fig. 2.13(c)] and

$$E_1 = (m_c/m_h)\,E_g \quad (2.22b)$$

for the CHCC-P process. The two expressions for E_1 are calculated by finding the heavy-hole kinetic energy at the wavevector k_i, where the final electron or hole state is at an energy E_g away from the band edge, so that, for example, in Eq. (2.22b) we have $E_1 = \hbar^2 k_i^2/2m_h$, where k_i is determined from $E_g = \hbar^2 k_i^2/2m_c$. Combining Eqs. (2.21) and (2.22), we see that $C_P(T)$ is approximately proportional to m_h^2 if we assume $\hbar\omega$ to be small compared with E_1. This then predicts a reduction in the phonon-assisted Auger rate if the heavy-hole mass is reduced and all else remains constant, but the expected reduction is much weaker than for the direct processes, so Auger recombination could still remain a significant loss mechanism in strained-layer 1.5-μm lasers.

To check that this is indeed the case, we have measured the variation of spontaneous emission efficiency as a function of current in both unstrained and strained QW 1.5-μm lasers (Braithwaite et al., 1995). The spontaneous emission was collected through a narrow window etched in the substrate electrode and displayed on an optical spectrum analyzer. By this method we were able to determine L, the integrated spontaneous emission from the well material as a function of the total current I through the device. Figure 2.15 shows a plot of I/L, which is proportional to the reciprocal of the spontaneous emission efficiency, against $L^{1/2}$, which we can assume is proportional to the carrier concentration n for (a) an unstrained laser and (b) a laser with 1% compressive strain. As can be seen, a good straight line is achieved over most of the range of L. The straight lines obtained show that $I \propto L^{3/2}$; so that, with $L \propto n^2$ in the Boltzmann

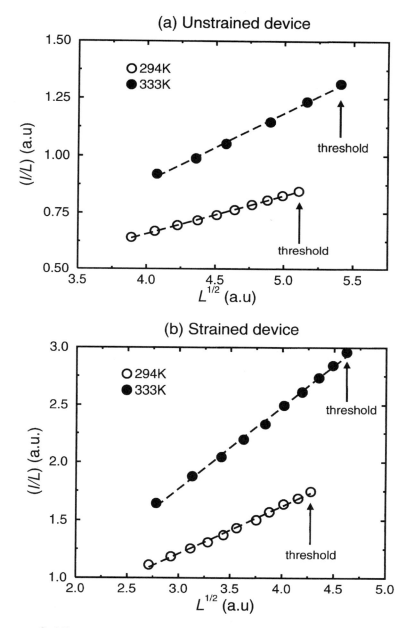

Figure 2.15: Variation of the inverse of the efficiency I/L against $L^{1/2}$ for (a) an unstrained device and (b) a device with 1% compressive strain. I is the total current though the device, and L is the integrated spontaneous emission from the active region observed though a small window etched in the substrate electrode.

approximation, we can conclude that to a good approximation in both our devices, $I \propto n^3$. Since this is typical for an Auger process, which involves three carriers, we take this as strong evidence that the current density in the strained as well as the unstrained devices is dominated by Auger recombination.

Another important measure of the relative strength of Auger recombination is obtained from the variation of threshold current with pressure. This is shown for several 1.5-μm lasers in Fig. 2.16, which, for comparison,

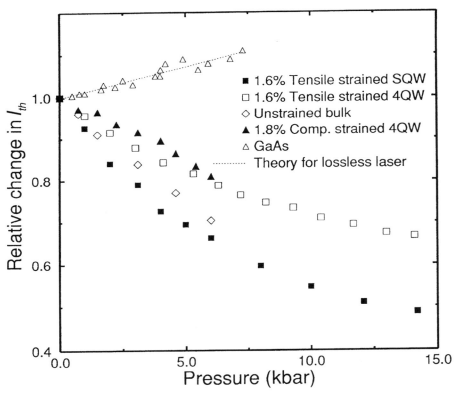

Figure 2.16: The pressure dependence of the threshold current density I_{th} for a bulk unstrained laser, a compressively strained 4QW laser, and tensiley strained SQW and 4QW lasers, all emitting at about 1.5 μm, compared with the predicted pressure dependence of an ideal laser and that of GaAs lasers. The threshold current density decreases rapidly with pressure in all the long-wavelength devices, indicating that Auger recombination controls the current flow in all cases.

2.5 Long-wavelength lasers

also shows some typical results for GaAs lasers (Adams *et al.*, 1989, 1993). In a loss-free laser, one would expect the threshold current to increase with pressure as the bandgap and mode density increase. This is shown by the dotted line in Fig. 2.16 and, as can be seen, is closely followed by the behavior of the GaAs lasers. By contrast, all the 1.5-μm devices show a considerable decrease in threshold current with increasing pressure, including the compressively and the tensiley strained lasers. This decrease can be understood if the current in the 1.5-μm devices is dominated by Auger recombination, which decreases with increasing bandgap. Although there is some difference between the different types of 1.5-μm devices, basically the trend is the same, indicating that Auger recombination controls the current flow in all the long-wavelength lasers.

Since more than 80% of the current at threshold is estimated to be due to Auger recombination, a study of threshold current density J_{th} as a function of strain gives a very direct measure of the effect of strain on the Auger recombination rate. Figure 2.17 shows a compilation by Thijs et al. (1994) of threshold current densities per QW deduced for infinite-cavity-length 1.5-μm lasers versus the strain in the InGaAs(P) QWs, using data reported in the literature. As can be seen, J_{th} is reduced by a factor of about 2 to 3 at either 1.5% compressive or tensile strain. Since the reduction in radiative current calculated in Fig. 2.8 is small on the scale of Fig. 2.17, the results indicate that the Auger current is itself reduced by a factor of about 2 to 3 by the strain. These reductions are of very considerable practical importance but are far less than originally predicted by the simple parabolic model (Adams, 1986). This is because strain primarily affects the bands near the zone center, having less effect at the larger wavevectors associated with Auger transitions (Silver et al., 1997).

2.5.4 The influence of strain on temperature sensitivity

The temperature sensitivity of a semiconductor laser is usually described by the T_0 parameter, where T_0 is related to the temperature dependence of the threshold current $I_{th}(T)$ by

$$1/T_0 = d/dT\,[\ln(I_{th})] \qquad (2.23)$$

Values of T_0 for long-wavelength lasers are typically in the range $T_0 = 55 \pm 20$ K.

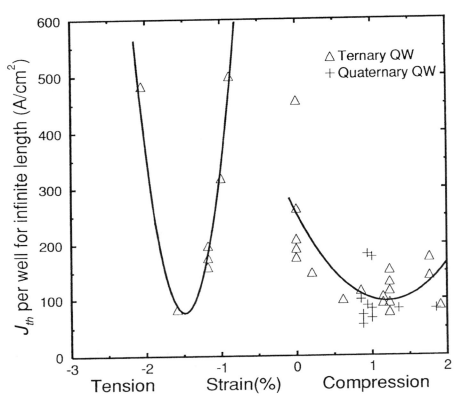

Figure 2.17: Summary of threshold current densities per QW deduced for infinite-cavity-length 1.5-μm lasers versus the strain in the InGaAs(P) QWs, using data reported in the literature. The solid lines serve as a guide to the eye (from Thijs, 1994, with permission). (© 1994 IEEE)

The temperature sensitivity of strained 1.5-μm lasers is better than that of bulk and lattice-matched QW devices, but not by a significant factor. Thijs et al. (1994) reported an improvement in T_0 values in both tensile and compressive lasers at room temperature, with the maximum T_0 values approaching 100 K in compressive structures. The performance of lattice-matched and compressive 1.5-μm lasers degrades rapidly at higher temperatures, whereas the best tensile lasers continue to operate to 140°C.

Further analysis of the integrated spontaneous emission L from the windows in the electrodes (Braithwaite et al., 1995) allowed us to investigate the relative influence of the radiative and nonradiative current paths on the temperature sensitivity of 1.5-μm lasers. Figure 2.18 shows at two temperatures how L increases sublinearly with current up to threshold, where the carrier density in the well becomes pinned and L saturates. The vertical dotted lines indicate the threshold current as determined from the radiation from the facet. They agree well with the point at which L becomes pinned and correspond to a T_0 of 60 K in both cases. However, the variation with temperature of L at threshold, which is directly proportional to the radiative threshold current, corresponds to values of T_0 of 300 and 265 K for strained and unstrained lasers, respectively. Theoretically, one would expect a T_0 of about 300 K, so the results show that the radiative current is behaving completely normally and there is no anomalous temperature dependence of the gain in these devices. However the threshold current is dominated by Auger recombination, as discussed earlier, which will therefore play a dominant role in the temperature sensitivity of the laser.

We have shown theoretically that T_0 close to 100 K is the best that can be achieved at room temperature in lasers where Auger recombination is the dominant contribution to the threshold current (O'Reilly and Silver, 1993). We also have been able to explain the better performance achieved in tensile lasers compared with compressive lasers at high temperature.

We assumed that Auger is an activated process, with the Auger coefficient $C(T)$ given by Eq. (2.19). For direct Auger processes, the activation energy E_a can be quite large (~70 meV). Phonon-assisted processes are much less temperature dependent and therefore have smaller values of E_a (~20–30 meV at room temperature). The threshold carrier density n_{th} increases linearly with temperature in an ideal quantum well (Ghiti et al., 1992), where only the lowest conduction and valence subbands are occupied. Any nonideal factors that degrade the differential gain, such as carriers occupying higher subbands, carrier spillover into the barrier, or IVBA, will increase the temperature dependence beyond linearity. We represent these effects on n_{th} as

$$n_{th} \propto T^{1+x} \qquad (2.24)$$

where $x = 0$ in an ideal quantum well, as described above. It has been shown experimentally that the differential gain degrades rapidly with temperature in lattice-matched and compressively strained InGaAsP lasers with very low T_0 values (~40 K) (O'Gorman et al., 1992; Zou et al., 1993). The threshold carrier density n_{th} consequently increases supralin-

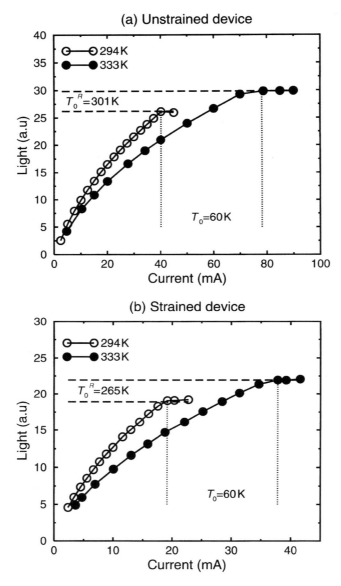

Figure 2.18: Variation of the window emission L versus current I at two temperatures for (a) an unstrained device and (b) a device containing 1% compressive strain. The vertical dotted lines show the current at which the device reached threshold as determined by the radiation from the end facet and correspond to a T_0 of 60 K in each case. The increase with temperature of the saturated value of L beyond threshold as shown by the horizontal dashed lines correspond to a T_0 for the radiative current of 301 and 265 K for the unstrained and strained devices, respectively.

2.5 Long-wavelength lasers

early in the devices investigated. Auger recombination nevertheless remains the dominant current path in such devices (Zou et al., 1993), so we approximate the threshold current by Eq. (2.18):

$$J_{th} \propto Cn_{th}^3$$

and substituting Eqs. (2.19) and (2.24) in Eqs.(2.18) and (2.23) gives

$$T_0 = \frac{T}{3 + 3x + E_a/kT} \tag{2.25}$$

This puts an upper limit on the T_0 value because the best case we can expect is that $x = 0$ and C is independent of temperature ($E_a = 0$) so that at room temperature

$$T_0 = \frac{T}{3} = 100 \text{ K} \tag{2.26}$$

This estimate agrees well with the best values reported and suggests that this is indeed an upper limit for T_0 in QW lasers dominated by Auger recombination. Therefore, the low T_0 values (60–100 K) can be explained by just a combination of Auger recombination with the theoretically predicted dependence of gain on temperature. However, the very low T_0 values (~40 K) observed in many InGaAsP QW lasers may be associated with a strong temperature dependence of the threshold carrier density n_{th} (Zou et al., 1993), while the Auger coefficient C has a much weaker temperature dependence (Fuchs et al., 1993) than was originally expected (Adams, 1986).

To investigate this point further, we calculated the threshold carrier density for InGaAs/InGaAsP/InP lasers operating at 1.5 μm with both tensile and compressive wells (O'Reilly and Silver, 1993). The gain as a function of carrier density was calculated using the 6 × 6 Luttinger-Kohn Hamiltonian for the valence states (Luttinger and Kohn, 1955; O'Reilly, 1989), while the conduction band states include the effect of carrier spillover into the barrier region by a self-consistent Poisson/Schrödinger calculation (Barrau et al., 1992; Silver and O'Reilly, 1994). The structures considered were based on the laser structures reported by Thijs et al. (1994) and consist of four $In_{1-x}Ga_xAs$ quantum wells between 170-Å $In_{0.2}Ga_{0.8}As_{0.45}P_{0.55}$ barriers with the well width adjusted to emit at 1.5 μm.

Figure 2.19 shows the calculated carrier densities at (a) transparency and (b) threshold for 1.2% compressive and tensile InGaAs/InGaAsP quantum well lasers. The compressive wells are 25 Å wide, while the tensile

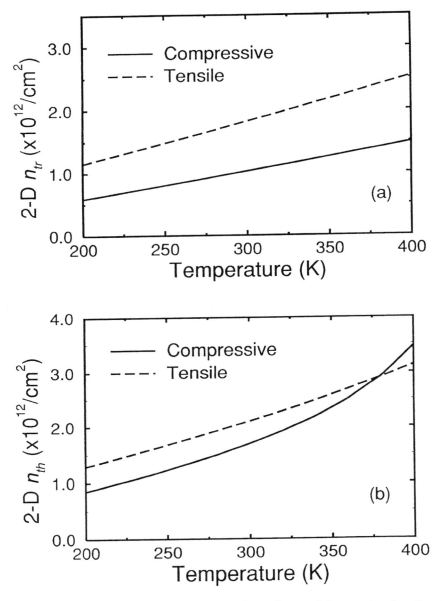

Figure 2.19: Calculated temperature dependence of the carrier density at (a) transparency and (b) threshold for 1.2% compressive and tensile InGaAs/InGaAsP quantum well lasers. Note that the compressive threshold carrier density increases rapidly above room temperature because of gain saturation effects, while the tensile structure shows an improved temperature sensitivity due to the lower threshold gain required for lasing.

wells are 110 Å wide. Because of the reduced optical confinement factor, the threshold material gain in the compressive laser needs to be much higher (~3000 cm^{-1}) than in the tensile structure (~500 cm^{-1}).

The temperature dependence of the transparency carrier density for both cases is close to the linear variation predicted for an ideal QW structure. The threshold carrier density n_{th} in the compressive structure, however, is strongly temperature dependent, due to the degradation of the differential gain at the high gains required for lasing. This rapid increase in n_{th} is responsible for the poor performance of these compressive lasers at high temperature, in agreement with the analyses of others (O'Gorman et al., 1992; Zou et al., 1993). Because of the lower threshold gain, the temperature dependence of n_{th} for the tensile laser is much weaker, even when carrier spillover effects are included. This leads to a reduced temperature sensitivity compared with the compressive laser above room temperature and hence to a higher operating temperature for the tensile devices, in agreement with experiment.

It may be surprising initially to find that carrier spillover into the barriers has little influence on the temperature dependence of the threshold carrier density in the tensile layers. The spillover of electrons into the barriers does, however, increase the magnitude of the threshold carrier density and also reduces the magnitude of the differential gain in these devices (see Silver and O'Reilly, 1994; also see Chap. 1, by Zhao and Yariv). This may limit the use of tensile-strained InGaAs quantum wells with InGaAsP barriers for high-speed applications.

2.6 Linewidth, chirp, and high-speed modulation

We describe briefly here how the introduction of strain enhances the dynamic properties of semiconductor lasers. The high-speed characteristics are reviewed in detail by Nagarajan and Bowers in Chap. 3, while Zhao and Yariv discuss in Chap. 1 how state filling in QW structures can degrade the laser dynamics.

A change in the carrier concentration n in the active region of a laser leads to a change in the refractive index N and hence to a change in wavelength of the longitudinal modes. Thus noise fluctuations in the photon density and in n lead to an increase of the width of the laser line. Similarly, any imposed variations in n as the device is modulated lead to chirp. Both effects are particularly undesirable in 1.5-μm lasers used in dispersive long-distance communication systems. They are usually described in terms of the linewidth enhancement factor α, which is given by (Henry, 1982)

$$\alpha = \frac{-4\pi}{\lambda} \frac{dN/dn}{dg/dn} \quad (2.27)$$

Since the rate of change of gain with carrier concentration dg/dn is larger in strained-layer lasers, much smaller excursions in carrier concentration and hence refractive index result (Suemune et al., 1988; Ohtoshi and Chinone, 1989; Ghiti et al., 1989; Lau et al., 1989). This reduces the linewidth $\Delta \nu$ for a given power P as

$$\Delta \nu \approx \frac{1 + \alpha^2}{P} \quad (2.28)$$

and also the chirp width, which is proportional to $(1 + \alpha^2)^{1/2}$.

There is a wide scatter in the published data on linewidth enhancement factor in lattice-matched and strained QW laser structures, due in part to differences in measurement technique between different laboratories and also due to differences in the laser structures and material systems being studied (Zhao et al., 1993). Factors such as carrier occupation of higher subbands (Burt, 1984) or carrier spillover into the barrier region (Zhao et al., 1993) can significantly degrade the differential gain and increase the α value, as described further in Chap. 1. Comparative studies within a given laboratory, however, generally show that the linewidth enhancement factor is reduced in going from lattice-matched to strained QW laser structures (Hirayama et al., 1991; Tiemeijer et al., 1992; Schönfelder et al., 1993), typically by a factor of about two.

An improved differential gain was predicted also to lead to enhanced high-speed performance in strained-layer lasers, since the resonant frequency f_r depends on dg/dn and the output power P in a single-mode laser as (Bowers et al., 1986)

$$f_r \propto (P \, dg/dn)^{1/2} \quad (2.29)$$

so that a given relaxation frequency should be obtained for a lower output power in strained-layer lasers. This has indeed been demonstrated by a number of groups, particularly for InGaAs QW lasers operating at wavelengths in the neighbourhood of 1 μm (Offsey et al., 1990; Nagarajan et al., 1991; Ralston et al., 1993). The 3-dB modulation bandwidth of a laser

f_{max} depends, however, not just on f_r but also on damping processes within the laser due to such mechanisms as spectral hole burning (Agrawal, 1987; Ghiti and O'Reilly, 1990), carrier heating (Gomatam and deFonzo, 1990), and carrier transport through the core region (Rideout et al., 1991). It has been found generally to date that due to these processes there is little change in the damping factor K and the experimentally determined f_{max} in strained-layer structures compared with lattice-matched devices (Ralston et al., 1993; Meland et al., 1990; Morton et al., 1992).

The introduction of p-doping in the active region is also predicted to improve the differential gain dg/dn (Uomi, 1990). It has been demonstrated that the combination of p-doping and strain can in fact simultaneously increase dg/dn and decrease K, yielding very efficient high-speed modulation and the first semiconductor lasers to achieve a direct modulation bandwidth of 30 GHz under dc bias (Ralston et al., 1993). This combination of p-doping and strain appears to be the most promising route to maximize the bandwidth of semiconductor lasers.

2.7 Strained laser amplifiers

The signal emerging from an optical fiber is, in general, randomly polarized, and therefore, it is very desirable that any optical amplifier provides gain that is independent of the direction of polarization of the light. In a normal QW system, the quantum confinement brings the heavy-hole band to the top of the valence band, and light polarized in the TE mode is amplified more than light polarized in the TM mode. This problem can be overcome using strained-layer techniques. First, it is possible to grow the wells or the barriers with a small amount of tensile strain (Magari et al., 1990). This raises the light-hole band, and the strain can be adjusted to increase the TM gain until it is just equal to that of the TE gain. Unfortunately, it is found that the relative gains in the two modes are sensitive to the magnitude of the amplifier current, and so the gains are equal only over a very limited range.

Another more promising approach has been adopted by Tiemeijer et al (1993), who introduced both compressively and tensilely strained wells alternately into the active region. This provided gain to the TE and TM modes, respectively, and the structure could be adjusted to make them equal. Tiemeijer et al. concentrated on 1.3-μm devices because there is no suitable rare earth optical amplifier available at this wavelength. They found that a combination of four 45-Å-wide compressive wells and three 110-Å-wide tensile wells of InGaAsP each with 1% strain gave TE and TM gain within 1 dB of each other over a wide wavelength band and over an order of magnitude change in amplifier current. This structure is an excellent example of the new flexibility in device design that is afforded by the introduction of strained-layer techniques. Other applications of such a structure include two-polarization/two-frequency lasers (Mathur and Dapkus, 1992) and polarization control elements.

2.8 Conclusions

We have reviewed the manner in which the properties of semiconductor lasers can be considerably improved by the introduction of either compressive or tensile strain. The changes observed in visible lasers can be explained well in terms of the influence of strain on the band structure and gain processes. For long-wavelength (1.5-μm) lasers, high-pressure measurements of the quantum efficiency indicate that intervalence band absorption is effectively eliminated, whereas measurements of the threshold current show that Auger recombination has been reduced significantly in the devices investigated to date. Theoretical analysis, together with the measured pressure dependence of the threshold current, indicates that despite these improvements, the Auger process still remains the dominant recombination path; this behavior is reflected in the fact that the lasers remain temperature sensitive, with values of T_0 below 100 K, as predicted by a simple theoretical analysis. The increased rate of change of gain with carrier concentration in strained-layer lasers decreases the linewidth enhancement factor α, resulting in narrower linewidth and reduced chirp, while the modulation bandwidth is also enhanced in p-doped strained structures. We also have discussed briefly how the relaxation of the constraint of lattice matching allows new materials combinations and novel device applications, such as polarization-insensitive amplifiers that contain both tensiley and compressively strained QW layers. In summary, strained layers retain the benefits of lattice-matched QW structures and are also able to take advantage of new physical effects and materials combinations. Finally, since it is observed that 1.5-μm devices have excellent low-degradation properties, with predicted operating lifetimes well into the next century, we can safely conclude that strained-layer lasers are now here to stay!

Acknowledgments

This work was supported by the U.K. Engineering and Physical Sciences Research Council and by the European Community ESPRIT programme. Much of the chapter was written while A. R. Adams was on sabbatical leave at the University of Montpellier supported by the Ministere de la Recherche et de la Technologie (France) and while E. P. O'Reilly was at the Fraunhofer Institute, Freiburg, supported by the Alexander von Humboldt Foundation (German Federal Republic). We are grateful to our sabbatical hosts Prof. J. L. Robert and Dr. J. D. Ralston, respectively.

References

Adams, A. R. (1986). *Electron. Lett.,* **22**, 249.

Adams, A. R., Asada, M., Suematsu, Y., and Arai, S. (1980). *Jpn. J. Appl. Phys.* **19**, 621.

Adams, A. R., Heasman, K. C., and O'Reilly, E. P. (1989). *Band Structure Engineering in Semiconductor Microstructures NATO ASI Series B: Physics,* **189**, 279.

Adams, A. R., Hawley, M. J., O'Reilly, E. P., and Ring, W. S. (1993). *Jpn. J. Appl. Phys.,* **32** (Suppl. 32–1), 358.

Agrawal, G. P. (1987). *IEEE J. Quantum Electron.,* **23**, 860.

Ahn, D., and Chuang, S. (1988). *IEEE J. Quantum Electron.,* **QE-24**, 2400.

Altarelli, M., Ekenberg, U., and Fasolino, A. (1985). *Phys. Rev. [B],* **32**, 5138.

Andersson, T. G., Chen, Z. G., Kulakovskii, V. D., Uddin, A., and Vallin, J. T. (1987). *Appl. Phys. Lett.,* **51**, 752.

Asada, M., Adams, A. R., Stubkjaer, K. E., Suematsu, Y., Itaya, Y., and Arai, S. (1981). *IEEE J. Quantum Electron.,* **17**, 611.

Barrau, J., Amand, T., Brousseau, M., Simes, R. J., and Goldstein, L. (1992). *J. Appl. Phys.,* **71**, 5768.

Bastard, G., and Brum, J. A. (1986). *IEEE J. Quantum Electron.,* **QE-22**, 1625.

Beattie, A. R., and Landsberg, P. T. (1959). *Proc. R. Soc. [A],* **249**, 16.

Bernard, M. G. A., and Duraffourg, B. (1961). *Phys. Stat. Sol.,* **1**, 699.

Bour, D. P., Gilbert, D. B., Fabian, K. B., Bednarz, J. P., and Ettenberg, M. (1990). *IEEE Photon. Technol. Lett.,* **2**, 173.

Bowers, J. E., Hemenway, B. R., Gnauck, A. H., and Wilt, D. P. (1986). *IEEE J. Quantum Electron.,* **22**, 833.

Braithwaite, J., Silver, M., Wilkinson, V. A., O'Reilly, E. P., and Adams, A. R. (1995). *Appl. Phys. Lett.,* **67** 3546.

Broido, D. A., and Sham, L. J. (1985). *Phys. Rev. [B],* **31**, 888.

Burt, M. G. (1984). *Electron. Lett.,* **20**, 27.

Chang, Y. C., Chu, H. Y., and Chung, S. G. (1986). *Phys. Rev. [B],* **33**, 7364.

Chen, Y. H., Wilkinson, C. I., Woodhead, J., David, J. P. R., Button, C. C., and Robson, P. N. (1997) *IEEE Photon. Technol. Lett.,* **9**, 143.

Colak, S., Eppenga, R., and Schuurmans, M. F. H. (1987). *IEEE J. Quantum Electron.,* **QE-23**, 960.

Coleman, J. J. (1993). In *Quantum Well Lasers* (P. S. Zory, ed.), Chap. 8, p. 367, Academic Press, San Diego.

Corzine, S. W., Yan, R.-H., and Coldren, L. A. (1993). In *Quantum Well Lasers* (P. S. Zory, ed.), Chap. 1, p. 17, Academic Press, San Diego.

Dunstan, D. J., Kidd, P., Howard, L. K., and Dixon, R. H. (1991). *Appl. Phys. Lett.,* **59**, 3390.

Eppenga, R., Schuurmans, M. F. H., and Colak, S. (1987). *Phys. Rev. [B],* **36**, 1554.

Foreman, B. A. (1994). *Phys. Rev. [B],* **49** 1757.

Forstmann, G. G., Barth, F., Schweizer, H., Moser, M., Geng, C., Scholz, F., and O'Reilly, E. P. (1994). *Semicond. Sci. Technol.,* **9**, 1268.

Frank, F. C., and van der Merwe, J. H. (1949). *Proc. Roy. Soc. [A],* **198**, 216.

Fuchs, G., Hörer, J., Hangleiter, A., Härle, V., Scholz, F., Glew, R. W., and Goldstein, L. (1992). *Appl. Phys. Lett.,* **60**, 231.

Fuchs, G., Schiedel, C., Hangleiter, A., Härle, V., and Scholz, F. (1993). *Appl. Phys. Lett.,* **62**, 396.

Fujii, H., Ueno, Y., Gomyo, A., Endo, K., and Suzuki, T. (1992). *Appl. Phys. Lett.,* **61**, 737.

Ghiti, A., and O'Reilly, E. P. (1990). *Electron. Lett.,* **26**, 1978.

Ghiti, A., and O'Reilly, E. P. (1993). "Valence Band Engineering in Quantum Well Lasers," in *Quantum Well Lasers* (P. S. Zory, ed.), Chap. 7, pp. 329–366, Academic Press, San Diego.

Ghiti, A., O'Reilly, E. P., and Adams, A. R. (1989). *Electron. Lett.,* **25**, 821.

Ghiti, A., Batty, W., and O'Reilly, E. P. (1990). *Superlatt. Microstruct.,* **7**, 353.

Ghiti, A., Silver, M., and O'Reilly, E. P. (1992). *J. Appl. Phys.,* **71**, 4626.

Gomatam, B. N., and DeFonzo, A. P. (1990). *IEEE J. Quantum Electron.,* **26**, 1689.

Gonul, Besire (1996). Ph.D. thesis, University of Surrey.

Harnagel, G. L., Paoli, T. L., Thornton, R. L., Burnham, R. D., and Smith, D. L. (1985). *Appl. Phys. Lett.,* **46**, 118.

Harrison, W. A. (1980). *Electronic Structure and the Properties of Solids,* Freeman, San Francisco.

Haug, A. (1987). *Appl. Phys. [B],* **44**, 151.

Haug, A. (1990a). *Electron. Lett.,* **26**, 1415.

Haug, A. (1990b). *Appl. Phys. [A],* **51**, 354.

Hausser, S., Fuchs, G., Hangleiter, A., Streubel, K., and Tsang, W. T. (1990). *Appl. Phys. Lett.,* **56**, 913.

Hawley, M. J., Adams, A. R., Silver, M., O'Reilly, E. P., and Valster, A. (1993). *IEEE J. Quantum Electron.,* **29**, 1885.

Hayakawa, T., Suyama, T., Takahashi, K., Kondo, M., Yamamoto, S., and Hijikata, T. (1988a). *Appl. Phys. Lett.,* **52**, 339.

Hayakawa, T., Suyama, T., Takahashi, K., Kondo, M., Yamamoto, S., and Hijikata, T. (1988b). *J. Appl. Phys.,* **64**, 297.

Henry, C. H. (1982). *IEEE J. Quantum Electron.,* **18**, 259.

Hirayama, Y., Morinaga, M., Tanimura, M., Onomura, M., Funemizu, M., Kushibe, M., Suzuki, N., and Nakamura, M. (1991). *Electron. Lett.,* **27**, 241.

Jaros, M. (1985). *Rep. Prog. Phys.,* **48**, 1091.

Joindot, I., and Beylat, J. L. (1993). *Electron. Lett.,* **29**, 604.

Jones, G., and O'Reilly, E. P. (1993). *IEEE J. Quantum Electron.,* **29**, 1344.

Jones, G., Smith, A. D., O'Reilly, E. P., Silver, M., Briggs, A. T. R., Fice, M. J., Adams, A. R., Greene, P. D., Scarrott, K., and Vranic, A., (1998), *IEEE J. Quantum Electron.,* **34**, 822.

Kasper, E. (1986). *Surf. Sci.,* **174**, 630.

Kobayashi, K., Kawata, S., Gamyo, A., Hino, A., Hino, I., Suzuki, T. (1985). *Electron. Lett.,* **21**, 931.

Krijn, M. P. C. M., t'Hooft, G. W., Boermans, M. J. B., Thijs, P. J. A., van Dongen, T., Binsma, J. J. M., Tiemeijer, L. F., and van der Poel, C. J. (1992). *Appl. Phys. Lett.,* **61**, 1772.

Lancefield, D., Batty, W., Crookes, C. G., O'Reilly, E. P., Adams, A. R., Homewood, K. P., Sundaram, G., Nicholas, R. J., Emeny, M., and Whitehouse, C. R. (1990). *Surf. Sci.,* **229**, 122.

Lau, K. Y., Xin, S., Wang, W. I., Bar-Chaim, N., and Mittelstein, M. (1989). *Appl. Phys. Lett.,* **55**, 1173.

Laurich, B. K., Elcess, K., Fonstad, C. G., Beery, J. G., Mailhiot, C., and Smith, D. L. (1989). *Phys. Rev. Lett.,* **62**, 649.

Los, J., Fasolino, A., and Catellani, A. (1996). *Phys. Rev. [B],* **53**, 4630.

Lui, W. W., Yamanaka, T., Yoshikuni, Y., Seki, S., and Yokoyama, K. (1993). *Appl. Phys. Lett.,* **63**, 1616.

Luttinger, J. M. (1956). *Phys. Rev.,* **102**, 1030.

Luttinger, J. M., and Kohn, W. (1955). *Phys. Rev.,* **97**, 869.

Magari, K., Okamoto, M., Yasaka, H., Sato, K., Noguchi, Y., and Mikami, O. (1990). *IEEE Photon. Technol. Lett.,* **2**, 556.

Mailhiot, C., and Smith, D. L. (1987). *Phys. Rev. [B],* **35**, 1242.

Mascarenhas, A., Kurtz, S., Kibbler, A., and Olson, J. M. (1989). *Phys. Rev. Lett.,* **63**, 2108.

Mathur, A., and Dapkus, P. D. (1992). *Appl. Phys. Lett.,* **61**, 2845.

Matthews, J. W., and Blakeslee, A. E. (1974). *J. Cryst. Growth,* **27**, 118; Matthews, J. W., and Blakeslee, A. E., (1975). *J. Cryst. Growth,* **29**, 273; Matthews, J. W., and Blakeslee, A. E. (1976). *J. Cryst. Growth,* **32**, 265.

Meland, E., Holmstrom, R., Schlafer, J., Lauer, R. B., and Powazinik, B. (1990). *Electron. Lett.,* **26**, 1827.

Meney, A. T., O'Reilly, E. P., and Ebeling, K. J. (1995). *Electron. Lett.,* **31**, 461.

Mollenkamp, L. W., Eppenga, R., t'Hooft, G. W., Dawson, P., Foxon, C. T., and Moore, K. (1988). *Phys. Rev. [B],* **38**, 4314.

Morton, P. A., Logan, R. A., Tanbun-Ek, T., Sciortino, P. F., Jr., Sergent, A. M., Montgomery, R. K., and Lee, B. T. (1992). *Electron. Lett.,* **28**, 2156.

Nagarajan, R., Fukushima, T., Bowers, J. E., Geels, R. S., and Coldren, L. A. (1991). *Appl. Phys. Lett.,* **58**, 2326.

Ninno, D., Gell, M. A., and Jaros, M. (1986). *J. Phys. [C],* **19**, 3845.

Offsey, S. D., Schaff, W. J., Tasker, P. J., and Eastman, L. F. (1990). *IEEE Photon. Technol. Lett.,* **2**, 9.

O'Gorman, J., Levi, A. F. J., Schmitt-Rink, S., Tanbun-Ek, T., Coblentz, D. L., and Logan, R. A. (1992). *Appl. Phys. Lett.,* **60**, 157.

Ohtoshi, T., and Chinone, N. (1989). *IEEE Photon. Technol. Lett.,* **1**, 117.

O'Reilly, E. P. (1989). *Semicond. Sci. Technol.,* **4**, 121.

O'Reilly, E. P., and Meney, A. T. (1995). *Phys. Rev. [B],* **51**, 7566.

O'Reilly, E. P., and Silver, M. (1993). *Appl. Phys. Lett.,* **63**, 3318.

O'Reilly, E. P., Jones, G., Ghiti, A., and Adams, A. R. (1991). *Electron. Lett.,* **27**, 1417.

Osbourn, G. C. (1986). *IEEE J. Quantum Electron.*, **22**, 1677.

Pabla, A. S., Woodhead, J., Khoo, E. A., Grey, R., David, J. P. R. and Reas, G. J. (1996). *Appl. Phys. Lett.*, **68**, 1595.

People, R., and Bean, J. C. (1985). *Appl. Phys. Lett.*, **47**, 322.

Ralston, J. D., Weisser, S., Esquivias, I., Larkins, E. C., Rosenzweig, J., Tasker, P. J., and Fleissner, J. (1993). *IEEE J. Quantum Electron.*, **29**, 1648.

Rideout, W., Sharfin, W. F., Koteles, E. S., Vassell, M. O., and Elman, B. (1991). *IEEE Photon. Technol. Lett.*, **3**, 784.

Ring, W. S., Adams, A. R., Thijs, P. J. A., and Van Dongen, T. (1992). *Electron. Lett.*, **28**, 569.

Sale, T. E., (1995) "Vertical Cavity Surface Emitting Lasers", John Wiley & Sons Inc., Chichester

Schirber, J. E., Fritz, I. J., and Dawson, L. R. (1985). *Appl. Phys. Lett.*, **46**, 187.

Schönfelder, A., Weisser, S., Ralston, J. D., and Rosenzweig, J. (1993). *Electron. Lett.*, **29**, 1685.

Schulman, J. N., and Chang, Y. C. (1985), *Phys. Rev. [B]*, **31**, 2056.

Seki, S., Yamanaki, T., Lui, W., Yoshikuni, Y., and Yokoyama, K. (1994). *IEEE J. Quantum Electron.*, **30**, 500.

Shanabrook, B. V., Glembocki, O. J., Broido, D. A., and Wang, W. I. (1989). *Superlatt. Microstruct.*, **5**, 503.

Silver, M., and O'Reilly, E. P. (1994). *IEEE J. Quantum Electron.*, **30**, 547.

Silver, M., O'Reilly, E. P. and Adams, A. R.(1997), *IEEE J. Quantum Electron.*, **33**, 1557.

Smith, A. D., Briggs, A. T. R., Vranic, A., and Scarrott, K. (1993). Extended abstract, European Workshop on MOVPE, Malmö, Sweden.

Suemune, I., Coldren, L. A., Yamanishi, M., and Kan, Y. (1988). *Appl. Phys. Lett.*, **53**, 1378.

Sugawara, M. (1992). *Appl. Phys. Lett.*, **60**, 1842.

Sun, D., Towe, E., Ostdiek, P. H., Grantham, J. W., and Vansuch, G. J. (1995). *IEEE J. Select. Top. Quantum Electron.*, **1**, 674.

Taylor, R. I., Abram, R. A., Burt, M. G., and Smith, C. (1985). *IEEE Proc. J. Optoelectron.*, **132**, 364.

Thijs, P. J. A. (1992). In *Proceedings of the 13th IEEE International Semiconductor Laser Conference*, Takamatsu, Japan, p. 2.

Thijs, P. J. A., Binsma, J. J. M., Tiemeijer, L. F., and van Dongen, T. (1991). *Proceedings of the European Conference on Optical Communication*, p. 31.

Thijs, P. J. A., Binsma, J. J. M., Tiemeijer, L. F., Kuindersma, P. I., and van Dongen, T. (1992). *J. Microelectron. Eng.,* **18**, 57.

Thijs, P. J. A., Tiemeijer, L. F., Binsma, J. J. M., and van Dongen, T. (1994). *IEEE J. Quantum Electron.,* **QE-30**, 477.

Tiemeijer, L. F., Thijs, P. J. A., Binsma, J. J. M., and van Dongen, T. (1992). *Appl. Phys. Lett.,* **60**, 2466.

Tiemeijer, L. F., Thijs, P. J. A., van Dongen, T., Slootweg, R. W. M., van der Heijden, J. J. M., Binsma, J. J. M., and Krijn, M. P. C. M. (1993). *Appl. Phys. Lett.,* **62**, 826.

Ueno, Y. (1993). *Appl. Phys. Lett.,* **62**, 553.

Uomi, K. (1990), *Jpn. J. Appl. Phys.,* **29**, 81.

Valster, A., van der Poel, C. J., Finke, C. J., and Boermans, M. J. R. (1992). In *Proceedings of the 13th IEEE International Semiconductor Laser Conference,* Takamatsu, Japan, p. 152.

Viña, L., Muñoz, L., Mestres, N., Koteles, E. S., Ghiti, A., O'Reilly, E. P., Bertolet, D. C., and Lau K. M. (1993). *Phys. Rev. [B],* **47**, 13926.

Voisin, P. (1988). *Quantum Wells and Superlattices in Optoelectronic Devices and Integrated Optics,* SPIE Vol. 861. Bellingham, WA: SPIE, p. 88.

Wang, M. C., Kash, K., Zah, C. E., Bhat, R., and Chuang, S. L. (1993). *Appl. Phys. Lett.,* **62**, 166.

Waters, R. G., and Bertaska, R. K. (1988). *Appl. Phys. Lett.,* **52**, 179.

Woodhead, J. (1995). Private communication.

Yablonovitch, E., and Kane, E. O. (1986). *J. Lightwave Technol.,* **4**, 504.

Yariv, A. (1989). *Quantum Electronics,* 3d. ed., p. 253. New York: John Wiley and Sons.

Zah, C. E., Bhat, R., Pathak, B., Caneau, C., Favire, F. J., Andreakis, N. C., Hwang, D. M., Koza, M. A., Chen, C. Y., and Lee, T. P. (1991). *Electron. Lett.,* **27**, 1414.

Zhao, B., Chen, T. R., Wu, S., Zhuang, Y. H., Yamada, Y., and Yariv, A. (1993). *Appl. Phys. Lett.,* **62**, 1591.

Zou, Y., Osinski, J. S., Grodzinski, P., Dapkus, P. D., Rideout, W., Sharfin, W. F., and Crawford, F. D. (1993). *Appl. Phys. Lett.,* **62**, 175.

Chapter 3

High-Speed Lasers

Radhakrishnan Nagarajan
SDL, Inc., San Jose, CA

John E. Bowers
Department of Electrical and Computer Engineering,
University of California, Santa Barbara, CA

3.1 Introduction

High-speed semiconductor lasers are an integral part in the implementation of high-bit-rate optical communications systems. They are compact, rugged, reliable, long-lived, and relatively inexpensive sources of coherent light, and due to the very low attenuation window that exists in the silica-based optical fiber at 1.55 μm and the zero dispersion point at 1.3 μm, they have become the mainstay of optical fiber communication systems. Semiconductor lasers can be easily amplitude or frequency modulated up to very high frequencies by directly varying the drive current at the required data rates. They also can be fabricated together with the drive electronics of the transmitter circuitry on the same wafer to form optoelectronic integrated circuits. Advances in the optical fiber amplifier technology also have made the semiconductor laser–based systems more economically attractive for long-haul high-bit-rate communications applications.

Over the years, the field of high-speed semiconductor lasers has seen a number of advances. In the 1980s, the realization that the device parasitics were the practical limit to higher and higher modulation rates led to the development of low parasitic device structures, and this resulted in large increases in the direct modulation bandwidth of semiconductor lasers. Toward the end of that decade, using advanced crystal growth techniques such as molecular beam epitaxy (MBE) and metal organic chemical vapor deposition (MOCVD), quantum well lasers became a reality. Lasers with quantum well active regions have larger differential gain compared with their bulk counterparts, and the flexibility to vary the amount of quantum confinement and the emission wavelength by changing the composition and the width of the quantum wells and the barrier regions resulted in additional degrees of freedom to design semiconductor lasers with higher and higher speeds. Although initially the expectations to realize ultra-high-speed devices were plentiful, the mere replacement of the bulk active regions with quantum wells did not result in automatic increases in modulation speeds. Soon, it became evident that the phenomena of carrier transport, capture, and emission are the more dominant limits to high-bit-rate operation in the quantum well lasers or in lasers generally with one or more degrees of quantum confinement in the active region.

This chapter will review the present level of understanding of the high-speed properties of semiconductor lasers. High-speed semiconductor lasers in both the GaAs and the InP systems have made great strides over years. The emphasis here will be on the principles of high-speed operation. Although InP-based lasers have become technologically important for optical communications system applications, device structures and experimental data will be drawn from both material systems to illustrate the points being discussed. There have been several review-style papers and book chapters on the high-speed properties of semiconductor lasers written in the past that are complementary to the present effort (Tucker, 1985; Lau and Yariv, 1985b; Agrawal and Dutta, 1986; Su and Lanzisera, 1986; Bowers, 1987, 1989; Olshansky et al., 1987; Bowers and Pollack, 1988). There have been a number of developments since these contributions, the most notable being that of carrier transport effects on the high-speed properties of quantum well lasers (Rideout et al., 1991; Nagarajan et al., 1991c; Nagarajan, 1994). Theoretical treatment and experimental evidence for these effects will be presented.

This chapter is organized into three main sections. The first deals with the rate equation description of the carrier and photon dynamics in semiconductor lasers. Small signal analysis of the amplitude modulation, relative intensity noise, and frequency modulation are presented. As far as possible, analytical expressions are derived to present the physics of high-speed lasers at a more intuitive level. The second section deals in detail with some of the concepts derived in the first and applies the principles to device design. Concepts of linear gain and nonlinear gain suppression are discussed, experimental evidence and optimization of carrier transport processes are presented, and the effects of device size and structure on the high-speed properties are explored. The third section is on the large signal modulation in semiconductor lasers and the related topic of short pulse generation via mode locking and gain switching.

3.2 Laser dynamics

A set of two linear coupled rate equations, one for the carrier density in the active area and the other for the photon density in the laser cavity, are commonly used to analyze the dynamic behavior of semiconductor lasers. These equations essentially keep track of the injected carriers and the emitted photons and describe the laser operation in a basic manner. Historical perspective of this form of analysis has been presented by Lau and Yariv (1985b) and Bowers (1989) and will not be dealt with here. We will use a set of three rate equations, one for the carrier density in the current confinement region as well, instead of the traditional set of two, to analyze the dynamics of a semiconductor laser under current injection. The additional equation will include the effects of carrier transport in the separate confinement heterostructure (SCH) region typically used in quantum well lasers for carrier and optical confinement.

There are a number of simplifications involved in arriving at the set of rate equations used for the analysis here. These approximations are such that an accurate yet physically intuitive picture of the laser dynamics can be obtained without having to resort to large numerical models for the solution of the problem. We assume that the lateral and transverse dimensions of the active layer are such that only one optical mode is supported in both these directions. Only the optical confinement factor in the transverse direction is considered, and in the lateral direction it is taken to be unity. For simplicity, we also have considered only one longitudinal mode. The Fabry Perot (FP) lasers can in principle support multiple longitudinal modes, and this assumption is strictly valid for the technologically significant single-frequency lasers that have frequency-selective distributed feedback (DFB) structures as an integral part of the laser cavity.

The variation in carrier density across the active area and the associated effects of lateral carrier diffusion on the modulation properties have been neglected. This is a very good approximation in high-speed lasers that have active regions that are sufficiently narrow in the lateral dimension. Carrier density variation in the longitudinal direction is not significant for lasers with facet reflectivity greater than 20% (Lau and Yariv, 1985b). Semiconductor lasers with cleaved facets have greater than 30% facet reflectivity and are thus not affected. Another factor that may lead to spatial inhomogenieties are the phenomena that occur on the time scale of the cavity transit time. This transit time is given by the photon lifetime, and any high-speed phenomena whose time scale approaches this limit cannot be described adequately by these sets of time and spatially averaged rate equations.

3.2.1 Rate equations

The rate equations for the carrier density in the quantum well (N_W) and the SCH layer (N_B) and the photon density in the cavity (S) are written as in Eqs. (3.1) to (3.3). A drawing of the laser structure modeled by these equations is given in Fig. 3.1. The quantum well, SCH, and cladding regions are clearly indicated in this figure.

$$\frac{dN_B}{dt} = \frac{I}{eV_{SCH}} - \frac{N_B}{\tau_r} - \frac{N_B}{\tau_{nb}} + \frac{N_W(V_W/V_{SCH})}{\tau_e} \tag{3.1}$$

$$\frac{dN_W}{dt} = \frac{N_B(V_{SCH}/V_W)}{\tau_r} - \frac{N_W}{\tau_n} - \frac{N_W}{\tau_{nr}} - \frac{N_W}{\tau_e} - \frac{v_g G(N_W)S}{1 + \varepsilon S} \tag{3.2}$$

$$\frac{dS}{dt} = \frac{\Gamma v_g G(N_W)S}{1 + \varepsilon S} - \frac{S}{\tau_p} + \Gamma \beta \frac{N_W}{\tau_n} \tag{3.3}$$

where τ_n is the radiative recombination lifetime in the quantum well, τ_{nr} is the nonradiative recombination lifetime in the quantum well, τ_{nb} is the total recombination lifetime in the SCH region, τ_p is the photon lifetime, τ_r is the carrier transport time across the SCH to the quantum well active

3.2 Laser dynamics

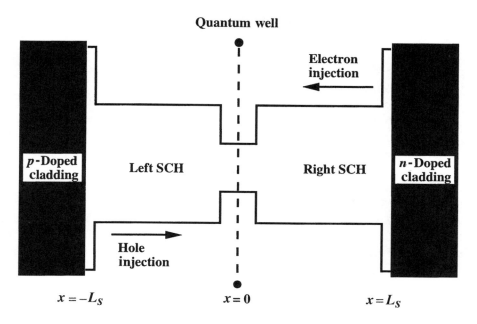

Figure 3.1: Schematic diagram of a single quantum well laser with a separate confinement heterostructure used in the carrier transport model.

region, τ_e is the carrier thermionic emission time from the quantum well into the barrier and SCH layers, Γ is the transverse optical confinement factor, β is the spontaneous emission feedback factor (the fraction of spontaneous emission that couples into the lasing mode), $G(N_W)$ is the carrier density dependent gain, v_g is the mode velocity, ε is the intrinsic gain compression factor, V_W is the volume of the quantum well, and V_{SCH} is the volume of the SCH. The spontaneous emission feedback term does not have a significant effect on the modulation response except at low power levels, where it contributes to additional damping. It will be neglected in the analysis of the laser dynamics. The nonradiative recombination τ_{nr} in the active region and carrier recombination in the SCH region τ_{nb} also have been neglected in the dynamic analysis because their contributions are not significant. They will be included later in the analysis of the effects of carrier transport on the effective linewidth enhancement factor. The optical gain $G(N_W)$ is a function of both the electron and hole carrier density within the quantum well. The electron and hole densities are in general not equal to one another in the case of nonuniform carrier injection in the quantum well lasers. We have neglected this effect here in the simplified analysis. Later we present results from a numerical model that highlight the nonuniform carrier distribution in the quantum well active region. The photon lifetime is the inverse of the total cavity loss rate and is written as $1/[v_g(\alpha_i + \alpha_m)]$, where α_i is the internal loss and α_m is the mirror loss given by $(1/2L_c) \ln [1/(R_1 R_2)]$. The front and rear facet reflectivities are given by R_1 and R_2, respectively, and L_c is the cavity length.

In the rate equations, the terms with τ_e represent the loss of carriers from the quantum well and the gain of carriers by the SCH layer, and the terms with τ_r vice versa. There are two components to τ_r; one is the transport time across the SCH, which is largely classical in nature, and the other is the quantum mechanical local capture time at the quantum well. The finite capture time of the quantum well is small (<1 ps) for the holes, which primarily contributes to the magnitude of τ_r. It is implicitly a part of τ_r, and we have not included the effects of carrier capture time separately. Models with an additional equation to describe the unconfined carriers in the active area and the effects of the local capture time also have been proposed (Tessler et al., 1992a, 1992b; Kan et al., 1992). Experimental evidence indicates that in practical laser structures at room temperature, hole transport across the SCH may indeed be the major speed-limiting factor (Nagarajan et al, 1993b). The magnitudes of τ_r and τ_e and the expressions to compute them will be derived and treated in detail later on in this section. The carriers in the quantum well are two dimensionally confined and the ones in the SCH layer have bulklike properties. The rate equations have been written in terms of volume carrier densities. The volume ratios V_{SCH}/V_W and V_W/V_{SCH} have been used in the rate equations to account for the fact that N_B and N_W are normalized with respect to two different volumes.

3.2.2 Small-signal amplitude modulation

The small-signal solution of the rate equations is obtained by first making the following substitution: $I = I_o + ie^{j\omega t}$, $N_B = N_{Bo} + n_B e^{j\omega t}$, $N_W = N_{Wo} + n_w e^{j\omega t}$, $S = S_o + se^{j\omega t}$, and $G = G_o + g_o n_W e^{j\omega t}$. The small-signal quantities i, n_B, n_W, and s, with the $e^{j\omega t}$, factor, have a sinusoidal time dependence with angular frequency ω. Optical gain varies linearly with carrier density in bulk active areas but exhibits a saturation behavior at

3.2 Laser dynamics

high carrier densities in quantum well lasers. Although $G(N_W)$ is a sublinear function of the carrier density in lower-dimensional active areas, it can always be linearized about some steady-state operating point, where g_o is the differential gain at that steady-state carrier density. For the small-signal carrier density variation n_W, g_o is a constant and is determined by the carrier density in the quantum well, which is clamped at the steady-state value of N_{Wo}. The g_o value is then different for different values of N_{Wo}, and this accounts for the saturation of gain at high carrier densities in quantum well lasers.

After the small-signal quantities are substituted into the rate equations [Eqs. (3.1) to (3.3)], the steady-state quantities are set to zero. The resulting small-signal equations are obtained by retaining only the linear terms in the carrier and photon densities, i.e., the terms with ω angular frequency variation. The higher-order terms with 2ω, 3ω, and higher-order variations give rise to harmonic distortion, which will not be considered here.

$$j\omega n_B = \frac{i}{eV_{SCH}} - \frac{n_B}{\tau_r} + \frac{n_W(V_W/V_{SCH})}{\tau_e} \qquad (3.4)$$

$$j\omega n_W = \frac{n_B(V_{SCH}/V_W)}{\tau_r} - \frac{n_W}{\tau_n} - \frac{n_W}{\tau_e} - \frac{v_g g_o S_o}{1 + \varepsilon S_o} n_W - \frac{v_g G_o}{(1 + \varepsilon S_o)^2} s \qquad (3.5)$$

$$j\omega s = \frac{\Gamma v_g g_o S_o}{1 + \varepsilon S_o} n_W + \frac{s}{\tau_p(1 + \varepsilon S_o)} - \frac{s}{\tau_p} \qquad (3.6)$$

The steady-state solution to the photon density equation, neglecting the term involving the spontaneous emission factor (which does not have a significant effect on the modulation response), gives the basic gain-loss relationship in the laser cavity; $(\Gamma v_g G_o)/(1 + \varepsilon S_o) = 1/\tau_p$. This set of small-signal equations can be reduced by eliminating n_B and n_W to give a relationship between the modulating current i and the optical output $s(\omega)$. The modulation response is given by $|M(\omega)|$, where $M(\omega)=s(\omega)/(i/e)$. The complete $M(\omega)$ function is (Nagarajan et al., 1992d)

$$M(\omega) = \left(\frac{\Gamma v_g g_o S_o}{V_W}\right) \frac{1}{A_0 + jA_1\omega - A_2\omega^2 - jA_3\omega^3} \qquad (3.7)$$

$$A_0 = \frac{v_g g_o S_o}{\tau_p}\left(1 + \frac{\varepsilon}{v_g g_o \tau_n}\right) \qquad (3.8)$$

$$A_1 = v_g g_o S_o \left(1 + \frac{\tau_r}{\tau_p}\right) + \frac{\varepsilon S_o}{\tau_p}\left(1 + \frac{\tau_r}{\tau_e} + \frac{\tau_r}{\tau_n}\right) + \frac{1}{\tau_n}(1 + \varepsilon S_o) \quad (3.9)$$

$$A_2 = (1 + \varepsilon S_o)\left(1 + \frac{\tau_r}{\tau_e} + \frac{\tau_r}{\tau_n}\right) + v_g g_o S_o \tau_r + \frac{\varepsilon S_o \tau_r}{\tau_p} \quad (3.10)$$

$$A_3 = \tau_r(1 + \varepsilon S_o) \quad (3.11)$$

The denominator of the exact expression is a third-order polynomial and can be solved numerically. The exact expression for $M(\omega)$ can be rewritten as (Nagarajan et al., 1992a)

$$M(\omega) = \left(\frac{1}{1 + j\omega\tau_r}\right) \frac{\dfrac{v_g g_o S_o}{qV_W}}{\left\{j\omega\left[1 + \left(\dfrac{\tau_r}{1+j\omega\tau_r}\right)\dfrac{1}{\tau_e}\right] + \dfrac{v_g g_o S_o}{1+\varepsilon S_o}\right\}\left[j\omega(1 + \varepsilon S_o) + \dfrac{\varepsilon S_o}{\tau_p}\right] + \dfrac{v_g g_o S_o}{\tau_p(1 + \varepsilon S_o)}} \quad (3.12)$$

$M(\omega)$ can be simplified by replacing the term within the parentheses in the first part of denominator by τ_r, i.e., by assuming that $|\omega\tau_r| \ll 1$. This is typically the case in the frequency region, where high-speed laser performance is limited by the carrier transport time across the SCH. With this replacement, $M(\omega)$ is written as (Nagarajan et al., 1992a)

$$M(\omega) = \left(\frac{1}{1+j\omega\tau_r}\right)\frac{A}{\omega_r^2 - \omega^2 + j\omega\gamma} \quad (3.13)$$

$$A = \frac{\Gamma(v_g g_o/\chi)S_o}{V_W(1+\varepsilon S_o)} \quad (3.14)$$

$$\omega_r^2 = \frac{(v_g g_o/\chi)S_o}{\tau_p(1 + \varepsilon S_o)}\left(1 + \frac{\varepsilon}{v_g g_o \tau_n}\right) \quad (3.15)$$

$$\gamma = \frac{(v_g g_o/\chi)S_o}{(1 + \varepsilon S_o)} + \frac{\varepsilon S_o/\tau_p}{(1 + \varepsilon S_o)} + \frac{1}{\chi\tau_n} \quad (3.16)$$

where a transport factor, $\chi = 1 + (\tau_r/\tau_e)$, has been introduced into the equations. The original expressions for γ and ω_r^2 are recoverable in the limit of $\chi = 1$, i.e., in the limit of negligible transport effects (Bowers, 1989).

From these analytic expressions, a number of effects of carrier transport in high-speed semiconductor lasers are evident. First, there is a low-frequency rolloff in the modulation response *only* due to the transport time across the SCH. This is a parasitic-like rolloff that would significantly limit the -3-dB modulation bandwidth for large τ_r and is *indistinguishable* from parasitics.

Second, the effective or dynamic differential gain has been reduced to g_o/χ. This reduction is *not* responsible for the rolloff above and is present even in the absence of a significant rolloff. This results in a reduction of the resonance frequency.

Third, the effective bimolecular recombination lifetime has been increased to $\chi\tau_n$ (Nagarajan et al., 1992d). Like the reduction in differential gain g_o above, this is a dynamic increase in τ_n and does not influence the static properties. This increase in τ_n is relevant to the case of switching in semiconductor lasers from below threshold, where the turn-on delay is linearly proportional to the carrier lifetime, and any adverse effect of carrier transport will increase this delay significantly. We will deal with this in detail in Sect. 3.4.

The gain compression factor remains unmodified. This is in contrast to the work by Rideout et al. (1991), where a well barrier holeburning model has been used to explain the degradation of high-speed properties via an enhancement in the nonlinear gain suppression factor ε.

From these analytic expressions, a relationship for the K factor can be derived. The K factor was originally defined as $\gamma = K f_r^2$ (Olshansky et al., 1987) to characterize the slope of the linear dependence of the damping rate on the square of the resonance frequency. However, experimentally, a dc offset is always observed, and therefore, a more accurate definition is $\gamma = K f_r^2 + \gamma_o$ (Nagarajan et al., 1992a). The maximum possible modulation bandwidth in a semiconductor laser is determined by the K factor. In principle, if one could drive the laser to arbitrarily high output power levels without the detrimental effects of device heating, then the maximum possible -3-dB bandwidth, in the absence of device parasitics and carrier transport rolloff, is derived by first setting the amplitude modulation response function to $1/\sqrt{2}$:

$$M(\omega) = \left| \frac{M(0)}{\omega_r^2 - \omega^2 + j\omega\gamma} \right| = \frac{1}{\sqrt{2}} \quad (3.17)$$

$$\omega^4 + (\gamma^2 - 2\omega_r^2)\omega^2 - \omega_r^4 = 0 \quad (3.18)$$

Using the linear dependence of ω_r^2 and γ on the photon density S_o, the expression for the maximum -3-dB bandwidth ω_{max} can derived by evaluating $d\omega/dS_o = 0$. The result is

$$\omega_{max}^2 = \frac{\omega_r^4}{[\chi(\gamma - \gamma_o) - \omega_r^2]} \tag{3.19}$$

Substituting this back into the original expression, we obtain

$$\gamma^2 = 2\omega_{max}^2 + \gamma_o\left(\frac{\gamma^2}{\omega_r^2}\right)(\gamma - \gamma_o) \tag{3.20}$$

Using this expression and the definition of the K factor,

$$\left(1 + \frac{\gamma_o}{Kf_{max}^2}\right)^2\left(1 - \frac{\gamma_o K}{4\pi^2}\right)f_{max}^2 = \frac{8\pi}{K^2} \tag{3.21}$$

$$f_{max} \approx \frac{2\sqrt{2}\pi}{K} + \frac{\gamma_o}{2\sqrt{2}\pi} \tag{3.22}$$

In deriving the final expression for f_{max}, we have assumed that $\gamma_o/(Kf_{max}^2)$ and $(\gamma_o K)/4\pi^2 \ll 1$. We then neglect the first term and take the first-order Taylor expansion of the second. In the limit of $\gamma_o = 0$, the original expression $f_{max} = (2\sqrt{2}\pi)/K$ as derived by Olshansky et al. (1987) is obtained.

Using the modified definition, the expressions for the K factor and γ_o are

$$K = 4\pi^2\left[\tau_p + \frac{\varepsilon}{(v_g g_o/\chi)}\right] \quad \gamma_o = \frac{1}{\chi\tau_n} \tag{3.23}$$

The intercept γ_o is a measure of the effective carrier recombination lifetime. Since τ_n has been modified to $\chi\tau_n$, this leads to a structure dependence of γ_o. Also, depending on the laser structure, the K factor can be significantly affected by carrier transport. The low-frequency rolloff caused by transport means that the maximum possible -3-dB modulation bandwidth in a laser may not even be determined by this reduced K factor. The transport delay is then the *intrinsic* limit to the maximum modulation bandwidth: $f_{max} = 1/(2\pi\tau_r)$ when the resonance frequency is higher than f_{max}.

3.2.3 Relative intensity noise

The relative intensity noise spectrum can be derived using the same formalism. The difference in this case is that the driving force is no longer the input current but rather the real part of the Langevin force F_r of the field due to the spontaneous emission, which is assumed to be uncorrelated white Gaussian noise (Kikuchi and Okoshi, 1985; Okoshi and Kikuchi, 1988). In general, there also will be a noise term associated with the carrier density fluctuations (Harder et al., 1982). The carrier density fluctuation term is not significant in the analysis of relative intensity noise in semiconductor lasers (Kikuchi and Okoshi, 1985). Including only the photon density fluctuations, the small-signal rate equations are (Nagarajan et al., 1992b)

$$j\omega n_B = -\frac{n_B}{\tau_r} + \frac{n_W(V_W/V_{SCH})}{\tau_e} \quad (3.24)$$

$$j\omega n_W = \frac{n_B(V_{SCH}/V_W)}{\tau_r} - \frac{n_W}{\tau_n} - \frac{n_W}{\tau_e} - \frac{v_g g_o S_o}{1 + \varepsilon S_o} n_W - \frac{v_g G_o}{(1 + \varepsilon S_o)^2} s \quad (3.25)$$

$$j\omega s = \frac{\Gamma v_g g_o S_o}{1 + \varepsilon S_o} n_W + \frac{s}{\tau_p(1 + \varepsilon S_o)} - \frac{s}{\tau_p} + F_r \quad (3.26)$$

The small-signal carrier density terms n_B and n_W are eliminated to obtain the expression for the Fourier component of the amplitude fluctuations $R(\omega)$ due to F_r. Without any simplifying assumptions, $R(\omega)$ is given by (Nagarajan et al., 1992b)

$$R(\omega) = F_r \frac{B_0 + jB_1\omega - B_2\omega^2}{A_0 + jA_1\omega - A_2\omega^2 - jA_3\omega^3} \quad (3.27)$$

$$B_0 = v_g g_o S_o + \frac{1}{\tau_n}(1 + \varepsilon S_o) \quad (3.28)$$

$$B_1 = (1 + \varepsilon S_o)\left(1 + \frac{\tau_r}{\tau_e} + \frac{\tau_r}{\tau_n}\right) + v_g g_o S_o \tau_r \quad (3.29)$$

$$B_2 = \tau_r(1 + \varepsilon S_o) \quad (3.30)$$

The denominator terms remain unchanged as in the expression for $M(\omega)$. The AM noise spectrum is defined as the power spectral density function $S_A(\omega)$ of the amplitude fluctuation, and it is written in terms of the Fourier transform of the amplitude fluctuations defined in the time interval $-T/2 < t < T/2$ (Okoshi and Kikuchi, 1988).

$$S_A(\omega) = \lim_{T \to \infty} \frac{1}{T} \langle R^*(\omega) R(\omega) \rangle \tag{3.31}$$

The relative intensity noise (*RIN*) is the ratio of the intensity fluctuation to the time-averaged intensity. To obtain the γ and ω_r in the *conventional* sense of their definitions, i.e., with respect to a second-order polynomial in the denominator, one would have to first compute the *RIN* spectrum using the expression given above and then numerically fit it to a conventional form of the expression and extract the relevant parameters (Ishikawa et al., 1992b).

Again, neglecting the frequency dependence of τ_r as in Eq. (3.12), the resulting expression for $R(\omega)$ is (Nagarajan et al., 1992b)

$$R(\omega) = F_r \frac{\gamma^* + j\omega}{\omega_r^2 - \omega^2 + j\omega\gamma} \tag{3.32}$$

$$\gamma^* = \frac{(v_g g_o/\chi) S_o}{1 + \varepsilon S_o} + \frac{1}{\chi \tau_n} \tag{3.33}$$

The other terms of the expression are the same as in the modulation response function. The γ and γ^* terms are almost identical at sufficiently low powers and small ε. The analytic expression for *RIN* spectrum can be written easily in this case as

$$RIN(f) = \frac{4}{\pi} (\delta f)_{ST} \frac{f^2 + (\gamma^*/2\pi)^2}{(f_r^2 - f^2)^2 + (\gamma/2\pi)^2 f^2} \tag{3.34}$$

where $(\delta f)_{ST}$ is the intrinsic Schawlow-Townes linewidth that is due to the spontaneous emission events (Okoshi and Kikuchi, 1988).

The additional low-frequency rolloff present in the expression for the modulation response is absent here. For large values of τ_r, the modulation bandwidth could be severely reduced, although the K factor values as determined from the noise measurements, which are affected by χ alone and not the low-frequency rolloff, could still be optimistic. It is generally believed that the noise spectrum measurement is a *parasitic-free* means of determining the actual or potential modulation performance of a laser. In the presence of significant transport effects, this is no longer true.

3.2 Laser dynamics

Figure 3.2 shows the maximum possible modulation bandwidth as a function of SCH width for a 300-μm cavity length $In_{0.2}Ga_{0.8}As$/GaAs single quantum well (SQW) laser (Nagarajan et al. 1992b). The dashed line shows the K factor limit, which is commonly inferred from noise spectra measurements. The bold line is the *real limit* determined by carrier transport across the SCH, i.e., the maximum modulation bandwidth that is possible in practice, despite an intrinsic high-frequency design, in a laser diode whose dynamic response is limited by carrier transport.

Herein lies the real danger of relying on K factors and *RIN* measurements to predict modulation bandwidths. A small K factor is indeed an indication of the good device and material quality, but one has to be careful as to the maximum modulation bandwidth limits implied by this quantity. One can truly use this as the limit only in devices where the carrier transport effects are minimal.

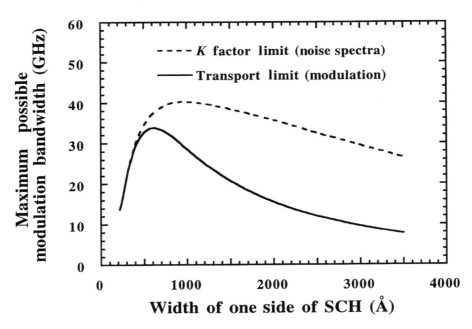

Figure 3.2: Comparison of maximum possible modulation bandwidth as a function of the width of SCH inferred from the modulation response and the relative intensity noise measurements for an $In_{0.2}Ga_{0.8}As/Al_{0.1}Ga_{0.9}As$ SQW laser.

3.2.4 Frequency modulation and chirping

The variation in carrier density under intensity modulation leads to a corresponding change in the refractive index in the laser cavity. This change in index causes a change in the lasing wavelength (or alternatively, the oscillation frequency) of the laser emission. This is commonly referred to as the *wavelength chirping* or the *frequency modulation (FM) response* of the semiconductor laser.

We start with the photon density rate equation to derive the magnitude of wavelength chirp under intensity modulation. Although a different form of gain suppression is used in our analysis, we essentially follow the method of Koch and Bowers (1984) and Koch and Linke (1986) for the derivation. Without assuming anything about the magnitude of the photon density variation and the corresponding change in gain under modulation, the following substitutions for the time-varying quantities—$G(t) \equiv G_o + \Delta G(t)$ and $S(t) \equiv S_o + \Delta S(t)$—are made where $G(t)$ is same as the $G(N_W)$ term we had used previously, but only the time dependence is considered here, and G_o is the steady-state gain at the laser operating point. The expansion for the $1/(1 + \varepsilon S)$ term is made to the first order. After the substitution for the time-varying and steady-state quantities are made in the photon density rate equation [Eq. (3.3)] and simplification, the resulting expression is

$$(1 + \varepsilon S)\frac{(dS/dt)}{S} + \varepsilon \frac{\Delta S(t)}{\tau_p} \approx \frac{(dS/dt)}{S} + \varepsilon \frac{\Delta S(t)}{\tau_p} = \Gamma v_g \, \Delta G(t)$$

(3.35)

The index of refraction in the laser cavity generally can be written in terms of its real and imaginary components, $n = n' + in''$. The gain G and the imaginary part of the index of refraction are related; $G = (4\pi/\lambda) n''$, where λ is the emission wavelength of the laser. In terms of the effective index, this is written as $\Gamma G = (4\pi/\lambda) n''_{\text{eff}}$, where Γ is the transverse optical confinement factor. The linewidth enhancement factor α (Henry, 1982) is the ratio of the variation of the real part of the index of refraction with carrier density to the imaginary part with carrier density: $(\partial n'/\partial N)/(\partial n''/\partial N)$. The linewidth enhancement factor also may be written in terms of the effective real and imaginary part of the index of refraction as $\alpha_{\text{eff}} = (\partial n'_{\text{eff}}/\partial N)/(\partial n''_{\text{eff}}/\partial N)$. The definition of the linewidth enhancement factor α in terms of the effective index quantities is consistent because both n'_{eff} and n''_{eff} are evaluated with respect to the same mode

3.2 Laser dynamics

overlap integral. Using this modified definition of the linewidth enhancement factor, the change in the effective (modal) gain and the real part of the index of refraction can be written in terms of α: $\Gamma G(t) = (4\pi/\lambda)[n'_{\text{eff}}(t)/\alpha]$. Substituting this in Eq. (3.35) gives

$$\Delta n'_{\text{eff}}(t) = \frac{\alpha}{4\pi}\frac{\lambda}{v_g}\left[\frac{d\ln(S)}{dt} + \varepsilon\frac{\Delta S(t)}{\tau_p}\right] \quad (3.36)$$

The change in the effective index $\Delta n'_{\text{eff}}$ and the group index n'_{geff} are related to the first order as $\Delta n'_{\text{eff}} = -n'_{\text{geff}}(\Delta\nu/\nu)$, where $\Delta\nu$ is the frequency deviation as the result of the change in the real part of the index of refraction $\Delta n'_{\text{eff}}$. This is then substituted in Eq. (3.36). The resulting expression for the time dependence of the frequency deviation is

$$\Delta\nu(t) = -\frac{\alpha}{4\pi}\left\{\frac{d\ln[P(t)]}{dt} + \frac{\varepsilon\Gamma}{V_W h\nu}\left(\frac{\alpha_m + \alpha_i}{\alpha_m}\right)P(t)\right\} \quad (3.37)$$

where $h\nu$ is the photon energy, and $P(t)$ is the optical power. Although we had assumed a $1/(1 + \varepsilon S)$ form of gain suppression, the expression for frequency deviation is the same as the one originally derived by Koch and Linke (1986).

Since the derivation of chirp under current modulation involves only the photon density rate equation, the carrier transport times that enter the equations for the carrier density in the SCH and the quantum well regions do not directly affect wavelength chirping. On the other hand, the *effective* index term that enters the definition of α_{eff} is directly affected by carrier transport. The additional contribution to wavelength chirping due to carrier transport can be derived in terms of α_{eff} for the whole laser structure consisting of the active layer and the SCH region. Thus the change in the effective gain and the real part of the index of refraction are related via an effective α that is true only for the particular laser structure, and hence, the α parameter is structure dependent (Nagarajan and Bowers, 1993a). We have called this the *effective* α. Although the α as originally derived is purely a material parameter, the α_{eff} defined here is affected by the structure-dependent carrier transport parameters.

The derivative of the effective index (n'_{eff}) of the mode with respect to the steady-state carrier density (N_{Wo}) in the quantum well, to the first order, is

$$\frac{\partial n'_{\text{eff}}}{\partial N_{Wo}} = \Gamma\frac{\partial n'}{\partial N_{Wo}} + (1-\Gamma)\frac{\partial N_{Bo}}{\partial N_{Wo}}\frac{\partial n'}{\partial N_{Bo}} \quad (3.38)$$

where the quantity of interest is the variation of the carrier density in the SCH with respect to the carrier density in the quantum well, $\partial N_{Bo}/\partial N_{Wo}$. In general, the index contribution from the confinement regions should be weighted by the appropriate mode overlap integral. Since the optical mode outside the active area lies mostly within these regions, the $(1 - \Gamma)$ term is used instead. To compute an expression for $\partial N_{Bo}/\partial N_{Wo}$, we have to look at the steady-state solution of the rate equation for N_B, which can be written as

$$\left(\frac{1}{\tau_r} + \frac{1}{\tau_{nb}}\right) V_{SCH} N_{Bo} = \frac{I_o}{e} + \frac{V_W N_{Wo}}{\tau_e} \qquad (3.39)$$

Taking the derivative of this equation with respect to I_o,

$$\frac{\partial(V_{SCH} N_{Bo}/\tau_b)}{\partial(I_o/e)} = \frac{1 + \frac{\partial(V_W N_{Wo}/\tau_n)}{\partial(I_o/e)}\left(\frac{\tau_n}{\tau_e}\right)}{1 + \frac{\tau_b}{\tau_r}} \qquad (3.40)$$

Typically τ_n is much larger than τ_e in the range where the transport effects are severe enough to cause a noticeable change in α_{eff}, and hence the second term in the numerator of the right-hand side of the equation is much larger than unity. With this assumption, the following expression may be written for $\partial N_{Bo}/\partial N_{Wo}$:

$$\frac{\partial N_{Bo}}{\partial N_{Wo}} \approx \left(\frac{V_W}{V_{SCH}}\right)\left(\frac{\tau_r}{\tau_e}\right)\left(\frac{\tau_b}{\tau_b + \tau_r}\right) = \left(\frac{V_W}{V_{SCH}}\right)(\chi - 1)\left(\frac{\tau_b}{\tau_b + \tau_r}\right) \qquad (3.41)$$

Using this expression for $\partial N_{Bo}/\partial N_{Wo}$, the final expression for α_{eff} is (Nagarajan and Bowers, 1993a),

$$\alpha_{\text{eff}} = \frac{4\pi}{\lambda g_o}\left[\frac{\partial n}{\partial N_{wo}} + \left(\frac{1}{\Gamma} - 1\right)\left(\frac{V_W}{V_{SCH}}\right)(\chi - 1)\left(\frac{\tau_b}{\tau_b + \tau_r}\right)\frac{\partial n}{\partial N_{bo}}\right] \qquad (3.42)$$

where g_o is the differential gain, which may be carrier density dependent. An interesting feature of the expression for α_{eff} is the geometric factor $[(1/\Gamma)-1](V_W/V_{SCH})$. In multiple quantum well lasers, the optical confinement factor scales linearly with the number of quantum wells (this is a very good approximation for a small number of wells), and so does the volume of the active region V_W. Thus, for a given SCH width, and in the absence of complications due to carrier transport in multiple quantum

3.2 Laser dynamics

well structures, the α_{eff} in multiple quantum well lasers will be lower due to the higher differential gain at equivalent threshold gain levels.

The effective linewidth enhancement factor can be written in terms of an intrinsic component from the index variation in the active area and a transport-related component from the index variation in the SCH region, where both components are with respect to the differential gain in the active area

$$\alpha_{\text{eff}} = \alpha_{\text{quantum well}} + \left(\frac{1}{\Gamma} - 1\right)\left(\frac{V_W}{V_{SCH}}\right)(\chi - 1)\left(\frac{\tau_b}{\tau_b + \tau_r}\right)\alpha_{SCH} \qquad (3.43)$$

In the absence of carrier transport effects, i.e., in the limit of τ_r being negligibly short or τ_e being significantly long, α_{eff} reaches its minimum value. In this limit, the wavelength chirping is solely governed by the index changes in the quantum well active area. The expression for the α_{eff} that has been derived here can be used directly in the Eq. (3.37) derived for $\Delta \nu(t)$ to compute the total wavelength chirp under high-speed current modulation of the quantum well laser structures.

3.2.5 Carrier transport times

Electron and hole transport from the doped cladding layers to the quantum well consists of two parts (Polland et al., 1988; Morin et al., 1991). First is the transport across the SCH. This is governed by the classical current continuity equations that describe the diffusion, recombination, and, in the presence of any electric field, drift of carriers across the SCH. The second part is the carrier capture by the quantum well. This is a quantum mechanical problem that has to take into account the relevant dynamics of the phonon scattering mechanism that mediates this capture. This scattering process is a function of the initial and final state wavefunctions, the coupling strength of the transition, and the phonon dispersion in the material. In addition to carrier-phonon scattering, carrier-carrier scattering also contributes to carrier capture (Preisel et al., 1994). In this analysis, as shown in Fig. 3.1, the SCH width is taken to be the distance between the quantum well active area and the point from which the claddings are being doped instead of just being the region sandwiched between the quantum well and the cladding. This distance is commonly referred to as the *doping offset* from the active during epitaxial growth. For carrier transport, it is the width of the undoped region between the quantum well and the cladding layer that is of significance, and in most cases it is the width of the SCH.

The next transport mechanism that is significant in devices operating at room temperature is the thermally activated carrier escape from the quantum well or thermionic emission. Although this process degrades the overall carrier capture efficiency of a SQW structure, it is essential for carrier transport between quantum wells in a multiple quantum well (MQW) structure. Another transport process of interest in an MQW system is tunneling between the quantum wells. Thermionic emission is a strong function of barrier height, while tunneling is sensitive to both barrier height and width. The barriers in an MQW structure should be designed such that the quantum wells efficiently capture and contain the carriers for laser action in a two-dimensionally confined system without adversely sacrificing the transport (leading to carrier trapping) across the structure. There have been predictions that in a biased multiple quantum well laser diode carrier transport by the resonant tunneling mechanism may be switched off due to band bending in the quantum well region (Tessler and Eisenstein, 1993b).

For small-signal modulation experiments, the laser diode is forward biased, and a small microwave signal is imposed on this dc bias. The response of the laser is then measured using a high-speed photodiode. Since the modulating signal is very small, the device is essentially at constant forward bias. We will consider large-signal modulation in Sect. 3.4. Further, we will only consider an SCH that is not compositionally graded, and this excludes any built-in fields in the SCH to aid carrier transport.

Under normal operation, the semiconductor laser is essentially a forward-biased p-i-n diode, with the claddings doped p and n and the SCH region nominally undoped. The electrons are injected from the one side and the holes from the other and across the SCH. The important difference is the quantum well in the middle of the SCH where the injected carriers recombine. Although the carrier injection is from the opposite ends of the SCH, any physical separation of the two types of charges across the quantum well layer would lead to very large electric fields between them. The laser normally operates under high forward injection, where the carrier density levels are about 10^{18} cm^{-3}, and solving the Poisson equation in one dimension, assuming that the SCH layers on the right- and left-hand sides of the quantum well are uniformly pumped with electrons and holes, respectively, leads to electric fields in excess of 10^6 V/cm be-

SCH transport time

The laser structure can then be analyzed like a heavily forward biased p-i-n diode (Howard and Johnson, 1965). The equations for the electron and hole current densities, including both the drift and diffusion components, are given by

$$J_n = eD_n \left(\frac{enE}{kT} + \frac{\partial n}{\partial x} \right) \qquad J_p = eD_p \left(\frac{epE}{kT} - \frac{\partial p}{\partial x} \right) \qquad (3.44)$$

In this expression we have used the Einstein relation $D/\mu = kT/e$.

The current continuity conditions are

$$\frac{\partial n}{\partial t} = \frac{1}{e} \frac{\partial J_n}{\partial x} - U(n, p) \qquad \frac{\partial p}{\partial t} = -\frac{1}{e} \frac{\partial J_p}{\partial x} - U(n, p) \qquad (3.45)$$

where $U(n, p)$ is the net recombination rate. Assuming high injection conditions, $n \approx p$, and charge neutrality, $\partial E/\partial x = 0$, the expressions in Eq. (3.45) can be combined. The electric field term E can be eliminated to give the following equation under steady-state conditions (i.e., $\partial n/\partial t = \partial p/\partial t = 0$):

$$\frac{d^2 n}{dx^2} - \frac{D_n + D_p}{2 D_n D_p} U(n) = 0 \qquad (3.46)$$

This simplification does not mean that the dynamic carrier effects have been eliminated from the charge transport equations. The steady-state carrier density distribution profile is derived first, and then the small signal response of this distribution is computed to determine the dynamic properties of the device under current modulation. In Eq. (3.46), if an ambipolar diffusion coefficient $D_a = 2 D_n D_p / (D_n + D_p)$ is introduced and the recombination rate U is determined by an ambipolar lifetime τ_a, then Eq. (3.46) modifies to

$$\frac{d^2 n}{dx^2} - \frac{n}{L_{ao}^2} = 0 \qquad (3.47)$$

where $L_{ao} = \sqrt{D_a \tau_a}$ is the ambipolar diffusion length. Generally in the III-V compounds used to fabricate semiconductor laser diodes, $D_n \gg D_p$, and this leads to an ambipolar diffusion coefficient $D_a \approx 2D_p$. In other words, the carrier transport proceeds as if it is purely diffusion, but with an effective diffusion coefficient that is twice the normal hole diffusion coefficient. It *must be emphasized* that in this derivation we have only assumed that $\partial E/\partial x = 0$, *not* $E = 0$, and in fact, $E \neq 0$. The ambipolar type carrier transport is the *result* of a nonzero electric field and is not an a priori assumption. Subpicosecond luminescence spectroscopy experiments done in graded and ungraded SCH GaAs/Al$_x$Ga$_{1-x}$. As quantum well structures have revealed evidence for ambipolar transport of carriers in structures similar to the one that is being modeled here (Deveaud et al., 1987; Blom et al., 1990).

The steady-state carrier distribution in the SCH region is obtained by solving the ambipolar diffusion equation [Eq. (3.47)] with appropriate boundary conditions. The general solution to this second-order differential equation is

$$n(x) = C_1 e^{+x/L_{ao}} + C_2 e^{-x/L_{ao}} \qquad (3.48)$$

where C_1 and C_2 are constants to be determined from the relevant boundary conditions. At the edge of the SCH ($x = -L_S$), a constant supply of carriers is established by the dc bias current flowing into the SCH. For a dc current of I_S, the required boundary condition is

$$\frac{I_S}{A} = -eD_a \left.\frac{dn}{dx}\right|_{x=-L_S} \qquad (3.49)$$

where A is the cross-sectional area of the laser diode. The second boundary condition is that at $x = 0$, i.e., at the position of the quantum well, the carrier density is given by the steady-state carrier density in the well N_{Wo}. The carrier density in the quantum well is determined by the steady-state carrier-photon dynamics within the quantum well. This is determined from the rate equations for the carrier and photon densities. In laser structures with very narrow SCH regions, the electric fields at the SCH-cladding and SCH–quantum well interfaces may significantly influence the carrier distributions. We will present experimental evidence later in this section that indicates that ambipolar transport is indeed the carrier transport mode in common high-speed quantum well structures.

3.2 Laser dynamics

The coefficients C_1 and C_2 are then

$$C_1 = \frac{N_{Wo} e^{+L_S/L_{ao}} - \frac{I_S L_{ao}}{eD_a A}}{e^{+L_S/L_{ao}} + e^{-L_S/L_{ao}}} \qquad C_2 = \frac{N_{Wo} e^{-L_S/L_{ao}} + \frac{I_S L_{ao}}{eD_a A}}{e^{+L_S/L_{ao}} + e^{-L_S/L_{ao}}} \quad (3.50)$$

The current flowing into the quantum well under steady-state conditions I_W is given by

$$I_W = -eD_a A \left.\frac{dn}{dx}\right|_{x=0} = I_S \operatorname{sech}\left(\frac{L_s}{L_{ao}}\right) - eV_W\left(\frac{L_{ao}}{L_W}\right) \tanh\left(\frac{L_S}{L_{ao}}\right) \frac{N_{Wo}}{\tau_a} \quad (3.51)$$

where V_W is the volume of the quantum well and L_W is the width of the quantum well.

A differential SCH transport factor α_{SCH} analogous to the common-base current gain of a bipolar junction transistor (BJT) (Sze, 1981a) can be defined as

$$\alpha_{SCH} = \frac{\partial I_W}{\partial I_S} = \operatorname{sech}\left(\frac{L_s}{L_{ao}}\right) \quad (3.52)$$

The quantity of interest is the small-signal value of α_{SCH}. The expression for this can be derived by substituting L_{ao} by L_a in Eq. (3.52), which is given by

$$L_a = \sqrt{\frac{L_{ao}^2}{1 + j\omega\tau_a}} \quad (3.53)$$

The small-signal expression simplifies to

$$\alpha_{SCH,\text{ small-signal}} = \frac{1}{\cosh[(L_S^2/D_a\tau_a)^{1/2}\sqrt{1+j\omega\tau_a}]} \approx \frac{1}{1 + j\omega(L_S^2/2D_a)} \quad (3.54)$$

In the final expression on the right-hand side of Eq. (3.54), the width of one side of the SCH has been assumed to be much smaller than the ambipolar diffusion length, i.e., $L_S \ll L_{ao} (= \sqrt{D_a\tau_a})$. This is a valid assumption because the nonradiative lifetime and even the spontaneous recombination lifetime in the SCH layer are on the order of nanoseconds and are much longer than the SCH layer ambipolar transport time given by $\tau_r = L_S^2/2D_a$. This is similar to the base layer of a BJT, where the bandwidth limit due to the minority carrier transport time is called the *alpha cutoff frequency* and is given by $f_T = 1/2\pi\tau_r$. The transport time across the SCH has been directly determined from gain recovery measurements in quantum well optical amplifiers and has been shown to vary experimentally as the square of SCH width (Eisenstein et al., 1991; Weiss et al., 1992).

The expression for τ_r, derived above, also can be written as

$$\tau_r = \frac{L_S^2}{2D_a} = \frac{L_S^2}{2}\left(\frac{D_n + D_p}{2D_n D_p}\right) = \frac{1}{2}\left(\frac{L_S^2}{2D_p} + \frac{L_S^2}{2D_n}\right) = \frac{\tau_{r,\text{holes}} + \tau_{r,\text{electrons}}}{2} \tag{3.55}$$

This means that the total ambipolar diffusion time can be viewed *as an average* of the individual hole and electron diffusion times. This is a rather useful way of conceptualizing the ambipolar carrier transport in a quantum well structure.

Thermionic emission time

The thermionic emission time is derived using the standard thermionic emission theory (Sze, 1981b). Assuming that the carriers in the barriers have bulklike properties and obey Boltzmann statistics, the thermionic emission lifetime τ_e from a quantum well is (Schneider and Klitzing, 1988)

$$\tau_e = \sqrt{\frac{2\pi m^* L_W^2}{kT}} \exp\left(\frac{E_B}{kT}\right) \tag{3.56}$$

where E_B is the effective barrier height, m^* is the effective carrier mass, k is the Boltzmann constant, and T is temperature in kelvins. Thermionic emission is a sensitive function of barrier height and temperature.

Tunneling time

In an MQW structure, there are a number of wells each separated by barriers of energy height E_B and thickness L_B. In this case the electron and hole wavefunctions are no longer completely localized within the individual quantum wells, and for any finite L_B there will be coupling between the wells. For a system of two symmetric wells, the interwell coupling would remove the energy degeneracy that exists in the limit of infinite L_B. The lowest odd (E_1) and even (E_2) bound states of such a two-well system are separated by some small energy, $\Delta E = E_1 - E_2$ ($E_1 > E_2$) (Kroemer and Okamoto, 1984; Bastard, 1988). The overall linear superposition wavefunction of this system corresponds to an electron or a hole oscillating between the wells at a frequency given by $\Delta E/h$, where h is the Planck constant. The tunneling time, defined as the one-half period of the oscillation, is given by $\tau_t = h/2\Delta E$. Although this result has been derived for a two-well system, it will generally be true for an MQW system in the limit of weak interwell coupling (Kroemer and Okamoto, 1984).

Capture time

Calculating the carrier capture time by the quantum well involves computation of the probability per unit time that a carrier in some initial state in the SCH emits a longitudinal optical (LO) phonon and ends up in some final state within the quantum well as dictated by energy and momentum conservation conditions. Early theoretical works based on semiclassical treatment argued that the quantum well width had to be on the order of or larger than the LO phonon scattering limited electron mean free path (~ 100Å) for the quantum well to efficiently capture carriers and act as a center for radiative recombination (Shichijo et al., 1978; Tang et al., 1982).

Quantum mechanical calculations have relaxed this condition but have predicted strong oscillations in the capture time as a function of the quantum well width. As the well width increases, first, the states within the quantum well become more tightly bound, and second, the virtual states above the quantum well become bound to the well. These result in the final energy states within the quantum well moving in and out of the reach of any state in the SCH separated by a phonon energy, from which a carrier can scatter (Brum and Bastard, 1986; Brum et al., 1986; Babiker and Ridley, 1986). Theoretically calculated capture times in SQW structures have been found to oscillate between 10 ps and 1 ns for quantum wells with an ungraded SCH (Brum and Bastard, 1986) and between 20 and 200 ps for quantum wells with a linearly graded SCH (Brum et al., 1986).

Later calculations done for MQW structures have predicted carrier capture times in the range of 1 ps, closer to the experimentally observed values (Blom et al., 1991). This calculation is an adaptation of the previous one for the SQW structure. The difference is that the increase in the number of wells results in an increase in the total number of final states to which carriers may scatter and a corresponding decrease in the total capture time. The capture time has in fact been calculated to be nearly inversely proportional to the number of wells, with all other structural parameters held constant (Blom et al., 1991). As noted previously, the capture time is a function of the quantum well width, and in addition, it is also sensitive to the choice of SCH width (also applicable to the SQW case), barrier width, and barrier height.

Experimentally, the carrier capture times have been measured by subpicosecond luminescence up-conversion spectroscopy (Deveaud et al., 1988; Morin et al., 1991; Kersting et al., 1992). Using this technique, the time-resolved barrier and quantum well luminescences are measured, and the effective electron and hole capture times are extracted from these measurements. Since the luminescence intensity is proportional to the number of both electrons and holes, the barrier luminescence is expected to decay with the faster of the two effective capture times, and quantum well luminescence is expected to rise with the slower of the two (Deveaud et al., 1988). The rise in the quantum well luminescence is not only influenced by the carrier transfer from the barrier layers but also by the carrier relaxation time within the quantum wells. This introduces additional complications to the interpretation of the quantum well luminescence rise time (Kersting et al., 1992). The quantum mechanical calculations indicate that the hole capture times are smaller than the electron capture times because of their larger effective mass. Faster hole capture times have been measured directly in experimental setups where it has been possible to distinguish between the two carrier types (Kersting et al., 1992).

The experimentally measured effective capture times are given in Table 3.1. The hole capture times are smaller than the electron capture times. This is due to the larger density of states in the valence band of the quantum well to which holes can scatter to from the SCH. The magnitudes of both the electron and hole local capture times are small, within a picosecond. The carrier capture times reported by Deveaud et al. (1988), Morin et al. (1991), and Kersting et al. (1992) were determined from ultrafast pump-probe-type measurements, and those reported by Hirayama et al. (1992a) were extracted using extensive modeling from the measurements of the spontaneous emission spectra above threshold in semiconductor lasers. Although the carrier density in the quantum well is clamped above threshold, the carrier density in the SCH continues to increase above threshold due to the finite carrier capture time. Hence, by

3.2 Laser dynamics

Reference	Material System	Hole Capture Time (ps)	Electron Capture Time (ps)
Deveaud et al., 1988	GaAs/AlGaAs	~0.65	~1.2
Morin et al., 1991	InGaAs/InP	<0.3	<1.0
Kersting et al., 1992	InGaAs/InP	~0.2	~0.8
Hirayama et al., 1992a	InGaAs/InP	0.2–0.25	

Table 3.1: Experimentally determined local electron and hole capture times in GaAs and InP quantum well systems.

measuring the spontaneous emission spectra due to barrier recombination above threshold, the carrier capture time may be estimated. Although this is somewhat indirect, it agrees well with the capture times determined by more direct methods. This is another indication of the significance of carrier transport times in the operation of quantum well semiconductor lasers. Experiments observing the theoretically predicted oscillations in the carrier capture time with QW width at low temperatures (8 K) (Blom et al., 1993) have been reported.

The use of a linearly or parabolically graded index (L-GRIN or P-GRIN) SCH layer has been shown experimentally to enhance the carrier collection efficiency of the quantum well (Polland et al., 1988; Morin et al., 1991). In the GaAs/Al$_x$Ga$_{1-x}$As system, this grading is achieved during crystal growth by continually varying the Al mole fraction x to introduce a quasi-electric field within the SCH. At low temperatures (6 K), quantum wells with the L-GRINSCH have been shown to exhibit almost 100% collection efficiency, while those with an ungraded SCH show only about 50% (Polland et al., 1988). Time-resolved photoluminescence experiments have shown that the decay time of the luminescence from an L-GRINSCH (2 ps) can be as much as 11 times faster than the ungraded SCH (22 ps) luminescence at low temperatures (80 K) (Morin et al., 1991). The performance of the P-GRINSCH quantum well is in between these two limits, probably due to the electric field going to zero at the quantum well interface. The capture and escape times are also greatly influenced by the energy-level separation of the electronic states in the SCH relative to the quantum well. The grading strategy adopted for the SCH has a profound effect on the availability and energy-level separation of these states.

At low temperatures, the main transport limit in the L-GRINSCH is the carrier capture by the quantum well. At room temperature, the drift time in the L-GRINSCH becomes important compared with the quantum well capture time. This is in contrast to the ungraded SCH, where the variation in transport time is essentially due to the variation in mobility with temperature. Thus at room temperature the L-GRINSCH has been predicted to be only twice as fast as the ungraded SCH (Morin et al., 1991). This would not lead to any improvement in the high-speed performance of L-GRINSCH lasers because the width of the L-GRINSCH needed to maximize the optical confinement factor is about twice that for an ungraded SCH, and for any shorter widths, the confinement factor decreases rather rapidly. The variation in the optical confinement factor with the SCH dimensions for different energy profiles is given in Fig. 3.3. This comparison of the various optical confinement schemes also was reported by Streifer et al., (1983). Further, some later experiments, in contrast with the preceding ones, have determined that the gain recovery times, which are a measure of the carrier transport/dwell times in the SCH, in L-GRINSCH quantum well optical amplifiers ($\sim$14 ps) are actually about 3 to 4 times larger than in amplifiers with ungraded SCH ($\sim$4 ps) of comparable width (Weiss et al., 1992). Thus at room temperature there may not be any advantage in using an L-GRINSCH for the design of a high-speed quantum well laser.

3.3 High-speed laser design

There are number of aspects to the design of high-speed semiconductor lasers. From the small-signal analysis of Sect. 3.2, specifically using the expressions for the resonance frequency, the K factor, and the effective linewidth enhancement factor, to design a high-speed laser device with low chirp, one has a number of parameters to optimize. The process is as follows ($\uparrow$ means to increase and $\downarrow$ means to decrease):

Differential gain g_o $\uparrow$

Carrier transport time across the confinement heterostructure τ_r $\downarrow$

3.3 *High-speed laser design* 203

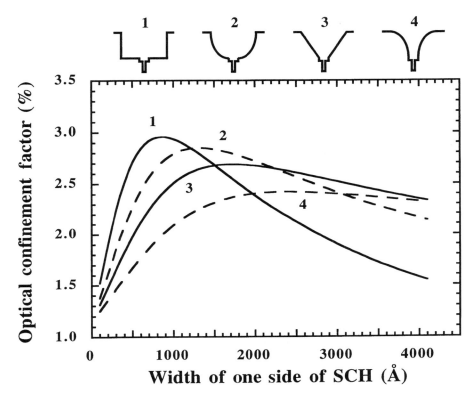

Figure 3.3: Variation of optical confinement factor with SCH width for different energy profiles. All the profiles have a 50-Å GaAs smoothing layer adjacent to the 80-Å $In_{0.2}Ga_{0.8}As$ single quantum well. The SCH is graded from $Al_{0.6}Ga_{0.4}As$ to $Al_{0.1}Ga_{0.9}As$ on either side of the smoothing layer over the distance plotted as the independent variable.

> Carrier escape time from the active area into the confinement heterostructure τ_e (not necessarily time for multiple quantum well active region) ↑
>
> Gain compression factor ε ↓
>
> Device size (photon lifetime τ_p) ↓
>
> Photon density S_o ↑
>
> Device parasitics ↓

In this section we will deal with these factors in detail and outline the compromises that have to be made in the final design.

3.3.1 Differential gain

To understand the role of differential gain and how it may be optimized, let us first look at the gain function in a semiconductor laser. In the early years of semiconductor lasers, the band tail model with the assumption that the momentum is not conserved in a optical transition was used extensively to compute the gain function. This was a good physical model for the early devices, where the active layers often were heavily doped. Doping the active area leads to the formation of recombination bands (and hence the name *band tails*) below the band edge, and the initial and final states of the optical transition had different momentum due to scattering by the charged impurity sites. A good review of the band tail model may be found in Casey and Panish (1978). In modern-day laser diodes, the active area is usually undoped to minimize the free carrier absorption losses. The band tail model is no longer physically correct for this case. Another model based on strict momentum conservation and relaxation broadening, derived using the density matrix analysis, was proposed. In this model, the experimental observation of the existence of band tails (in the spontaneous emission spectra) and smooth gain profiles (especially in quantum well and wire lasers), even for the case of undoped active regions, has been attributed to intraband relaxation processes (Nishimura and Nishimura, 1973; Zee, 1978; Yamada and Suematsu, 1979; Kazarinov et al., 1982; Asada and Suematsu, 1985). Over the years, the relaxation broadening model has been widely accepted, and we will use it for the gain analysis here.

Using the relaxation broadening model, the optical gain as a function of the emission energy E_i is written as (Asada et al., 1984)

$$g(E_i) = \frac{\pi \hbar e^2}{m_o^2 N_r \varepsilon_o c E_i} \int M^2 \rho_r(E)[f_c(E) - f_v(E)] L(E) dE \qquad (3.57)$$

where m_o is the free electron mass, N_r is the refractive index of the gain medium, ε_o is the permittivity of free space, c is the velocity of light, M^2 is the momentum matrix element that gives the relative strength of the various optical transitions, ρ_r is the reduced density of states function, f_c and f_v are the Fermi occupation probabilities in the conduction and valence bands, respectively, and $L(E)$ is the gain broadening function.

We will consider the band structure, for the computation of gain, in the parabolic band approximation without the complications of the band-mixing effects in the valence bands. The effects of band mixing are important when considering active regions that are composed of low-dimensional materials like the quantum wells, which in addition also may be coherently strained. One of the main effects of the band-mixing phenomena is the modification of the valence band effective masses (Suemune, 1991b). This can be included, as we will, in the parabolic band approximation by replacing the effective mass values that appear in the expressions for the gain and the density of states with those which have been specifically computed (Ridley, 1990) or experimentally determined (Jones et al., 1989) for these cases. In this approximation, the reduced density of states function for the various types of active regions may be written as

Bulk: $\quad \dfrac{4}{\pi^2}\left(\dfrac{m_r}{2\hbar^2}\right)^{3/2} \sqrt{E - E_{\text{bulk}}}$ (3.58)

Quantum well: $\quad \displaystyle\sum_{n=1}^{\infty} \dfrac{1}{\pi}\left(\dfrac{m_r}{\hbar^2}\right) H(E - E_{\text{bulk}} - E_{cn} - E_{vn})$ (3.59)

Quantum wire: $\quad \displaystyle\sum_{n=1}^{\infty} \dfrac{1}{\pi}\left(\dfrac{2m_r}{\hbar_2}\right)^{1/2} \dfrac{1}{\sqrt{E - E_{\text{bulk}} - E_{cnz} - E_{vnz} - E_{cny} - E_{vny}}}$ (3.60)

where $H(E)$ is the unitary Heaviside function, E_{bulk} is the bandgap of the equivalent bulk material, E_{cn} and E_{vn} are the quantized subband energy levels in the conduction and valence bands, respectively, and m_r is the reduced effective mass, which is written in terms of the conduction band and valence band effective masses, m_c and m_v, as $1/m_r = 1/m_c + 1/m_v$. For the quantum well and wire cases, the density of states functions are additionally summed over the various quantized subbands. The z and y subscripts for the quantum wire are used to distinguish between the two Cartesian coordinate directions along which the carriers have been confined.

The Fermi functions f_c and f_v are determined by the respective conduction and valence band quasi-Fermi levels μ_c and μ_v, which in turn are related to the electron n and hole p carrier densities in the active area. The relations for the quasi-Fermi levels are written as

$$\begin{aligned} n &= \int_0^{\infty} \rho_c(E) \dfrac{1}{1 + \exp[(E - \mu_c)/kT]} dE \\ p &= \int_0^{\infty} \rho_v(E - E_g) \dfrac{1}{1 + \exp[(E - E_g - \mu_v)/kT]} dE \end{aligned} \quad (3.61)$$

where ρ_c and ρ_v are the density of states in the conduction and valence bands, respectively. The zero point of the energy band diagram is taken with respect to the conduction band, so the energies that appear in the computation for μ_v are offset by the bandgap energy E_g. The value for E_g is the *effective* bandgap in the sense that it also includes all the shifts in the band edge due to carrier quantization in the quantum wells and wires and is equal to E_{bulk} in the case of bulk active regions. In semiconductor lasers fabricated these days, the active areas are generally undoped, and hence the condition of charge neutrality is assumed to hold, i.e., $n = p$.

Once the quasi-Fermi levels are known, the Fermi occupation probabilities used in the equation for optical gain [Eq. (3.57)], f_c and f_v are written assuming that the momentum is conserved (the photon carries very little momentum, so its contribution is ignored) during the optical transition—the so-called k conservation principle, which states that the initial (conduction band) and final (valence band) states of the optical transition have the same k value. The f_c and f_v functions are written as

$$f_c(E) = \frac{1}{1 + \exp[(m_r/m_c)E - \mu_c]}$$
$$f_v(E) = \frac{1}{1 + \exp[-(m_r/m_v)(E - E_g) - \mu_v]}$$
(3.62)

where E_g is again the *effective* bandgap.

Using the $\boldsymbol{k \cdot p}$ method, the momentum matrix element for the case of the bulk active area is written as (Kane, 1957)

$$M^2 = \frac{m_o}{6}\left(\frac{m_o}{m^*} - 1\right)\frac{E_g(E_g + \Delta_o)}{E_g + \frac{2}{3}\Delta_o}$$
(3.63)

where E_g is the original bulk bandgap, Δ_o is the spilt-off band energy, and m^* is not the effective electron mass m_c, as it has been commonly assumed, but a modified quantity that also includes the effects of higher- or lower-lying bands in addition to the usual valence and split-off bands (Hermann and Weisbuch, 1984; Yan et al., 1990). There are further modifications to this basic form of M^2 in active areas with quantum wells or wires. They lead to interesting properties, such as TE mode of oscillation being favored over the TM in lattice-matched (Yamanishi and Suemune, 1984) and compressively strained quantum well lasers and TM favored over TE for tensile strain in the active area. In quantum wire lasers, the matrix elements are also a function of the direction of propagation of the electric field vector (Asada et al., 1985). Although these properties of matrix elements play a role in active-layer design, we will not consider them here because they are not numerically large prefactors.

The broadening function $L(E)$ has been a source of much controversy. Using the density matrix analysis, this can be written simply as a Lorentzian

$$L(E) = \frac{1}{\pi} \frac{\hbar/\tau_{in}}{(E - E_i)^2 + (\hbar/\tau_{in})^2} \qquad (3.64)$$

where τ_{in} is the intraband scattering time, and the $1/\pi$ factor has been used to normalize the Lorentzian lineshape function when the integration is done for the gain function $g(E)$ in the energy space $E(0, \infty)$. The intraband scattering time τ_{in} is commonly known as the *dephasing time* T_2 in the density matrix analysis. In the early models for the lineshape function, a constant value was assumed for τ_{in}. This leads to the appearance of gradual tailing in the spontaneous emission spectrum and the unphysical absorption regions in the gain spectrum at photon energies below the bandgap. Mathematically speaking, this is the result of the weak convergence of the Lorentzian lineshape used for the gain-broadening function. More elaborate treatments of the intraband relaxation process including various forms of scattering mechanisms have managed to eliminate these shortcomings (Landsberg and Robbins, 1985; Yamanishi and Lee, 1987; Asada, 1989; Kucharska and Robbins, 1990; Ohtoshi and Yamanishi, 1991). A common result of these analyses is that τ_{in} itself is energy-dependent, leading to a faster converging lineshape function whose rate of convergence is typically between a Lorentzian and a Gaussian. Interface roughness in the active region can introduce additional broadening of the energy levels. This is especially severe in lower-dimensional active areas like the quantum wires, which have small fabrication tolerances (Zarem et al., 1989). The magnitude of τ_{in} is important because it affects both the linear gain [$g(E)\uparrow$ as $\tau_{in}\uparrow$] and the nonlinear gain (from spectral hole burning arguments) ($\varepsilon\uparrow$ as $\tau_{in}\uparrow$). From the expression for the K factor [Eq. (3.23)], it is the ratio of the nonlinear to the linear gain that needs to be minimized. For the purposes of gain computation here, we have assumed a constant value of 100 fs for τ_{in}.

Quantum size effects and strain

The use of quantum wells, wires, or even boxes (quantum confinement in all three spatial dimensions) instead of bulk material for the active area of semiconductor lasers was predicted to increase the differential gain dramatically (Asada et al., 1986; Arakawa and Yariv, 1986). Using the expression for gain in Eq. (3.57), it can be seen that except for the minor variations in the magnitude of momentum matrix element M^2, the major influence on the gain function comes from the $\rho_r(E)[f_c(E) - f_v(E)]$ term, which is a product of the reduced density of states and the difference in the conduction band and valence band Fermi occupation probabilities. The enhancement in the differential gain is the direct result of the modification of the density of states function in the lower-dimensional materials.

Figure 3.4 shows the gain and differential gain plotted as a function of carrier density for bulk, quantum well, and quantum wire active areas. Although the effective masses are somewhat modified in the quantum well and wire, we have assumed $m_e = 0.067m_o$, $m_h = 0.5m_o$, and $m^* = 0.056m_o$ for all three cases. The quantum well has a steplike density of states, and in the quantum wire, the density of states goes to infinity at the band edge. In practice, the band-edge density of states in the quantum

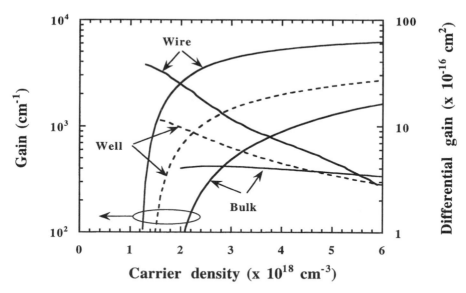

Figure 3.4: Comparison of gain and differential gain as a function of carrier density for bulk, quantum well, and quantum wire active areas. The quantum well is taken to be 100 Å thick and the quantum wire with a square cross section of 100 Å on the sides. The calculations are for the GaAs/AlGaAs system. The effective masses used in the calculations are in the text.

3.3 High-speed laser design

wires does not go to infinity and is affected by inhomogeneous broadening. In both these cases, unlike the bulk case, where the density of states is zero at the band edge, the enhancement in the density of states function results in a much larger gain at lower carrier densities. This leads to lower-transparency carrier densities and overall much reduced threshold current densities as well. The differential gain, which is the slope of the gain–carrier density relationship dg/dn, is also very much enhanced in the case of quantized active regions, but unfortunately, it is also a sensitive function of the carrier density, i.e., a function of threshold gain itself. The dg/dn curve is fairly flat and is independent of the carrier density for the case of bulk active area.

Figure 3.4 has been computed for a single quantized level in a single quantum well and wire. This results in a rapid gain saturation with carrier density, which is a major drawback of the lower-dimensional active areas. In practice, there may be additional quantized levels to provide gain, but the phenomenon of gain saturation with a single quantum well or wire is still sufficiently severe. To overcome this, multiple quantum wells or wires may been used. Figure 3.5 illustrates the principle with multiple quantum well (MQW) lasers. Starting with the gain function for a single quantum well (SQW), the gain and carrier density are multiplied by the number of wells. The carrier density at transparency increases proportionately to the number of quantum wells, but for lasers that have a high threshold gain, MQW structures offer a valuable tradeoff in increased differential gain. The bands in Fig. 3.5 show the regions of operation for minimum threshold current for various ranges of threshold gain. Maximum differential gain is always obtained at the minimum carrier density point irrespective of the active layer type or size.

In the III–V group of direct bandgap materials generally used to make semiconductor lasers, there is a large asymmetry between the conduction band electron mass and the dominant valence band heavy hole mass, which is typically about 7 times heavier. The lasing condition is satisfied once the separation of the quasi-Fermi levels exceeds the bandgap, i.e., $\mu_c - \mu_v > E_g$, of the material (Bernard and Duraffourg, 1961), and the optimal way to achieve this is in a material where the conduction and valence bands are mirror reflections of one another and have low density of states such that the electron and hole distributions become degenerate at low injected carrier densities (Adams, 1986; Yablonovitch and Kane, 1986). The triple degeneracy at the valence band edge is broken in the bulk material by the spin-orbit coupling that leads to the formation of

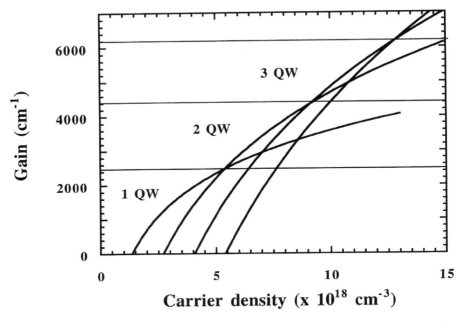

Figure 3.5: Gain–carrier density relationship for 1, 2, and 3 quantum wells. The calculation is first done for 1 quantum well, and the result is then scaled to obtain the family of curves. For a given threshold gain, the horizontal regions define the optimal range of operation for the minimum threshold carrier density. The material parameters are the same as the quantum well case in Fig. 3.4.

the split-off band. The large mass of the heavy hole band results from the interaction between the remaining double degeneracy. There are a number of ways to remove this degeneracy, and one of them is quantum confinement, as in the quantum wells. This can be further enhanced by the use of compressive strain, which moves the light-hole band edge further below the heavy-hole band edge such that the topmost valence band is now more parabolic with a *light-hole character*. In the case of tensile strain, the light-hole band edge moves up relative to the heavy-hole band and at sufficiently high levels of strain becomes the topmost band with a low effective mass.

To illustrate the effects of incorporation of strain on differential gain, we again use the expression for gain in Eq. (3.57). We have used the experimental effective masses determined for the strained $In_{0.2}Ga_{0.8}As/$

3.3 High-speed laser design

GaAs system for our calculations: $m_e = 0.071m_o$ and $m_h = 0.086m_o$ (Jones et al., 1989). Figure 3.6 shows the two distinct effects of the inclusion of strain, i.e., the drastic reduction in the transparency carrier density and the increase in the differential gain (Suemune 1991a; Lester et al., 1991b). Large enhancements in the modulation bandwidth were predicted based on this increase in differential gain for strain-incorporated structures (Suemune et al., 1988). Figure 3.6 also points out that the gain for strained structures also saturates more rapidly than for the lattice-matched case. This saturation is easily overcome with MQW structures or when there is more than one quantized level present.

In the GaAs system, strained active layers are typically obtained using $In_xGa_{1-x}As$ as the active area. For increasing values of In mole fraction, the active region becomes more compressively strained. In the InP system, $In_xGa_{1-x}As$ is lattice matched for $x = 0.53$; for $x > 0.53$, the active area is under compression, and for $x < 0.53$, the active area is

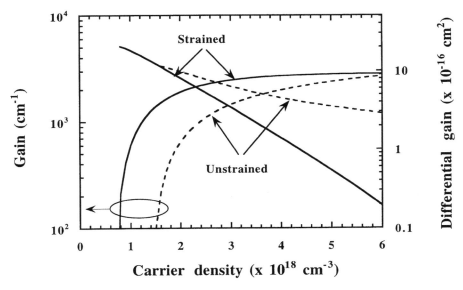

Figure 3.6: Comparison of gain and differential gain for strained and lattice matched active areas. The gain is enhanced and the transparency carrier density reduced for the case of strain in the active region. The calculations compare the strained SQW InGaAs/GaAs and unstrained GaAs/AlGaAs systems.

under tension. It is also possible to obtain quartenaries, $In_xGa_{1-x}As_yP_{1-y}$, with both types of strains in the InP system. Although low-threshold InP lasers have been demonstrated for compressive and tensile strained active areas, both theoretical (Corzine and Coldren, 1991) and experimental (Fukushima et al., 1991a, 1993a) work suggests that lasers with compressively strained active regions may be better for high-speed performance.

Experimentally, there is a maximum thickness, called the *critical thickness*, to which lattice-mismatched materials may be grown, for a given amount of strain, before it becomes thermodynamically favorable for this mismatch to be accommodated via the formation of dislocations. This thickness is typically on the order of 100s of angstroms. This means that typically it is only the quantum well lasers that are coherently strained as well. In the InP system, where materials with both compressive and tensile strain can be grown, it is possible to beat this critical thickness limit by some form of strain compensation (Miller et al., 1991; Seltzer et al., 1991).

Presently, the largest modulation bandwidths in both the InP and GaAs systems have been obtained with strained MQW structures. A modulation bandwidth of more than 40 GHz has been demonstrated for $In_{0.35}Ga_{0.65}As$/GaAs compressively strained quantum well lasers with four quantum wells (Weisser et al., 1992, 1996; Ralston et al., 1994), and InGaAs/InP lasers comprising seven compressively strained quantum wells with maximum bandwidths of 25 GHz also have been fabricated (Morton et al., 1992).

Quantum well or wire lasers have very small active areas, and unlike the bulk lasers with much larger active areas, they require a separate confinement heterostructure (SCH) to maximize the transverse optical confinement factor. In the preceding section we saw the effects of carrier transport across the SCH region on the dynamic properties of semiconductor lasers. As the carrier density rises in the active region, there is an increasing probability that the SCH region also gets significantly populated. Due to the much larger density of states present in the SCH region, this process of bandfilling or carrier overflow causes a proportionately smaller increase in the quasi-Fermi levels for the same level of carrier injection. From the gain equation, it can be seen that the $[f_c(E) - f_v(E)]$ function is correspondingly reduced, leading to lower levels of gain for the same injected carrier densities.

3.3 High-speed laser design

The bandfilling function or the movement of the quasi-Fermi levels under carrier injection, including the effects of the SCH regions, may be computed as

$$n = \frac{L_{SCH}}{L_{QW}} \int_{E_b}^{\infty} \rho_{c,SCH}(E - E_b) \frac{1}{1 + \exp[(E - \mu_c)/kT]} dE \\ + \int_{0}^{\infty} \rho_{c,QW}(E) \frac{1}{1 + \exp[(E - \mu_c)/kT]} dE \qquad (3.65)$$

where E_b is the energy barrier in conduction band, L_{SCH} is the width of the SCH region, L_{QW} is the width of the quantum well active area, and $\rho_{c,SCH}$ and $\rho_{c,QW}$ are the conduction band density of states in the SCH and quantum well regions, respectively. The L_{SCH}/L_{QW} is included to account for the larger number of states (volumetrically) available in the SCH compared with the quantum well. The calculation is similarly carried out for the valence band to determine μ_v. These are then used as input to the gain equation. The bandfilling function is calculated assuming that the carrier distributions in the SCH regions are uniform.

Figure 3.7 shows the effect of carrier overflow on the gain function. The SCH region is assumed to be 2000 Å wide and has a total band offset of 350 meV, of which 60% is across the conduction band. The effective masses in the SCH regions have been assumed to be the same as those in the active area. The total gain, as well as the differential gain, is reduced as a result of carrier overflow from the active region. From Fig. 3.7 it also can be seen that the conduction band overflow is a more sensitive function of the carrier density. Although initially the hole overflow is larger than the electron overflow due to the larger density of states, it is also not very sensitive to carrier density for the same reason and remains at low values throughout. The carrier occupation of the confinement regions, as well as the higher-lying bands like the L and X valleys, has a profound effect on the static properties, like the threshold current, as well as the dynamic properties of quantum well lasers (Nagle et al., 1986; Chinn et al., 1988; Nagarajan et al., 1989; Blood et. al., 1989; Zhao et al., 1991). The effect of bandfilling is dealt with in detail in Chapter 1 of this book.

p-Doping

Doping the active area with an acceptor impurity has been shown to increase the differential gain and improve the dynamic performance of

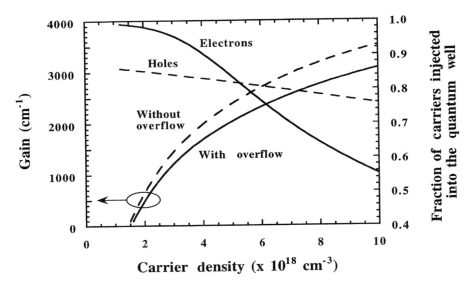

Figure 3.7: Effect of bandfilling and carrier overflow on the gain–carrier density relationship. Due to the lower mass and a smaller density of states in the current confinement regions, electron overflow is more significant than hole overflow. The SCH region is assumed to be 2000 Å wide and has a total band offset of 350 meV, of which 60% is across the conduction band.

semiconductor lasers (Su and Lanzisera, 1984; Uomi et al., 1990; Aoki et al., 1990; Sugano et al., 1990; Zah et al., 1990; Cheng et al., 1991; Lealman et al., 1992). Although there are some modifications to the band structure, like the formation of band tails, from the gain function it can be seen that doping the active region affects the $[f_c(E) - f_v(E)]$ function primarily.

Figure 3.8 shows the effect of both p and n doping of the active region. Intuitively, any form of doping leads to some initial separation of the quasi-Fermi levels, thereby reducing the level of carrier injection required to reach threshold. This is reflected in the reduction of the transparency carrier density for both cases but more so for the case of n doping (Vahala and Zah, 1988; Shank et al., 1992; Yamamoto et al., 1994). This is due to the lower conduction band effective mass and the resulting smaller density of states. The differential gain is higher only for the case p doping, and it is lower in the case n doping (Vahala and Zah, 1988). The reduction in differential gain in the case of n doping is not as puzzling as it may seem. The n doping pushes the initial electron Fermi level into the conduction

3.3 High-speed laser design

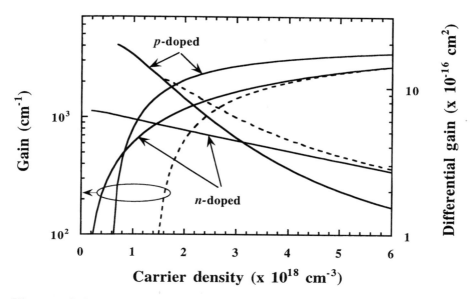

Figure 3.8: Modification of the gain function by p and n doping of the active region. n doping the active area leads to lower carrier density at transparency, and p doping it leads to enhanced differential gain. The material parameters are the same as the quantum well case in Fig. 3.4.

band, where the electron density of states is larger. Thus any increment of carriers under injection does not cause as rapid a separation of the Fermi levels (and hence as large an increase in gain) as at the band edge. At sufficiently high carrier densities, the differential gain enhancement is no longer present due to gain saturation. This is a real problem, because doping the active area generally leads to increased levels of internal loss, thus requiring much larger gain at threshold (Cheng et al., 1991). The use of a quantum well active area allows for a modulation doping scheme where the barriers and/or the SCH are selectively p doped (Uomi et al., 1990). This allows the acceptor sites to be separated from the active region. This limits the increase in free carrier absorption loss that can result from directly doping the active region. Selectively doping the barriers also prevents the complications of band tail formations that can result from doping the active region.

Some of the largest bandwidth lasers demonstrated to date have p-doped active regions. In addition to the compressively strained MQW structures, the lasers with the largest modulation bandwidths at 1-μm wavelength (InGaAs/GaAs) are modulation p doped with Be (Weisser et al., 1992; 1996; Ralston et al., 1994) and at 1.55-μm wavelength (InP) are completely p doped with Zn (Morton et al., 1992). Modulation bandwidths as high as 22 GHz have been demonstrated in Zn p-doped bulk active region lasers at 1.3 μm (Huang et al., 1992).

Wavelength detuning

This method of differential gain enhancement is applicable to laser structures where the lasing wavelength can be set independently of the peak wavelength of the material gain, as in the distributed feedback (DFB) lasers, where the lasing wavelength is determined by the internal grating structure that provides the feedback. This wavelength need not coincide with the peak of the gain spectrum as long as there is sufficient gain in the spectral region where the lasing occurs.

The principle of differential gain enhancement using wavelength detuning is illustrated in Fig. 3.9, where the gain and the differential gain are plotted as a function of wavelength for a quantum well active area. It can be seen that the differential gain is higher on the shorter-wavelength side of the gain peak. Negative detuning of the lasing wavelength from the gain peak is employed in DFB structures to take advantage of this differential gain enhancement (Kamite et al., 1987; Nishimoto et al., 1987). The fastest bulk (Uomi et al., 1989) and quantum well (Morton et al., 1993, 1994) DFB lasers operating at 1.5-μm wavelength are negatively detuned.

Another interesting point in Fig. 3.9 is that there is a maximum in the spectral variation of the differential gain. Although intuitively it will be best for the high-speed devices to operate at that point, unfortunately, it occurs in the region of negative gain for the regular undoped quantum well lasers. As also illustrated in the same figure, the wavelength at which the maximum in the differential gain occurs can be shifted into the region of positive gain by p-doping the active region. The use of strain will shift it further toward the gain peak. What is not obvious here is that as the dg/dn term passes through the maximum, the dn'/dn, i.e., the variation in the refractive index with carrier density, which is related to the differential gain via the Kramers-Kronig relationship, will pass through zero. The linewidth enhancement parameter, which is the ratio of the index variation with carrier density to the gain variation with

3.3 High-speed laser design

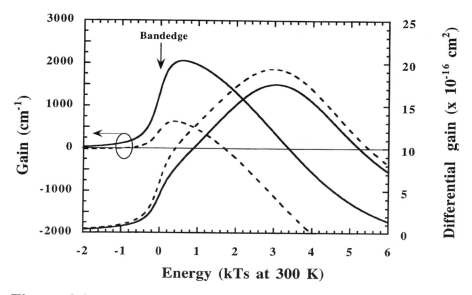

Figure 3.9: Spectral variation of gain and differential gain. The differential gain reaches a maximum typically on the shorter wavelength side of the peak of the gain spectrum. The position of this maximum may be tailored to occur in the region of positive gain by p doping the active region. The material parameters are the same as in Fig. 3.8. The dashed curve is for the undoped case, and the bold curve is for the p-doped case.

carrier density, will be zero at this wavelength (Yamanaka et al., 1992). Thus it is possible to fabricate lasers with ultralow chirp under modulation using wavelength detuning. Lasers with α parameters less than unity in the wavelength region with positive material gain have been demonstrated by Kano et al. (1993) by detuning the lasing mode in p modulation doped strained multiple quantum well InGaAsP/InP lasers.

3.3.2 Optimization of carrier transport parameters

In Sect. 3.2 we had included the effects of the carrier transport time across the separate confinement heterostructure (SCH) and the carrier emission time out of the active area into the SCH in the rate equation analysis of laser dynamics. An SCH layer is typically used together with active areas of very small widths, as in quantum well lasers, to maximize the transverse optical confinement factor. This region is usually absent in bulk laser designs, which have much wider active areas. The SCH layer, which is optimized for both optical and carrier confinement, is inserted between the normally doped cladding regions and the quantum well active area.

The SCH region is nominally undoped to reduce the internal loss. This leads to carrier transport delays in the SCH that translates into a parasitic-like roll-off in the modulation-response curves. The bandgap energy chosen for the SCH region determines the confinement energy for the carriers in the active area. The carrier escape rate is an exponential function of this energy, and a small barrier results in a degradation of the differential gain via static bandfilling effects and dynamic carrier emission events. The transport effects are more complicated in a multiple quantum well laser structure, where one has to also consider the transport mechanisms between the various wells. Carrier trapping can lead to severe localization of carriers within certain wells. The resulting nonuniform pumping of the quantum wells will degrade the high-speed response of multiple quantum well lasers.

Carrier transport time

From the analysis of Sect. 3.2, the most dramatic effect of carrier transport is the low-frequency roll-off in the amplitude modulation response. To investigate the effects of varying the transport time τ_r on the modulation response, lasers were fabricated from three SQW $In_{0.2}Ga_{0.8}As/Al_{0.1}Ga_{0.9}As$ samples with different SCH widths: sample A with 760-Å-wide SCH, sample B with 1500-Å-wide SCH, and sample C with 3000-Å-wide SCH. Figure 3.10 shows the experimental C.W. modulation response of samples A and C, respectively. The samples both have 300-μm-long cavities and 2.5-μm-wide ridges. Although the devices are identical except for the SCH width, the modulation response of sample C with the widest SCH region is completely damped at about half the output power level of sample A, and the bandwidth of sample A is six times that of sample C at higher power levels. The maximum C.W. modulation bandwidth for sample A is 18.1 GHz, which is the largest reported to date in SQW lasers (Nagarajan et al., 1992a). Such a roll-off in wide SCH but otherwise identical laser structures in the InP system also has been reported (Wright et al., 1992; Grabmaier et al., 1993).

The damped response of sample C shows all the effects predicted by the carrier transport model. The response is *similar* to one of a device that is limited by a low-frequency parasitic-like roll-off. In this case it

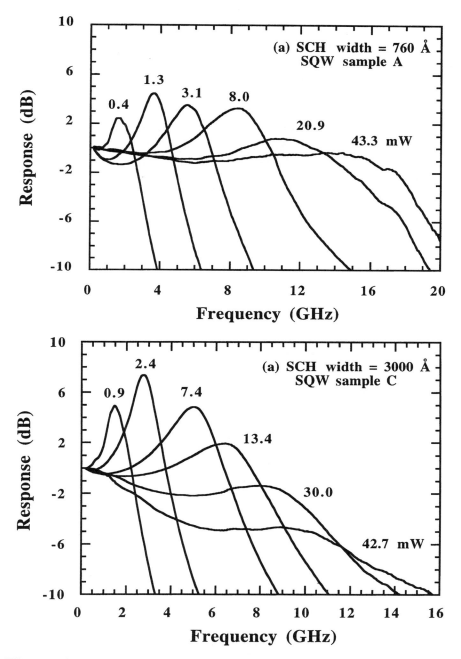

Figure 3.10: C.W. modulation response for (a) the narrow SCH and (b) the wide SCH sample SQW lasers.

cannot be attributed to device parasitics because both devices had been processed identically. To ascertain this, the parasitics, including the bond wire inductance, were extracted from the S_{11} parameter measured at the input port to the laser mount, and in addition, the series resistance was measured using the HP 4145B semiconductor parameter analyzer. The series resistance of the devices is 4 Ω. The roll-off frequency due to the device parasitics is about 25 GHz. From the fit to the low-frequency roll-off in the modulation response of sample C, the value of τ_r is determined to be about 54 ps for the 3000-Å-wide SCH, in agreement with the values calculated from published mobility values (Nagarajan et al., 1992d).

The variation of differential gain and gain compression factor ε with SCH width in these lasers is obtained from the modulation response as well as the relative intensity noise spectra. As shown in Fig. 3.11, the results from the modulation response agree with the data extracted from the noise spectra measurements. This indicates that except for the roll-

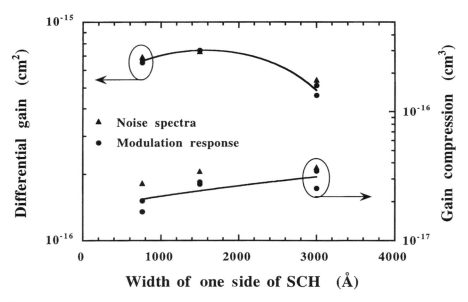

Figure 3.11: Differential gain and gain compression factor determined from the modulation response and noise spectra. They are in good agreement with one another. The measurements were done on SQW $In_{0.2}Ga_{0.8}As/GaAs$ lasers with 80-Å-wide quantum wells. The lines have been drawn to show the trend in the data.

3.3 High-speed laser design

off that places, an upper limit on the amplitude modulation bandwidth, the differential gain, and gain compression data can be extracted accurately from the noise spectra measurements. Figure 3.11 shows that the gain compression factor slightly increases for larger SCH widths, but given the scatter in the data, this variation of ε with SCH is not significant, and this variation alone cannot explain the severe low-frequency roll-off in sample C with the widest SCH region. The variation in the differential gain extracted from the modulation response and noise spectra measurements, which can be thought of as an *effective* differential gain, is in contrast to the results from the threshold gain measurements (derived from the broad area threshold current density measurements) in Fig. 3.12, which predicts an increasing differential gain with increasing width of the SCH layer. Sample C with the lowest threshold current density

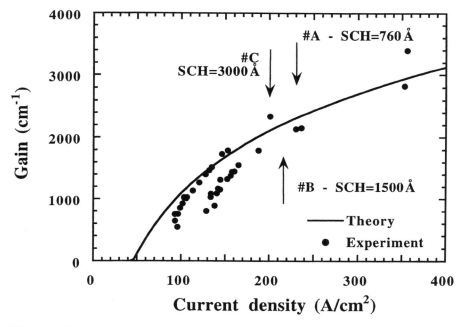

Figure 3.12: Theoretical (bold curve) and experimental (filled circles) variation of material gain with current density for an 80-Å $In_{0.2}Ga_{0.8}As$ quantum well. The experimental data are from the broad area measurements on samples A, B, and C. The arrows indicate the respective operation points of 300-μm cavity length devices.

has the lowest effective differential gain. This is again due to carrier transport, which reduces the differential gain from g_o to g_o/χ. This reduction in the differential gain is also responsible for the reduction in resonance frequency at high power levels in sample C compared with sample A [Fig. 3.10(a) compared with Fig. 3.10(b)].

Figure 3.13 shows the variation in the -3-dB modulation bandwidth with the square root of optical power for samples B and C. There is good agreement between the experiment and model over a large range of optical power, and the model accurately predicts the discontinuity, caused by carrier transport, in the -3-dB bandwidth for sample C. In the calculations, the value for ε is taken to be 1.5×10^{-17} cm^3 throughout. The bold line gives the exact solution [Eq. (3.12)], and the dashed line shows the analytic results [Eq. (3.13)]. For a SCH width of 1500 Å, both equations give identical results. At 3000 Å, the bandwidth values predicted by both

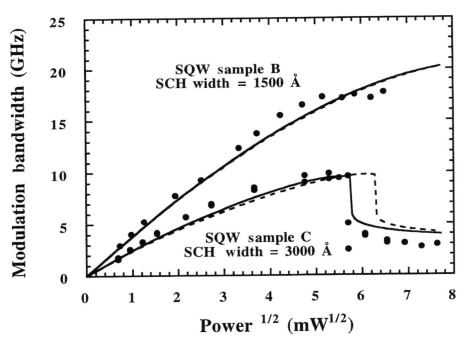

Figure 3.13: Comparison of the carrier transport model and experiment (filled circles) for the variation in modulation bandwidth with the square root of power. The dashed line is the result from the analytic solution, and the bold line that of the exact expression. Severe low-frequency roll-off due to carrier transport causes the discontinuity in the curve for sample C at higher power levels.

3.3 High-speed laser design

expressions are almost the same, except in the vicinity of the power level at which the low-frequency roll-off causes a *jump* in the curve. This shows that the assumption made to obtain the analytic expressions is valid even for the devices with the widest SCH layers.

Figure 3.14 shows the experimental variation in the -3-dB modulation bandwidth with SCH width at different power levels. Here again, the predictions of the analytic expression are very close to the results from the exact expression. The optimal SCH width also corresponds roughly to the point at which the optical confinement factor is a maximum. For a narrow SCH, the bandwidth drops off due to a decreasing optical confinement factor, resulting in a larger threshold gain and thus a lower differential gain. At larger SCH widths, the combination of a decreasing confinement factor and an increasing carrier transport time across the undoped regions of the SCH limits the modulation bandwidth. At suffi-

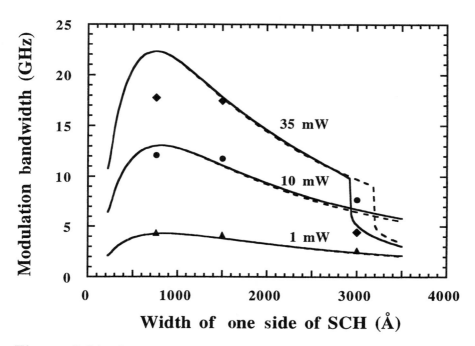

Figure 3.14: Comparison of the calculated and measured variation (filled circles) in modulation bandwidth with SCH width. The dashed line is the result from the analytic solution, and the bold line is that of the exact expression. The carrier transport effects dominate at sufficiently high power levels.

ciently high powers for wide SCH devices, the characteristic *drop* due to carrier transport appears in the modulation bandwidth curve.

Figure 3.15 shows the variation in α_{eff} with width of the SCH. For a given carrier confinement energy in the SCH, laser structures with wide SCH regions have a larger amount of wavelength chirping. This translates into higher FM efficiencies in lasers with wider SCH regions (Nagarajan and Bowers, 1993a). This trend has been observed experimentally (Wright et al., 1992; Yamazaki et al., 1993; Kitamura et al., 1993). The increase in the amount of chirp with the SCH width is due primarily to the increase in carrier transport time, which is proportional to the square of the SCH width, and the decrease in the optical confinement factor.

The experimental values for the *intrinsic* variation of index with carrier density in the quantum well active region ($\partial n/\partial N_{wo}$) and the SCH region ($\partial n/\partial N_{bo}$) are required as material parameter inputs to the analytic expressions for α_{eff} in Eq. (3.43). There have been a range of values between 10^{-20} and 10^{-22} cm^{-3} reported in literature for the GaAs/AlGaAs (Manning et

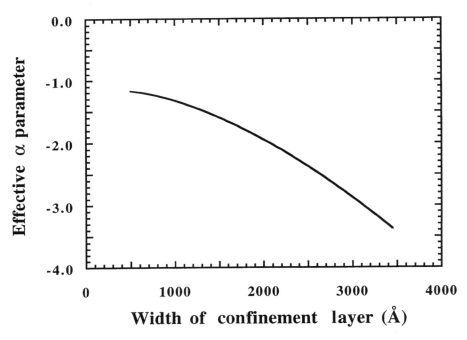

Figure 3.15: Variation of the effective α parameter as a function of the SCH width for an SQW In$_{0.2}$Ga$_{0.8}$As/Al$_x$Ga$_{1-x}$As laser structure. The calculation is done for 10% Al mole fraction in the SCH.

3.3 High-speed laser design

al., 1983; Dutta et al., 1984) and the InGaAs/GaAs systems (Rideout et al., 1990; Dutta et al., 1992), with the later results in the InGaAs/GaAs system (Dutta et al., 1992) being at lower end of the spectrum. In our case, the SCH is bulk AlGaAs, and the active region is a InGaAs quantum well. We have taken $\partial n/\partial N_{wo}$ (quantum well active region) to be 5×10^{-21} cm^{-3} and $\partial n/\partial N_{bo}$ (bulk SCH region) to be 5×10^{-20} cm^{-3}.

Carrier escape time

The experimental data for the dependence of the high-speed parameters on the thermionic emission time is obtained from two samples with different energy barriers in the SCH: sample D with Al$_{0.15}$Ga$_{0.85}$As SCH and sample E with GaAs SCH. Figure 3.16 shows the variation in resonance frequency with the square root of the power obtained from the modulation response data for the two cases considered here. The resonance frequency

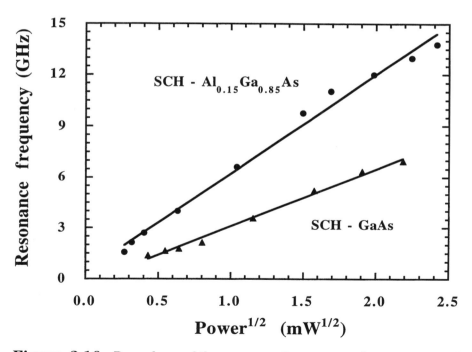

Figure 3.16: Dependence of the resonance frequency on the square root of power for the MQW samples with a higher (Al$_{0.15}$Ga$_{0.85}$As) and lower (GaAs) energy barrier in the SCH. The samples each have a 205-μm-long cavity and a 2-μm-wide ridge. The slopes are 5.8 GHz/mW$^{1/2}$ for the Al$_{0.15}$Ga$_{0.85}$As SCH and 3.4 GHz/mW$^{1/2}$ for the GaAs SCH.

for the MQW laser with $Al_{0.15}Ga_{0.85}As$ SCH (higher barrier and hence a longer thermionic emission time) is almost twice that of the laser with GaAs SCH at equivalent power levels.

The data for the comparison is summarized in Table 3.2 (Nagarajan et al., 1992c), where it can be seen that the differential gain for sample D is more than 2.5 times that of sample E. The gain compression coefficient ε for both cases is about the same. Both devices have a 205-μm-long cavity and a 2-μm-wide ridge. The larger differential gain in the case of sample D, with $Al_{0.15}Ga_{0.85}As$ SCH, is surprising considering that it had a larger threshold gain due to a larger internal loss and a smaller optical confinement factor. Consequently, due to gain saturation in quantum well lasers, it is *expected* that they have a lower differential gain. This result is consistent with the carrier transport model, which attributes part of this discrepancy to the reduction in the *effective* or *dynamic* differential gain in sample E due to an increase in χ caused by a reduction in τ_e. Further, the lower energy barrier and the larger density of states in SCH of sample E lead to a severe *carrier overflow*, thereby causing a much slower rise in the quasi-Fermi levels under carrier injection (Nagle et al., 1986; Nagarajan et al., 1989; Zhao et al., 1992). This reduces the differential gain even under *static* operating conditions.

The various carrier transport times in a quantum well laser are a sensitive function of temperature. As a further proof of validity of the present model, the noise spectra of the lasers at various temperatures between -65 and $+50°C$ were measured, and the temperature dependence of the K factor was extracted.

In semiconductor lasers, even in the absence of transport effects, the K factor would increase with temperature if the gain compression factor is temperature-independent. This is so because the gain, and thus the differential gain, decreases as the temperature increases. In the presence of transport effects, this increase would be more pronounced because the transport factor χ also increases with temperature, leading to a further reduction in the differential gain.

Figure 3.17(a) shows the variation in K factor with temperature for sample F, which is an SQW $In_{0.2}Ga_{0.8}As/Al_{0.1}Ga_{0.9}As$ laser with an SCH width of 900 Å, and Fig. 3.17(b) shows that of sample C, which is also an SQW $In_{0.2}Ga_{0.8}As/Al_{0.1}Ga_{0.9}As$ laser but with an SCH width of 3000 Å. The K factor for sample C is much more sensitive to temperature variations than that of sample F (Ishikawa et al., 1991a; Nagarajan et al., 1992d).

Results from the model for three different cases have been presented in Fig. 3.17 together with the experimental data. The first curve was computed

3.3 High-speed laser design

Sample	Structure	Cavity Length and Ridge Width	Threshold Gain	Differential Gain	Gain Compression Factor, ε
D	SCH–Al$_{0.15}$Ga$_{0.85}$As	205 μm, 2 μm	1440 cm^{-1}	8.6×10^{-16} cm^2	1.8×10^{-17} cm^3
E	SCH–GaAs	205 μm, 2 μm	1080 cm^{-1}	3.3×10^{-16} cm^2	1.3×10^{-17} cm^3

Table 3.2: Effect of bandfilling and thermionic emmission on the threshold gain, differential gain, and gain compression factor as measured in two otherwise identical devices with different carrier confinement energies in the SCH region.

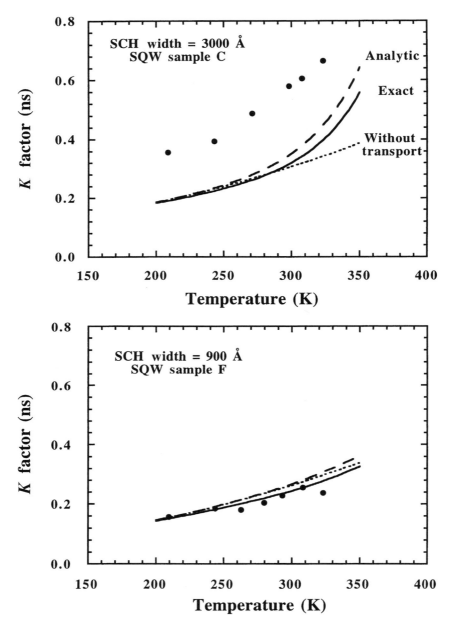

Figure 3.17: Temperature dependence of K factor for (a) a narrow SCH and (b) a wide SCH SQW $In_{0.2}Ga_{0.8}As/Al_{0.1}Ga_{0.9}As$ laser. A large contribution to the significant variation of K factor with temperature for wide SCH devices comes from the temperature dependence of the transport parameters.

3.3 High-speed laser design

using the analytic expression, the second using the exact solution, and the third using the analytic solution but neglecting transport effects and considering only the variation in differential gain with temperature. For the case of sample F, where the transport effects are not significant, all three curves lie close to one another and do not show much variation with temperature. For sample C, variations in differential gain with temperature alone are clearly insufficient to explain the K factor trend with temperature. From the Fig. 3.17(a) and (b) it can be seen that the analytic solution is in good agreement with the exact solution. There is some discrepancy in the actual K factor values predicted in the wide SCH case. The model assumes a constant value for the gain compression factor ε, but experimentally, there is some increase in the ε values for the wide SCH devices (see Fig. 3.11). This increase in ε will have to be taken into account for better absolute agreement between the model and the experiment.

The enhanced temperature sensitivity of the K factor in the wide SCH case is the result of the temperature dependence of the transport parameters. Figure 3.18 shows the calculated variation of τ_e, τ_r and χ

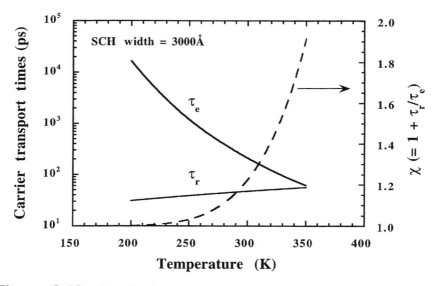

Figure 3.18: Calculated temperature dependence of the transport times and the transport factor χ for the wide SCH sample. As the temperature increases, the thermionic emission lifetime from the quantum well is considerably reduced, and the transport time across the SCH is somewhat enhanced, leading to a significant increase in the transport factor χ for wide SCH devices.

with temperature for sample C with the widest undoped SCH region. The thermionic emission time decreases exponentially with temperature and is the major contributor to the increase in χ. The SCH diffusion time is temperature-dependent via the temperature dependence of the ambipolar diffusion coefficient. Generally, the carrier mobilities decrease as the square of the temperature and the diffusion coefficient increases linearly with temperature, leading to a linear increase in τ_r with temperature. Both the decrease in τ_e and the increase in τ_r lead to an increase in χ and thus an enhanced reduction in K. This clearly demonstrates the significant role played by carrier transport in the modulation dynamics of semiconductor lasers.

The enhanced temperature dependence of K factor in laser structures where transport effects are dominant also have been observed in the InP system. Figures 3.19 and 3.20 show the temperature dependence of the K factor for 1.3- and 1.5-μm wavelength lasers (Ishikawa et al., 1992a).

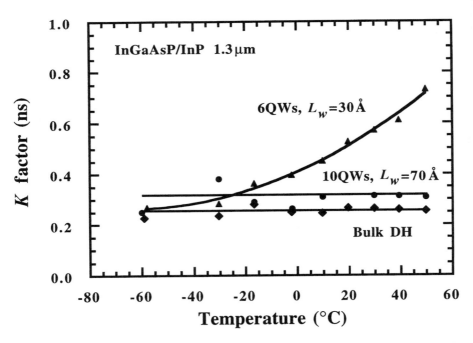

Figure 3.19: Temperature dependence of the K factor for 1.3-μm-wavelength InGaAsP/InP MQW lasers. The lines have been drawn to show the trend in the data.

3.3 High-speed laser design

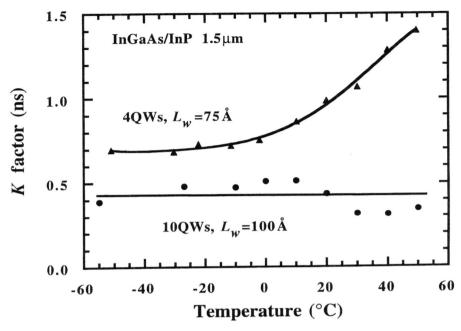

Figure 3.20: Temperature dependence of the K factor for 1.5-μm-wavelength InGaAs/InP MQW lasers. The lines have been drawn to show the trend in the data.

The lasers with fewer and narrower quantum wells have a large K factor at room temperature, and the temperature dependence is also more significant. The K factors are less sensitive to temperature for the lasers with a larger number and wider quantum wells. The K factor for the bulk double heterostructure (DH) sample is also insensitive to temperature.

Laser structures with lower carrier confinement energies in the SCH layers also have larger α_{eff} and hence larger amounts of wavelength chirping under current modulation (Nagarajan and Bowers, 1993a). As shown in Fig. 3.21, $\text{In}_{0.2}\text{Ga}_{0.8}\text{As}/\text{Al}_x\text{Ga}_{1-x}\text{As}$ lasers require Al mole fractions (x) of at least 15% for the wavelength chirping to reach its minimum value.

The variation of differential gain g_o with carrier confinement energy, which is required as an input for the computation of the results presented in Fig. 3.21, was theoretically calculated using more sophisticated models than discussed earlier, considering valence band mixing effects and the appropriate momentum matrix elements (Corzine et al., 1990). This calcu-

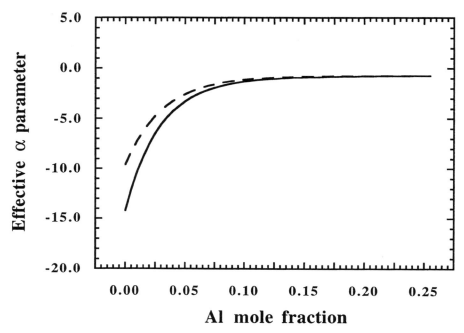

Figure 3.21: Variation of the effective α parameter with Al mole fraction in an InGaAs/AlGaAs SCH SQW laser structure. The calculation is done for a 1000-Å-wide SCH. The dashed curve shows the contribution due solely to the *static* bandfilling effects, and the solid curve is the sum of both the *static* bandfilling and the *dynamic* carrier transport effects.

lation also considered the carrier occupation in the SCH region under bandfilling and carrier overflow effects. The *static* differential gain increases as the confinement energy increases due to a reduction in the bandfilling or the carrier overflow effect. The dashed line in Fig. 3.21 shows the variation in the effective α parameter caused solely by the variation in the differential gain with confinement energy. This *could* be called the *structure dependent contribution* as opposed to the *purely transport component,* which is given by the difference between the bold and dashed curves.

In Fig. 3.15, the variation in α_{eff} with width of the SCH is almost all due to carrier transport. The carrier overflow is not a sensitive function of SCH width, at least not for the moderately high confinement energy (Al mole fraction of 10%) considered in Fig. 3.15, so the differential gain does not vary significantly with SCH width. The only time when there may be a significant variation in differential gain with SCH width is when the loss levels in the cladding regions are dominant (Fig. 15a in Nagarajan et al., 1992d). In this case, differential gain will be reduced at very small SCH widths due to a large gain required at threshold, and the curves in Fig. 3.15 would roll over in that region.

Multiple quantum well structures

There are additional complications due to carrier transport between the various quantum wells in the MQW system. Carrier transport across the MQW structure, i.e., between the quantum wells, is either via thermionic emission and then transport across the barrier and capture by the next quantum well or tunneling (τ_t) through the barrier. For a well-designed barrier, the barrier transport and capture times are negligible compared with the initial thermionic emission time from the quantum well. Thermionic emission and tunneling are competing processes, and the faster one will dominate. If a barrier transport time τ_b is defined, then

$$\frac{1}{\tau_b} = \frac{1}{\tau_e} + \frac{1}{\tau_t} \tag{3.66}$$

The barrier transport time τ_b corresponding to the electrons and holes will be different. The derivation of the ambipolar tunneling or emission times is more difficult. In the case where the dynamics of one carrier dominates, its tunneling or emission time may be taken to be the effective quantity with little effect on the final results. Figure 3.22 shows the variation in carrier transport times, at room temperature, for different barrier thicknesses for an $In_{0.2}Ga_{0.8}As$ quantum well and a GaAs barrier (Nagarajan et al., 1991c, 1992d). For sufficiently small widths (<50 Å), tunneling dominates the carrier transport across the barriers. For intermediate structures, hole transport is mainly by thermionic emission, while tunneling dominates electron transport. For thick barriers, the experimental tunneling times are somewhat larger than the calculated ones because the scattering in the barriers makes the tunneling process incoherent (Leo et al., 1990). This does not affect our model because thermionic emission and subsequent diffusion across the barrier dominate both electron and hole transport in this regime. The slope of the electron and hole transport times decreases for wider barriers, and this apparent saturation is due to the fact that the diffusion time is still a small part of the total

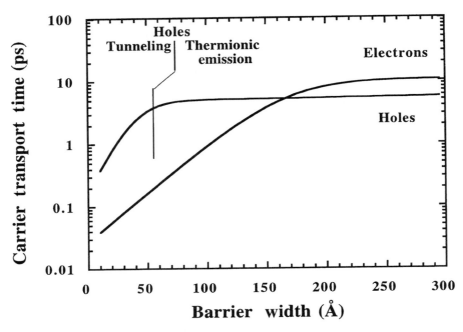

Figure 3.22: Variation of the electron and hole transport times between two 80-Å $In_{0.2}Ga_{0.8}As$ quantum wells with barrier thickness for a GaAs barrier.

transport time across the barrier. For very wide barriers, the diffusion time across the barrier dominates over the thermionic emission time, which is independent of the barrier width.

Our analysis so far has considered a single quantum well active area. The multiple quantum well structure also may be analyzed using the ambipolar current injection mechanism discussed previously. Figure 3.23 compares the carrier distribution in the SCH obtained using the exact self-consistent solution to the current continuity and Poisson's equations with the one obtained using the assumption of ambipolar current flow. Although the electron and hole injections are from the opposites of the SCH, the final carrier distribution is weighted toward the p cladding side due to the smaller mobility of the holes and follows the ambipolar carrier distribution very closely. This calculation is done for the case of an unstrained 10-quantum-well InGaAs/InP laser (Ishikawa et al., 1994) with barrier bandgap wavelength of 1.2 μm. The quantum wells are 100 Å thick.

3.3 High-speed laser design

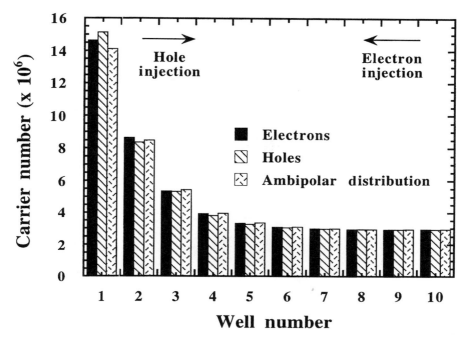

Figure 3.23: Numerically calculated distribution of electrons and holes in a 100-Å-wide 10 quantum well InGaAs laser with a barrier bandgap wavelength of 1.2 μm and barrier width of 100 Å. Results obtained using the ambipolar carrier transport assumption are also shown.

Although the ambipolar model may be used to accurately describe the spatial carrier density variation among the quantum wells in an MQW structure, the actual carrier density within the quantum well will be different because only part of the carriers are two dimensionally confined to the quantum well, whereas the rest remain unconfined in the three dimensional bulk like energy states above the quantum well. Figure 3.24 shows the distribution of the carriers (electrons and holes) between the confined and unconfined states. This calculation is again done for the same InGaAs/InP laser structure, where the holes are more effectively confined and the electron overflow is significant. The differential gain of the active region is determined by the confined carriers, and hence the spatial variation in the carrier density will lead to the degradation in the overall device performance (Tessler and Eisenstein, 1993a, 1994). Proper

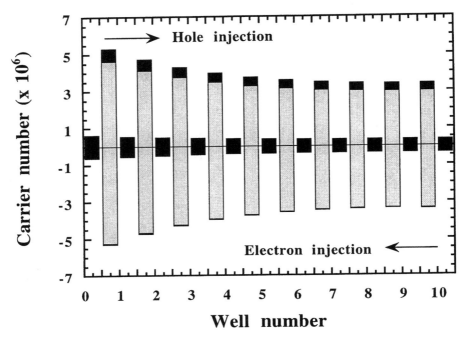

Figure 3.24: Calculated carrier (holes and electrons) profile as in the case of Fig. 3.23 but showing the distribution of the carriers between the two-dimensionally confined states in the quantum well and three-dimensionally confined states above the quantum well or in the barriers.

design of the SCH, barrier, and quantum well layers is needed to minimize these effects.

There are a number of differences between designing high-speed lasers in the InGaAs/GaAs/AlGaAs and the InGaAsP/InP systems. The first difference in the InP system is that the total change in the confinement energy that can be obtained in going from InP to InGaAs, under lattice-matched conditions, is only 0.75 eV compared with 1.59 eV in the bandgaps of GaAs and AlAs lasers. InGaAs has a smaller bandgap than GaAs, thus offering an even larger energy range. The refractive index difference obtainable for the same composition range is also larger in the InGaAs/GaAs system. These factors lead to more efficient design of the quantum confinement and waveguiding structures in the InGaAs/GaAs system as compared with the InGaAs/InP system.

The second difference in the InGaAs/InP system is that the conduction band offset is 40% and thus is smaller than the valence band offset. This, together with the fact that the energy range for the design of quantum structures in the InGaAs/InP system is small, leads to two problems. The first is that the electrons are not sufficiently confined to the quantum wells, and the second is that the holes are rather easily trapped in the quantum wells. Although this may seem like two different ways of stating the same problem, the implications are different. The first implies a small thermionic emission time, and thus a large χ, leading to a reduction in the effective differential gain. The second implies a large barrier transport time, leading to hole trapping, a large χ, and poor operation of an MQW structure. Since the carrier transport in this case is also ambipolar, the average of these two effects will govern the dynamics of lasers in this system.

One solution to the carrier transport problems encountered in long-wavelength, high-speed lasers in the InGaAs/InP system is to use InAlGaAs instead of InGaAsP for the barrier, SCH, and the cladding regions. The conduction band offset ratio in the InGaAs/InAlGaAs system is the same as in the InGaAs/GaAs system. InGaAs/InAlGaAs lasers operating at 1.55 μm have been shown to have a smaller K factor and a smaller gain compression factor compared with lasers in the InGaAs/InGaAsP system (Grabmaier et al., 1991). Figure 3.25 shows the calculated modulation bandwidth as a function of the undoped SCH width for InGaAsP and InAlGaAs SCH material with the same bandgap wavelength of 1.25 μm and width of 2000 Å (Ishikawa et al., 1992b). The thermionic emission time is longer for the InAlGaAs SCH due to the deep conduction band well, and it leads to overall better carrier confinement and consequently a larger modulation bandwidth.

In the carrier transport model, the effective transport time is computed for the undoped SCH region between the doped cladding and the quantum well active region. Keeping the SCH region undoped couples the problem of optimizing the optical confinement factor and the carrier transport time. Figure 3.25 shows how the SCH width and the carrier transport time may be independently optimized by doping the SCH region part way. The SCH width is optimized for maximum optical confinement factor, and the doping profile is adjusted to obtain the minimum transport time without increasing the internal loss, due to free carrier absorption, to unacceptable levels at small doping offsets. The increase in internal loss, and hence a reduction in the differential gain due to gain saturation,

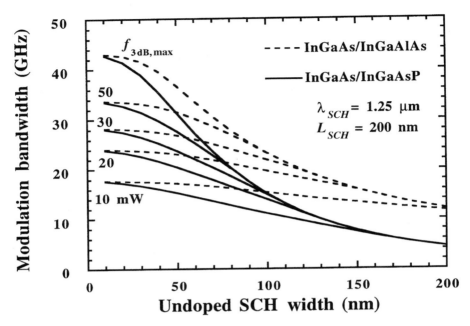

Figure 3.25: Calculated −3-dB modulation bandwidth as a function of the undoped SCH width. The influence of the thermionic emission time is varied by changing the SCH material system.

with decreasing doping offset from the quantum well active region has not been considered in this calculation.

We will not discuss all the tradeoffs required in the design of an optimized MQW laser structure in detail here. Instead, we will highlight the major reason that was stated previously for using more than one quantum well in the active region, i.e., the gain saturation in a single quantum well and the larger available differential gain with multiple quantum wells. This gain saturation with carrier density leads to an optimal cavity length for the maximum modulation bandwidth. This is quite unlike lasers with bulk active areas, where the differential gain is fairly insensitive to carrier density variations and shorter cavity lengths lead to larger modulation bandwidths due to the reduction in the photon lifetime. In quantum well lasers, the carrier density–dependent differential gain works in opposition to the effects of a reduced photon lifetime. These tradeoffs are illustrated in Fig. 3.26, which shows the experimental

3.3 High-speed laser design

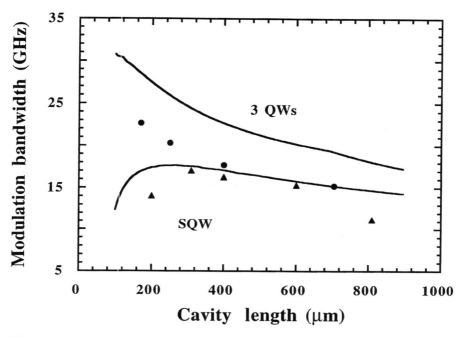

Figure 3.26: Comparison of the calculated (bold line) and measured (filled circles) variation in modulation bandwidth with cavity length for SQW and MQW lasers. The SQW lasers exhibit a broad maximum for the cavity-length dependence of modulation bandwidth, while MQW lasers are generally more sensitive to cavity-length variations.

and theoretical variations in modulation bandwith with cavity length for SQW and MQW lasers. The parasitics were more significant in this set of devices, and the RC product was about 14.5 ps. The effects of the parasitics have been deconvolved from the experimental data presented in Fig. 3.26. As the cavity length is reduced, the threshold gain increases, leading to a decrease in the differential gain. This is balanced by the reduction in photon lifetime, and this results in an optimal cavity length for the maximum modulation bandwidth. Generally, MQW lasers have a larger modulation bandwidth than SQW lasers, but the cavity length in the MQW case has to be sufficiently short to take full advantage of this. The optimal cavity length for the SQW case is 300 μm. For three or more quantum wells, the optimal cavity length is below 100 μm (Nagarajan et al., 1992d). It is difficult to cleave lasers below this limit, and one may have to resort to dry etching techniques to fabricate lasers with very short cavity lengths (Lester et al., 1991a). Further, lasers with short cavity lengths are more prone to thermal heating and may not perform as well as theoretically expected. To overcome this practical problem, one may have to use MQW lasers with nonoptimal cavity lengths.

3.3.3 Nonlinear gain

In addition to the gain saturation at high carrier densities, which is more pronounced in low-dimensional active regions, the gain also saturates at high photon densities, irrespective of the type of active area. In the rate equation analysis of semiconductor laser dynamics, this saturation phenomenon has been included via the gain compression factor ε. The form and the source of the gain suppression have been a source of controversy.

Using the third-order perturbation theory in the density matrix approach, the linear form of gain suppression is written as $g_{\text{eff}} = g(1 - \varepsilon S)$ (Yamada and Suematsu, 1979; Kazarinov et al., 1982; Asada and Suematsu, 1985; Agrawal, 1987), where S is the photon density in the laser cavity and g is the material gain that was derived previously also using the same density matrix method. This form has the drawback that g_{eff} becomes negative at sufficiently large photon densities. For ε, which is typically on the order of 10^{-17} cm^3, these photon densities are well within the operating range of the semiconductor lasers.

Using the theory of gain saturation in homogeneously broadened two-level laser systems (Yariv, 1989), Channin (1979) suggested the $g_{\text{eff}} = g/(1 + \varepsilon S)$ form of gain suppression for analyzing semiconductor laser dynamics. Although the experimental data for high-speed lasers is in better agreement with this form of gain suppression originally derived for a two-level laser system (Bowers, 1985), it does not strictly hold for semiconductor lasers, where light emission is the result of band-to-band transitions that occur between two continuums in the energy space. In the limit of $\varepsilon S \ll 1$, i.e., for small photon densities, both expressions for gain suppression are equivalent (the former is the first-order Taylor expansion of the latter).

Again using the density matrix analysis by modeling the laser gain medium as an ensemble of two-level systems with different transition frequencies, Agrawal (1988) proposed the $g_{\text{eff}} = g/\sqrt{1 + \varepsilon S}$ form as an approximate analytic expression for gain suppression. This expression does not have the drawbacks of the original one, and again, for small

3.3 High-speed laser design

photon densities, it is close to the original expression derived using the third-order perturbation analysis. In a more elaborate work, including the effects of non-Markovian type of intraband relaxation on the density matrix formalism, Tomita and Suzuki (1991) have shown that of the three *analytic* expressions that have been proposed, the $g_{\text{eff}} = g/(1 + \varepsilon S)$ form best agrees with the exact *numerical* solution of the problem. We have thus used this throughout our analysis here.

The source of this nonlinearity also has been under much debate. The two commonly accepted causes are spectral hole burning and transient carrier heating (Kesler and Ippen, 1987; Gomatam and DeFonzo, 1990; Willatzen et al., 1991). The relaxation broadening model for gain implies the spectral hole burning model for gain suppression. The third model attributes the nonlinear gain suppression to the feedback from the spatial dielectric grating that is induced by the electric field standing wave in the lasing cavity (Su, 1988). Later investigations in this area (Frankenberger and Schimpe, 1990; Eom et al., 1991) have largely eliminated this as a probable cause for gain suppression in semiconductor lasers.

Figure 3.27 shows the essential differences between spectral hole burning and transient carrier heating as the causes of nonlinear gain suppression in semiconductor lasers. In the spectral hole burning model, depletion of carriers in the spectral vicinity of the lasing mode, which generally coincides with the peak position of the gain function, causes a reduction in the gain. A larger photon density causes an increased amount of carriers to be depleted and hence a stronger suppression of gain. The magnitude of this so-called spectral hole will depend not only on the rate at which the carriers recombine but also on the rate at which they are replenished via intraband scattering. The carrier heating model makes use of the fact that the lasing energy is smaller than the separation of the quasi-Fermi levels, and hence the recombination process actually removes cold carriers from the conduction and valence bands. This leads to an effective rise in the carrier temperature and hence a reduction in the gain. A crucial difference in these two is that carrier heating component is negative across the entire gain spectrum; i.e., the gain is reduced and the absorption is increased (absorption is negative gain, so the same signs have opposite effects) by this mechanism. The spectral hole burning component is negative in the gain region and positive (i.e., it reduces the absorption by the creation instead of depletion of carriers) in the region to the short-wavelength side of the transparency point in the gain spectrum.

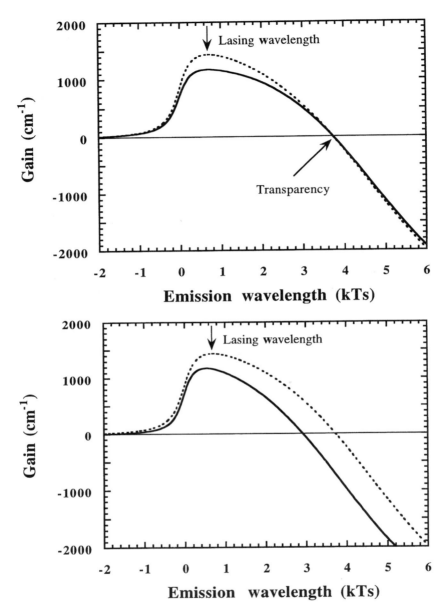

Figure 3.27: Illustration of (a) the spectral hole burning and (b) the transient carrier heating mechanisms proposed to explain the origin of the intrinsic gain reduction with increasing photon densities in semiconductor lasers. In both figures, the dashed curve is the original material gain, and the bold curve is the reduced (real) gain profile as affected by the respective gain-suppression mechanism.

Measurements in semiconductor lasers using direct modulation techniques (Frankenberger and Schimpe, 1992) and ultrafast pump-probe techniques (Mark and Mørk, 1992; Hall et al., 1992) have indicated that both spectral hole burning and transient carrier heating components contribute to nonlinear gain suspension.

Spectral hole burning

From the density matrix analysis of nonlinear gain and spectral hole burning arguments, the analytic expression for the gain compression factor can be written as (Yamada, 1983; Agrawal, 1987; Uomi and Chinone, 1989)

$$\varepsilon = \left(\frac{e^2}{m_o^2 \varepsilon_r E_i}\right) M^2 \left(\frac{\Gamma_3}{\Gamma_1^2}\right) 2(\tau_c \tau_v) \tag{3.67}$$

where ε_r is the dielectric constant of the material, and τ_c and τ_v are the intraband relaxation times in the conduction and valence bands, respectively. This expression has been derived the case for the bulk active region here and thus is smaller than that given by Uomi and Chinone (1989) by a factor of 1.5. This factor is the result of the enhancement in the momentum matrix elements, at the band edge, in the case of the quantum wells. The effective intraband relaxation time τ_{in} used to compute the gain function is related to τ_c and τ_v as $1/\tau_{in} = (1/\tau_c + 1/\tau_v)/2$. The (Γ_3/Γ_1^2) is a geometric factor that is defined as

$$\frac{\Gamma_3}{\Gamma_1^2} = \frac{d \int_{-d/2}^{+d/2} |\psi(x)|^4 \, dx}{\left(\int_{-d/2}^{+d/2} |\psi(x)|^2 \, dx\right)^2} \tag{3.68}$$

where d is the total width of the active region, $\psi(x)$ is the electric field distribution, in the transverse direction, of the optical mode in the laser cavity, and Γ_1 and Γ_3 are the first- and third-order optical confinement factors, respectively.

The conduction and valence band relaxation times are generally functions of the quantum well width, in addition to being also sensitive to the carrier density and the energy from the respective band edges (Asada, 1989). Since the intraband relaxation times are also part of the gain function, it is thus possible to optimize the effect of spectral hole burning on the high-speed performance of quantum well lasers by optimizing the

$\varepsilon/(dg/dn)$ ratio or the K factor (Uomi and Chinone, 1989). Although from the density matrix analysis the gain compression takes the form of $g_{\text{eff}} = g(1 - \varepsilon S)$, it is still a useful way to optimize high-speed quantum well structures.

Figure 3.28 shows the result of such an optimization assuming a constant value for τ_c of 200 fs and τ_v of 70 fs, which gives a consistent value of about 100 fs for τ_{in}. Since the intraband scattering times are not a very sensitive function of quantum well width (Asada, 1989), the minimum in the $\varepsilon/(dg/dn)$ function generally occurs in the region of maximum dg/dn. Figure 3.28 shows the variation in the K factor and the threshold carrier density with well width for a modal gain of 40 cm^{-1}. There is a minimum in the K factor at around well widths of 80 to 100 Å. This is consistent with a previous report of such an optimization process, which also considered the variation in intraband scattering times with quantum well width, for the InGaAs/InP quantum well structures, where a minimum in the K factor was computed in the same range of quantum well widths (Uomi and Chinone, 1989). Figure 3.28 shows that

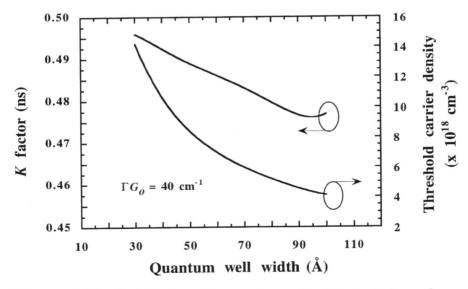

Figure 3.28: Optimization of the quantum well width for high-speed performance based on the spectral hole burning model by minimizing the ε/g_o ratio contribution to the total K factor. The details of the calculation are given in the text.

the threshold carrier density also increases as the well width is reduced. This calculation does not include carrier overflow, which will make this trend even more severe. Thus, for both high-speed and low-threshold current, quantum well widths between 80 and 100 Å are optimal. A more detailed analysis of spectral hole burning and its variation with quantum confinement or strain is given by Takahashi and Arakawa (1989) and Ghiti and O'Reilly (1990). These two works and others (Takahashi and Arakawa, 1991) predict an enhancement in the nonlinear gain suppression with increasing levels of quantum confinement. Although the trend with quantum confinement has been observed, the theoretically predicted enhancement of gain suppression in quantum well lasers has not been observed experimentally (Nagarajan et al., 1991b, 1992c). There also have been some experimental reports that p doping the active region leads to a reduction in the nonlinear gain compression factor (Aoki et al., 1990; Ralston et al., 1993). The reduction in ε has been attributed to a reduction in the intraband relaxation time as the result of doping. In these reports, this trend in ε was deduced from K factor measurements in high-speed lasers.

The experimental data for the gain compression factor presented previously in Fig. 3.11 shows that it is about the same magnitude as reported in lasers with bulk active areas and did not vary significantly with SCH width. The carrier transport model that was derived in Sect. 3.2 did not influence ε at all, although other models with enhancement in ε due to carrier transport also have been proposed (Rideout et al., 1991; Wu et al. 1992; Tessler et al., 1992b). Although the ε values quoted for lasers in the InGaAs/InP system (Uomi et al., 1991; Fukushima et al., 1991a, 1991b; Shimizu et al., 1991), on average, are higher than the ones that we have reported for the InGaAs/GaAs system (Nagarajan et al., 1991b), they are by no means anomalously high or enhanced in quantum well lasers as they were initially speculated to be (Sharfin et al., 1991). Experimental reports seem to indicate that ε is independent of the laser structure and is even unaffected by the inclusion of strain, compressive or tensile, in the quantum wells (Hirayama et al., 1991; Nagarajan et al., 1991b; Fukushima et al., 1991a; Shimizu et al., 1991). Figure 3.29 shows the experimental variation in the gain compression factor with cavity length for SQW and MQW lasers. The gain compression factor is lower for shorter-cavity lasers and lasers with a smaller number of wells. In this case, the lowest value of ε was for an SQW laser of 200-μm cavity length. This value of 9.98×10^{-18} cm^3 is about 5 to 6 times lower than the

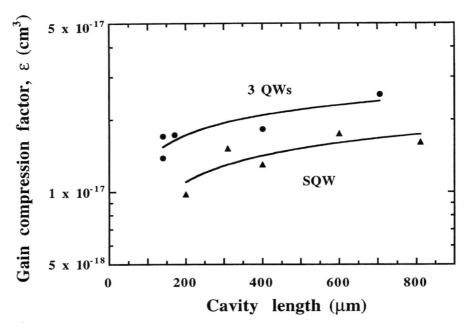

Figure 3.29: Variation of the gain compression factor with cavity length for MQW and SQW lasers. The data are for $In_{0.2}Ga_{0.8}As/GaAs$ quantum well lasers. The lines have been drawn to show the trend in the data.

values reported for MQW lasers operating at around 1.55-μm wavelength (Westbrook et al., 1989; Kitamura, 1989; Uomi et al., 1991; Fukushima, et al., 1991b). The higher values of gain compression for MQW lasers and lasers with longer cavity lengths indicate that this nonlinearity is enhanced by increasing the quantum confinement of the carriers.

Transient carrier heating

Analysis of gain suppression due to transient carrier heating is somewhat more complicated (Gomatam and DeFonzo, 1990; Willatzen et al., 1991). In addition to the rate equations for the carrier and photon densities, the analysis also requires separate equations for electron and hole energy densities, which are a function of the respective carrier temperatures. Willatzen et al. (1991) have derived an analytic expression for ε assuming

3.3 High-speed laser design

the $g_{\text{eff}} = g(1 - \varepsilon S)$ form of gain suppression. We will just quote the result here:

$$\varepsilon = \sum_{i=c,v}\left[-\frac{\partial g}{\partial T_i}\right]_N \left[\frac{\partial U_i}{\partial T_i}\right]_N^{-1} \tau_{\text{heat},i} v_g \left\{\left[\frac{\partial U_i}{\partial N_i}\right]_{T_i} + E_i\right\} \quad (3.69)$$

where T_i is the carrier temperature, U_i is the energy density, E_i is the carrier energy at the lasing wavelength, $\tau_{\text{heat},i}$ is the time constant with which the carrier energy density relaxes toward the equilibrium lattice energy density, v_g is the optical mode velocity in the laser cavity, and the summation is carried over the conduction and valence bands, $i = c, v$. The value for ε is positive because the $\partial g/\partial T_i$ term, i.e., the variation of gain with temperature, is usually negative. This term can be designed to be positive, in which case *negative* ε or gain enhancement with increasing optical power density is possible. Experimental observation and theoretical treatment of such an enhancement in vertical cavity surface emitting lasers were reported by Wang et al. (1993).

3.3.4 Photon density

From the expression for the resonance frequency derived in Eq. (3.15) one of the parameters required for wide modulation bandwidth is a large photon density in the laser cavity. The photon density S_o is related to the output optical power P_o and the bias current, I_o as

$$S_o = \left(\frac{1}{V/\Gamma}\right)\frac{1}{h w_g \alpha_m} P_o = \left(\frac{1}{V/\Gamma}\right)\frac{1}{e v_g (\alpha_m + \alpha_i)} \eta_i (I_o - I_{th}) \quad (3.70)$$

where P_o is the total output power (including both facets), I_{th} is the threshold current, η_i is the internal quantum efficiency, α_m is the mirror loss, α_i is the internal loss, V is the volume of the active region, and V/Γ is the volume of the optical mode.

From Eq. (3.70) it can be seen that a large output power does not necessarily imply a correspondingly large photon density within the optical cavity. Although a high output power capability in a semiconductor laser is highly desirable for large modulation bandwidths, it is the mode volume that largely determines the photon density within the laser cavity. Thus a large modulation bandwidth laser structure requires a tight optical confinement within the laser cavity, and lasers using structures with large optical cavities (LOCs) to increase the reliable output power capability will not be suitable for high-speed operation (Lau and Yariv, 1985a).

In distributed feedback (DFB) lasers that have internal phase-shift/adjustment regions (a total of $\lambda/4$ shift) for single-mode operation, the large photon density required for high-speed operation within the optical cavity leads to the spatial hole burning effect. In these DFB lasers, the photon density within the cavity is nonuniform and is much higher at the center of the cavity, where the phase shift region is located. The carrier density in this region is reduced due to stimulated recombination, leading to spatial variations in the refractive index. This degrades the single-mode behavior of the DFB lasers, and the side-mode suppression is reduced because the cavity is now capable of supporting multiple longitudinal modes (Soda et al., 1987). The magnitude of the spatial hole burning effect increases with the amount of phase shift incorporated and the magnitude of the coupling coefficient of the grating.

Spatial hole burning is undesirable for a number of reasons, the main one being that it limits the maximum useful output power from a DFB laser, thus limiting the maximum photon density in the laser cavity and the resonance frequency. The nonuniform photon density distribution also leads to nonuniform gain suppression (via ε, the gain compression factor) over the length of the cavity. This affects the overall modulation response of the DFB laser. The spatial hole burning effect also has been shown to limit the minimum attainable laser linewidth (Okai et al., 1989) and significantly modifies the wavelength chirping characteristics (Kinoshita and Matsumoto, 1988) in DFB lasers.

3.3.5 Device operating conditions

The original expression for the resonance frequency [Eq.(3.15)] was derived in terms of the photon density. Ignoring the carrier transport effects (which will only reduce the effective value of differential gain) and the minor terms in ε that have little effect on the analysis here, the equation for the resonance frequency may be rewritten in terms of the experimentally measured quantities such as the output power and the bias current as

$$\omega_r^2 = \frac{v_g g_o}{\tau_p} S_o = \left(\frac{1}{V/\Gamma}\right)\left(\frac{v_g g_o}{h\nu}\right)\frac{\alpha_m + \alpha_i}{\alpha_m} P_o = \left(\frac{1}{V/\Gamma}\right)\frac{v_g g_o \eta_i}{e}(I_o - I_{th}) \qquad (3.71)$$

3.3 High-speed laser design

Equation (3.71) is interesting because it indicates different optimization procedures for devices that are limited by the photon density within the laser cavity, output power, or thermal heating due to the bias current. Using the expression in terms of S_o for photon density–limited devices, reducing τ_p will lead to enhancements in the resonance frequency. This can be achieved by low-reflectivity facet coatings, and the assumption is with the reduced reflectivity and the corresponding decrease in the external differential efficiency that the bias current can be increased sufficiently to restore the original power level, if so required, without any detrimental effects.

However, most devices are current limited or limited by the thermal dissipation of the device structure. Using the expression in terms of I_o, η_i has to be increased and I_{th} has to be reduced to enhance the resonance frequency.

Although the expression for ω_r^2 in terms of S_o does not present the complete picture, it is the physical limitation, while the previous two are more like device limitations. This is the one that has been used over the years as a starting point for high-speed device optimization, and using this approach, reducing the photon lifetime or increasing the cavity loss rate, will lead to an enhancement in the resonance frequency. This can be achieved by using short-cavity lasers and/or antireflection coatings on the facets. This is only recommended if the device is neither power nor thermally limited. These constraints on the output power and the bias current can be met successfully to various degrees with modern crystal growth and device fabrication technology. One does not strictly operate in any one of these regimes, and as the device characteristics are modified, e.g., by facet coating, etc., one in fact moves from one to another. It is desirable to be in the photon density–limited regime and not to be limited either by the output power level or the device thermal dissipation limit. From these tradeoffs one can determine an optimal design point, i.e., facet reflectivity for a given cavity length, with respect to these parameters (Mar et al., 1990).

Given the constraint of having to keep the device size smaller than the modulating signal wavelength, one has to go to shorter cavity lengths to achieve higher speeds of operation. This will increase the cavity loss and may cause the device to become thermally limited. One can then employ some of the methods, such as facet coating, discussed earlier to compensate for this.

3.3.6 Device structures with low parasitics

In practice, the often dominant limit to the modulation response of laser diodes is the device parasitics. The simplest model that may be used to represent the laser diode is an inductance (due to the bond wire) in series with a parallel combination of the device capacitance and resistance. There could be additional capacitance in parallel before the series inductance due to packaging elements such as the bonding pads, etc. The intrinsic resistance of the laser diode given by $\eta k T/eI_o$, where I_o is the bias current and η is the ideality factor, is very small at the large bias currents (about 0.5 Ω for $I_o = 100$ mA and $\eta = 2$, which is typical in laser diodes) used in the high-speed modulation experiments that it appears to be essentially a short circuit. The major goals in any high-speed device design is then to minimize the series resistance and the parasitic capacitance. The effect of bond wire inductance may be eliminated in the measurements by the use of laser structures suitable for use with coplanar microwave probes (Offsey et al., 1990). In practice, the bond wire inductance is an important consideration in all packaged semiconductor lasers required for high-speed system applications (Bowers et al, 1986).

Figure 3.30 shows the two major contributions to the total device series resistance R_d; the first is the contact resistance R_c, and the second

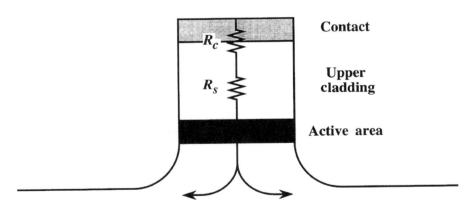

Figure 3.30: Major contributions to the total device series resistance in a semiconductor laser; the top contact resistance R_c and the upper cladding resistance R_s.

is the bulk resistance of the top cladding layer R_s. In the fabrication of a typical laser structure, a ridge/mesa is first etched partway down the upper cladding or all the way down past the active area for lateral mode and current confinement. The structure is then planarized either by regrowing current blocking layers to the sides of the mesa or using a thick layer of polyimide. In many cases a thin dielectric layer (SiO_x or SiN_x) is also used. An electrode is then deposited to contact the top of the original mesa for current injection. The upper cladding is usually heavily p doped (for n substrates), and in some cases where p substrates are used, it is n doped. The lower cladding region is not etched, and the contact to the substrate side is large (size of the laser chip is typically about 500 μm on one side and the cavity length on the other). Due to this, the resistance contributions from the bottom cladding and contact and the substrate are negligible.

The contact resistance is determined by the contact resistivity Ω_c of the metal-semiconductor interface, which is a function of the doping in the contact/cap layer of the laser structure and the metal sequence used as the electrode. A good contact resistivity is about $\Omega_c = 10^{-6}\ \Omega\ cm^2$. Let us consider a laser structure with a cavity length L_c of 150 μm, mesa width L_s of 1.5 μm, and upper cladding thickness L_h of 1 μm. Doping levels of $n = 10^{18}\ cm^{-3}$ are typical in the cladding regions, and the majority carrier mobility values are small at this large doping level. Let us assume a mobility value of $\mu = 100\ cm^2/V\ s$. Using these values, the total device resistance is

$$R_d = R_c + R_s = \frac{\Omega_c}{L_s L_c} + \frac{L_h}{en\mu L_s L_c} = 2.8 + 0.4 = 3.2\ \Omega \quad (3.72)$$

Device resistance of about 2 to 3 Ω is typical for high-speed laser diodes. For a given contact resistivity, the contact resistance is large for a smaller device geometry, but the parasitic device capacitance scales inversely with the area ($L_c L_s$ product), and hence the total RC product is unchanged with respect to the contact resistance component. The dominant contribution to R_d comes from R_s. This can be reduced by having narrower cladding regions (reducing L_h), but it will be at the expense of poorer optical mode confinement in the lateral direction, if the refractive index of the cladding region cannot be decreased adequately to compensate for the reduced thickness. The more commonly employed method is to increase the doping level in the cladding and use very heavily doped (1–5 $\times\ 10^{19}\ cm^{-3}$) graded and/or low-bandgap material to form the contact layer. Since a part of the optical mode is in the cladding regions, the main drawback of this scheme is the increased free carrier absorption loss. A number of material combinations for low contact resistance in devices based on III-V compounds have been reported in the literature, and details of many of these may be found in a review by Shen et al. (1992). Since the majority of these contacts are alloyed, optimization of the alloying temperature and time is also important to minimize the resistance at the metal-semiconductor interface.

Since the contribution of the lower cladding and substrate to the total resistance is minimal, the use of p-type substrates instead of the typical n-type has been suggested for fabricating lower-resistance devices. In this structure, the n-type material with the higher electron mobility will form the upper cladding, which has the largest contribution to the device resistance. In the GaAs system, where the cladding regions are composed of $Al_xGa_{1-x}As$ with large Al mole fractions (60% and above), the electron mobility saturates at an intrinsically low value slightly less than 200 cm^2/Vs because the Γ valley is no longer the minimum at these high Al compositions, and the electrons from X valley with the much larger effective mass dominate the conduction process (Saxena, 1981). This is only slightly larger than the hole mobility values (in the range of 100 to 200 cm^2/Vs in $Al_xGa_{1-x}As$ for similar Al mole fractions (Masu et al., 1983). The actual mobility values in the devices will be lower due to impurity scattering when the claddings are heavily doped. Thus there is no distinct advantage to the use of p-type substrates for laser diodes in the GaAs system.

In the InP system, there may be advantages in fabricating lasers on p-type substrates. Tsang and Liau (1987) have reported modulation bandwidths up to 16.4 GHz in mass-transported InP lasers emitting at 1.3 μm on p-type substrates. They also have suggested that with p-type substrates the devices need not be mounted junction down for better heat dissipation, a procedure that tends to increase the parasitic capacitance and is not desirable for high-speed operation.

The other major component of the device parasitics is the capacitance. Since the device resistance is constrained by the minimum cladding layer thickness required for good lateral optical confinement and the maximum doping in the cladding for acceptable levels of free carrier absorption loss, lowering the parasitic capacitance is the most effective way of controlling the RC product.

3.3 High-speed laser design

Figure 3.31 shows a number of low parasitic structures employed in the InP system to fabricate high-speed lasers. This is by no means a list of all the structures that have been reported to date, but a few chosen to illustrate the principles involved in such designs. The most straightforward of these methods is shown in Fig. 3.31(a), which illustrates the use of isolation channels to minimize the parasitic capacitance. A regular buried heterostructure (BH) laser with reverse-biased p-n junctions to minimize leakage currents is first grown. The parasitic capacitance of such a laser is large mainly due to the capacitance of the reverse-biased current-blocking layers. This capacitance is then reduced considerably by etching deep grooves on either side of the mesa close to the active region (Ishikawa et al., 1987; Kamite et al., 1987). In addition to the isolation channels, an air bridge may be used to contact the narrow ridge across

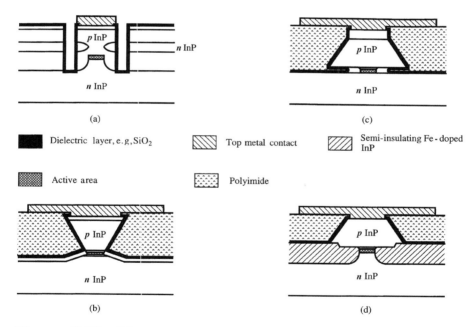

Figure 3.31: Schematic diagrams illustrating the principle behind the various low parasitic capacitance laser structures reported in technical literature to date: (a) use of isolation channels in a buried heterostructure configuration; (b) polyimide buried ridge waveguide geometry; (c) undercut mesa-type structures; and (d) Fe-doped semi-insulating InP-based structures.

the channels. This overcomes the problem of having to directly contact a fairly narrow ridge after the isolation etch. InGaAsP/InP DFB lasers operating at 1.3 μm utilizing this air bridge contact configuration with bandwidths in excess of 17 GHz have been reported (Chen et al., 1993).

The use of a polyimide layer to minimize the parasitic capacitance has become fairly common over the years. The polyimide is insulating and has the advantage that thick, dense layers can be easily spun on, cured, and planarized unlike the dielectric insulating layers, which do not planarize and can only be deposited in limited amounts. The simplest laser structure is one that uses a polyimide layer instead of regrowth around the mesa that is first etched. Figure 3.31(b) shows one such structure where some amount of initial regrowth has been used to provide lateral confinement for the optical mode and heat-dissipation path around the active region (Uomi et al., 1989). A good alternative would be to not etch past the active region at all and simply bury the ridge in a thick polyimide layer (Nagarajan et al., 1991a). Large modulation bandwidths have been demonstrated in both these structures. Alternatively, in these ridge waveguide lasers, an insulating dielectric layer, e.g., SiN, may be deposited followed by the deposition of a patterned top metal contact designed to reduce the parasitic capacitance. DFB lasers with bandwidths of 22 GHz have been demonstrated in these structures (Lu et al., 1993).

An inverted-V-type mesa or the "mushroom" laser structure is illustrated in Fig. 3.31(c). In this type of laser, the mesa width can be made independent of the active layer width by undercutting the active region after the mesa etch. There are etchants that will preferentially etch a V or an inverted V depending on the crystallographic orientation of the etch mask. The undercut mesa is then buried partway (to minimize the area of the semiconductor junction that will contribute to the parasitic capacitance) either by vapor-phase regrowth (Bowers et al., 1985) or by mass transport (Tsang and Liau, 1987). A polyimide layer can then also be used in addition to minimize the parasitic capacitance from the top contact metal. Another interesting undercut method is the self-aligned constricted mesa structure that has been used to fabricate DFB lasers. This method has an advantage that the undercut etching stops automatically at the InP interface used to bury the quaternary active region. Modulation bandwidths as high as 17 GHz have been demonstrated in these lasers (Hirayama et al., 1992b).

The use of an Fe-doped semi-insulating current-blocking layer is an attractive alternative for the fabrication of high-speed InP-based lasers.

3.3 High-speed laser design

Fe doping of InP renders it semi-insulating due to the formation of deep midgap impurity levels. An advantage of this method is that the regrowth step is also planarizing, and the use of a semiconductor burying layer results in a more symmetric optical mode profile, which results in a more efficient coupling of light into an optical fiber. Figure 3.31(d) shows one such structure where a polyimide layer also has been used to further reduce the parasitic capacitance (Bowers et al., 1987). Bandwidths as high as 22 GHz have been demonstrated at 1.3 μm in lasers with low-capacitance structures employing a buried crescent active region, semi-insulating current blocking layers, and a layer of polyimide beneath the top contacts (Huang et al., 1992).

A conventional laser structure has a vertical current injection path, and in this configuration, some minimum parasitic capacitance as dictated by the area of the active layer, with cavity length on one side and the active region width on the other, is unavoidable. The use of transverse junction structures where the current injection occurs laterally from the sides of the active region will minimize this residual parasitic capacitance considerably. These structures also have the advantage that both the n and p contacts are on the top side, and the lasers may be fabricated on semi-insulating substrates and easily integrated with other electronic devices.

Figure 3.32(a) shows one such implementation of the structure where regrowth has been used to form the lateral p and n layers. Capacitance

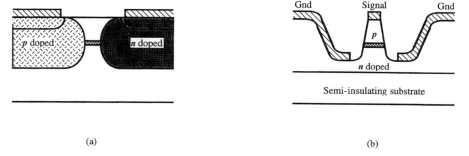

(a) (b)

Figure 3.32: Low parasitic device structures on semi-insulating substrates. The first is the lateral current injection structure with low parasitic capacitance, and the second is a structure with a top probe geometry commonly used with microwave coplanar probes to eliminate the parasitic effects of the bond wire inductance.

less than 50 fF (about 20 times smaller than for the conventional vertical injection structures) has been demonstrated in these lasers (Shimoyama et al., 1988). Lateral current injection (LCI) structures also can be formed by impurity-induced disordering. This results in a planar structure, and stray capacitance as low as 270 fF has been demonstrated in these lasers (Furuya et al., 1988). In conventional multiple quantum well lasers, where the vertical transport between the various quantum wells may be hindered by the large energy height and/or the width of the barriers, LCI structures have an advantage that the current injection is from the sides, resulting in a more uniform carrier distribution. Although the LCI structures have very low parasitic capacitance, the transport delay in the current injection, which now has to take place over the ridge/stripe width instead of just the SCH width, significantly affects the modulation properties. Large modulation bandwidths have not been presently reported in lasers with LCI structures.

An additional contribution to the total device parasitics comes from the bond wire inductance of the laser mount (Bowers et al., 1986). Strictly speaking, it is not a device parameter but has a significant influence on the modulation properties of the laser diode. The inductive parasitic component can be minimized by using short wires or meshes instead of wires to bond the device. Another technique is to use coplanar microwave probes to characterize the high-speed properties of these lasers. Special laser structures are required for this, and Fig. 3.32(b) shows the top probe geometry suitable for use with the microwave probes. InP MQW lasers at 1.3 μm with small signal bandwidths up to 20 GHz have been demonstrated using this structure (Lipsanen et al., 1992). InGaAs/GaAs lasers with modulation bandwidths of greater than 40 GHz are also of this geometry (Weisser et al., 1992; 1996; Ralston et al., 1994). Although the use of microwave probes eliminates the influence of bond wire inductance on the modulation properties of semiconductor lasers, it is always a source of concern in the finally packaged versions of these devices.

3.3.7 Device size and microwave propagation effects

The device analysis up to this point has assumed that the modulation current pumping the laser is uniform in amplitude and phase along the laser stripe. At very high frequencies, this assumption is invalid because the time delay from the feed point to the far end of the device can be a significant fraction of the modulation period. The distributed microwave

3.3 High-speed laser design

nature of the laser under these circumstances modifies the high-frequency current injection and the optical modulation response. High-speed laser structures that are on the order of 300 μm long should be treated as distributed electrical elements at frequencies of 25 GHz and greater (Tauber et al., 1994). This is important for quantum well lasers because they are often this long or longer in order to ensure low threshold current densities and high differential gain as well as good thermal properties and high cleave yields. The distributed effects clearly will be important in longer devices at higher modulation frequencies.

A picture of the microwave currents flowing in the device under high-frequency modulation is shown in Fig. 3.33. The transverse current across the active region pumps the structure, and the longitudinal currents are needed to satisfy the boundary conditions imposed on the propagating electromagnetic wave. The key point of this picture is that all these are dissipative currents, and they therefore contribute to microwave loss. Further, because the longitudinal currents are subject to skin effects in the conducting regions, the loss should increase monotonically with frequency. The loss that results from the injection of transverse current into the active region is bias-dependent and increases as the diode conductance increases. Above threshold, this transverse conductance is fixed, and the loss is set by the series resistance of the contacts and the cladding layers.

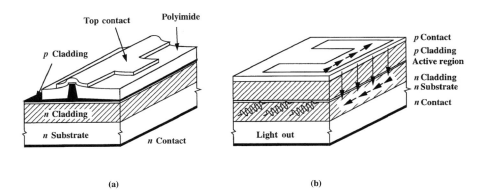

Figure 3.33: (a) High-speed ridge waveguide laser structure. (b) Conceptual picture of microwave currents flowing in laser under high frequency modulation. The horizontal arrows represent longitudinal currents, while the vertical arrows represent the small-signal injection current into the active region.

In order to characterize the microwave propagation, a calibrated, 0.045- to 40-GHz two-port S-parameter extraction on a HP 8510 network analyzer was performed on the laser. The laser used for the measurements reported here was an SQW ridge waveguide structure (Nagarajan et al., 1992a). The device length was 830 μm, and the microwave probes used to launch and receive the signal were separated by 600 μm along the laser stripe. Similar results were measured on other structures.

The measured loss per unit length and the phase velocity as a function of frequency are shown as solid lines in Fig. 3.34. The measurements were done for both zero bias and above-threshold operation. The striking features in the graphs are (1) the enormous value of loss, near 600 dB/cm at 40 GHz for the forward-biased case, and (2) the slow phase velocity, around one-tenth the speed of light in vacuum. The large loss has its origin in three sources. First, the thick n-doped cladding and substrate form a lossy ground plane for the transmission line. Because the skin depths in the semiconductor for the measured frequency range are significantly less than the 100-μm thickness of the n cladding and n substrate, essentially all the longitudinal current on the n side is carried in the poor-conductivity doped semiconductor and not in the metal contact. In contrast, the 1-μm-thick p-cladding layer is thinner than the semiconductor skin depths and thus the longitudinal p-side current will penetrate into the gold contact layer. The current should therefore be carried primarily in the high-conductivity gold, and the loss should be low if the gold layer is thick enough. However, the skin depth in gold at the frequencies we are interested in is greater than 4000 Å, while the p-side contact metal thickness in these devices is only 3500 Å. Therefore, a second source of loss is the thin p-side gold contact. The forward-biased junction is a third source of loss, because under forward bias a dissipative conduction path is opened that is not present for zero bias.

The slow-wave nature of the propagation is significant because the device lengths and frequencies at which these distributed effects are important are determined by the propagation velocity. Transmission lines fabricated on semiconductors often exhibit slow-wave behavior (Hasegawa et al., 1986; Liou and Lau, 1993). Further, large skin losses are always accompanied by an internal inductance $L_i = (2\pi f \sigma \delta)^{-1} = \sqrt{\mu/4\pi f\sigma}$, where σ is the conductivity, δ is the skin depth, f is the frequency, and μ is the magnetic permeability (Ramo et al., 1984). The wave velocity can be approximated by $v = \sqrt{1/LC}$, where L and C represent the inductance and capacitance per unit length, respectively, and the added inductance

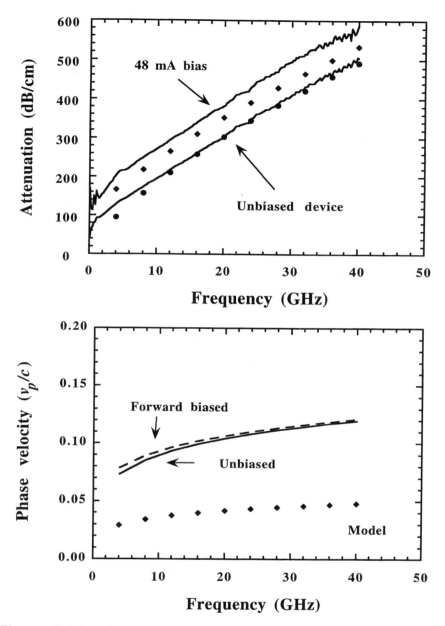

Figure 3.34: (a) Microwave attenuation per unit length versus frequency. (b) Phase velocity, normalized to the vacuum speed of light, versus frequency. The solid lines represent measured values, and the points represent calculations based on the circuit model of Fig. 3.35. The measurements were done on an SQW $In_{0.2}Ga_{0.8}As/GaAs$ laser.

can cause slow-wave propagation. Another point to note is that the frequency dependence of the internal inductance explains the dispersion seen in the phase-velocity graphs.

A distributed circuit model for the laser was constructed in order to obtain a first-order estimate of the propagation parameters. Figure 3.35 is a schematic of the model, and the data points on the attenuation and phase-velocity plots (Fig. 3.34) are calculations based on this model. The numerical values used in the model are per unit length values and are extracted from measured lumped-element values or are estimated from the device geometry. The model predicts the loss accurately and reveals the correct dispersion trend (a phase velocity that increases monotonically with frequency). The slow-wave nature of the propagation is predicted as well, although the phase velocity is underestimated. The discrepancy between theory and experiment is possibly caused by the fact that the distributed circuit models quasi-TEM propagation, which is not strictly valid for structures with large attenuation in the longitudinal direction. Another point to make is that the evanescent and higher-order modes excited at the feed point, which normally decay over a negligible length of standard transmission line, may be significant for short structures such as diode lasers. These modes are neglected in the distributed-circuit model.

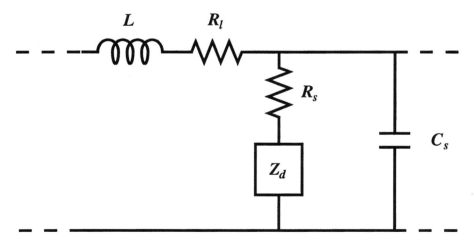

Figure 3.35: Distributed circuit model used to calculate microwave propagation parameters for the laser.

3.3 High-speed laser design

The primary conclusion to draw from these results is that the frequency-dependent loss results in a frequency-dependent current injection to the active region. As the frequency rises, the loss increases, which in turn causes a roll-off in the active region current injection. This can be accounted for by including in the small-signal response a roll-off term (Fig. 3.36) that can be calculated directly from the network measurements or approximated from the circuit model. The curves shown in Fig. 3.36 are extracted from the measured propagation parameters and show the square magnitude of the total injected current, normalized to dc, versus frequency, for three different device lengths. The -3-dB roll-off frequency occurs at approximately 25 GHz for the 300-μm-long device. In these plots, the device is fed at the center of the laser stripe. This is the optimal feeding condition because waves can propagate out in both directions from the center and maximally pump the device. An end-fed device should have a lower roll-off frequency because only a single forward wave can

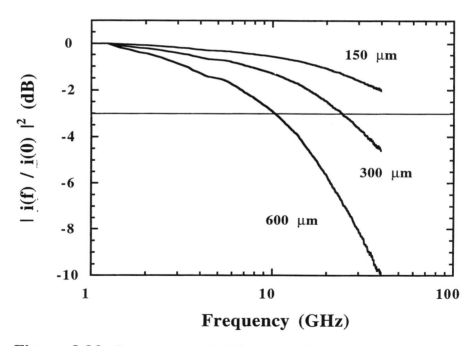

Figure 3.36: Square magnitude of the total small-signal current versus frequency for three different device lengths, normalized to dc. The injection current is calculated from the measured propagation parameters shown in Fig. 3.34.

propagate, and therefore, the fraction of the device that is pumped is smaller.

3.4 Large-signal modulation

Many practical applications of high-speed lasers generally involve some form of large-signal modulation. These are generally in the area of short optical pulse generation or data transmission using direct modulation techniques. Thus far we have used the small-signal performance as the criterion for measuring the high-speed properties of a semiconductor laser. In this section we will apply the principles discussed previously to the area of large-signal modulation.

In directly intensity-modulated optical fiber systems, the non-return to zero (NRZ) modulation format gives the maximum bit-rate performance. If B is the small-signal bandwidth of the laser, in linear systems, the maximum NRZ bit rate is approximately $2B$. Large-signal switching dynamics in a semiconductor laser are complicated. There is usually a time delay between application of the electrical signal and emission of the optical pulse. Considering also the finite rise and fall times during the switching process, and ignoring the undesirable effects of intersymbol interference at the laser diode transmitter, the maximum bit-rate performance has been predicted to be about $1.3B$ (Tucker et al., 1986).

Gain switching and mode locking are the most commonly used techniques for short pulse generation in semiconductor lasers. Short, large-signal optical pulses are needed in applications such as soliton generation and electro-optic sampling. In gain switching, each optical pulse is independent and builds up from spontaneous emission. The advantage is that the pulse is electronically triggerable, and the repetition rate and amplitude are variable. Mode locking consists of resonantly amplifying the pulse as it travels around the cavity. The pulse become shorter and shorter until it reaches a steady state, where the pulse-broadening mechanisms in the cavity balance the pulse-shortening mechanisms. The advantages of mode-locked pulses are shorter optical pulses with less timing jitter. The disadvantages of mode locking are fixed repetition rate and power. The repetition rate is tunable, but the timing of individual pulse timing is relatively fixed. The following two sections describe these two pulse-generation techniques.

3.4.1 Gain switching

The two primary structures used for gain switching are shown in Fig. 3.37. The simplest structure contains a single electrode that is driven by a short current pulse. The rise and fall times can be analyzed easily from the rate equations given earlier in the limit of very short electrical drive pulses. For this analysis, the photon density in Eqn. (3) is rewritten as

$$\frac{dS}{dt} = \Gamma v_g g(N_W - N_t) - \frac{S}{\tau_p} + \Gamma\beta\frac{N_W}{\tau_n} \quad (3.73)$$

In Eqn. (73), the linear approximation of the gain function has been used; $G(N_W) = g(N_W - N_t)$ where the gain has been linearized about the transparency carrier density N_t. The differential gain is given by g which is simply the slope of the linearized gain function. Assuming that gain compression is negligible for small photon densities during the rising edge of the pulse and neglecting the small contribution of spontaneous emission feedback into the lasing mode, Eqn. (73) can be written as

$$\frac{1}{S} dS = \frac{1}{\tau_r} = \Gamma v_g g(N_W - N_{th}) \, dt \quad (3.74)$$

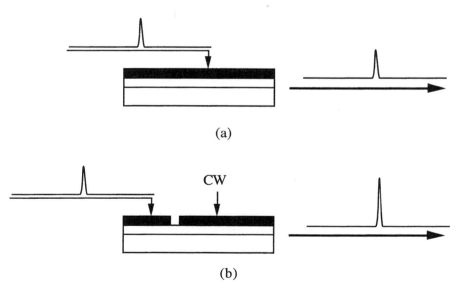

Figure 3.37: (a) Schematic of a simple gain-switched laser. (b) Q-switched laser where the primary energy storage is in the dc-driven section, and the modulated section is used to hold off lasing while the energy is built up.

where N_{th} is the carrier density at threshold ($N_{th} = N_t + 1/\Gamma v_g g \tau_p$) and τ_r is the rise time of the optical pulse (Downey, et al., 1987). Eqn. 74 has the following solution,

$$S = \begin{cases} S_0 & \text{for } t < 0 \\ S_0 e^{t/\tau_r} & \text{for } 0 < t < t_{on} \end{cases} \quad (3.75)$$

where the rise time is given approximately by

$$\frac{1}{\tau_r} = \Gamma v_g g \left(\frac{Q}{qV} + N_0 - N_{th} \right) \quad (3.76)$$

For short, high current pulses used in gain switching, the carrier density rises to a high value and remains constant while the optical density builds up and depletes the carrier density. In this case it is the total charge in the electrical pulse (Q) and not the shape that is important (Bowers, 1989). In Eqn. 76, N_0 is the dc carrier density and V is the volume of the active region. Clearly, the rise time can be quite short, just a few picoseconds if the charge in the pulse is very large. If the laser is pre-biased at threshold, the rise time is

$$\tau_r = \frac{qV}{v_g g \Gamma Q} \quad (3.77)$$

The fall time is

$$\tau_f = \frac{1}{v_g g \Gamma (N_{th} - N_f)} \quad (3.78)$$

where N_f (below N_{th}) is the final level to which the carrier density falls at the trailing edge of the optical pulse. The fall time is limited by the difference in the threshold and transparency carrier densities and is usually substantially longer, typically 10 ps. Note that the peak of a gain-switched pulse occurs when the carrier density is exactly equal to the threshold value. These results were verified for a 115-μm-long laser with a round-trip time of 3 ps. Figure 3.38 shows a short drive pulse, the integral of the drive pulse (proportional to the carrier density and gain in the absence of stimulated or spontaneous emission), and the measured laser output on linear and logarithmic scales. The exponential behavior is quite a good approximation over most of the pulse. Note that the fall time is about twice as long as the rise time.

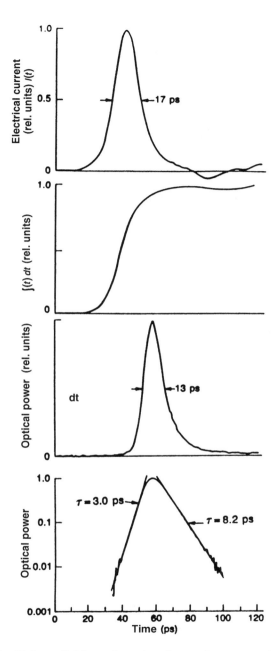

Figure 3.38: Gain switching of semiconductor lasers: (a) the current drive waveform; (b) the integral of the current waveform, which is the form of the carrier density and gain changes in the absence of emission; (c) light output (linear scale); (d) light output (log scale). This illustrates the exponential nature of the rise and fall of the intensity as well as the slower fall times.

The pulse energy obtainable is limited by the charge in the electrical drive pulse, which is limited by the peak voltage of the drive circuit. The preceding equation [Eq. (3.77)] shows that the shortest pulses with the largest peak intensities occur when the laser is prebiased at threshold. This energy storage can be greatly improved with the structure shown in Fig. 3.37(b). We will refer to this structure as a *Q-switched laser* in analogy with solid-state lasers. Here one section is biased into a high-gain region, but the short section is biased below threshold for the entire laser. A short pulse driving this section above transparency causes the energy of the entire cavity to be emitted. The increase in pulse energy is not as large as can be obtained in solid-state Q-switched lasers because the spontaneous carrier lifetime in the energy-storage medium is only 0.5 ns.

The jitter in timing of the output of large-signal-modulated semiconductor lasers results from two effects. One is a pattern effect resulting from the difference in carrier density and thus rise time and delay time that occurs after different numbers of ones or zeros. This can be minimized for return-to-zero formats by proper biasing. For any size drive signal and resulting optical output pulse, there is always a bias point where the carrier densities before and after the optical pulse are identical, and pattern effects in the pulse amplitude and delay do not occur. The other source of jitter (and the only source when the drive current is a simple repetitive-pulse stream) is due to buildup of the pulse from spontaneous emission. This case has been analyzed extensively by Bell et al. (1991).

3.4.2 Modulation without prebias

One of the important emerging applications of high-speed laser technology is in the area of short-distance–high-speed data links. These links may be between high-speed switching or computing systems, where data communication between the various processing elements is the bottleneck for achieving high-speed operation. Low-data-rate short links (<400s Mb/s) can be implemented using light-emitting diode and photodetector arrays connected via multimode fibers (Deimel et al., 1985; Brown et al., 1986). Higher-data-rate systems operating at a few gigabytes per second over longer distances need laser transmitter arrays and single-mode fiber channels (Dutta et al., 1992). In these high-speed applications, the laser transmitters are typically designed to be driven directly by buffered electronic

3.4 Large-signal modulation

logic circuits, and low power consumption is essential for the implementation of high-speed parallel data buses that are typically 64 to 128 bits wide in modern-day computers. One proposed solution for such an application is the use of ultra-low-threshold-current lasers and optical switching (from below threshold) without any prebias on the laser diodes. The light output extinguishes once the input electrical signal goes to zero but does not turn on immediately after the application of the next positive going pulse. This leads to partial eye closure, and the turn-on delay becomes an important consideration to ensure the largest eye opening. This also determines the highest data rate that can be employed in such systems.

The expression for the time delay τ_d between application of the electrical pulse and onset of optical emission as first derived by Konnerth and Lanza (1964) in the absence of stimulated emission is given by

$$\tau_d = \tau_n \ln\left(\frac{I}{I - I_{th}}\right) \quad (3.79)$$

where τ_n is the carrier lifetime, which has been assumed to be a constant and independent of carrier density, I_{th} is the laser threshold current, and I is the final bias current level (usually the peak current of the input laser pulse). The turn-on delay can be reduced by having a small prebias I_o on the laser. In this case, the expression for τ_d in Eq. (3.79) modifies to (Dyment et al., 1972)

$$\tau_d = \tau_n \ln\left(\frac{I - I_o}{I - I_{th}}\right) \quad (3.80)$$

From Eq. (3.80) it can be seen that the delay τ_d can be reduced to very small values as I_o approaches I_{th}. Lasers with very low threshold currents would again be very useful here to reduce the prebias level to limit power consumption.

In semiconductor lasers where the bimolecular recombination mechanism is dominant, the radiative recombination rate is usually written as Bn^2 for $n = p$, where B is taken to be a constant. The bimolecular recombination constant B is also carrier-density-dependent, but this is generally not very significant. The carrier-density-dependent component is equivalent to the Auger recombination component, which is written as Cn^3. The Auger recombination coefficient is significant in long-wavelength InP lasers. Considering only the quadratic bimolecular radiative recombination term, the delay for the onset of laser emission with prebias can be written as (Chinone et al., 1974; Agrawal and Dutta, 1986)

$$\tau_d = \tau_n\,(n_{th})\sqrt{\frac{I_{th}}{I}}\left(\tanh^{-1}\sqrt{\frac{I_{th}}{I}} - \tanh^{-1}\sqrt{\frac{I_o}{I}}\right) \quad (3.81)$$

The turn-on delay for switching in the semiconductor laser is essentially determined by the carrier recombination lifetime. The carrier recombination lifetime decreases with increasing carrier density, and the value of τ_n extracted from fitting the experimental data to the first two expressions [Eqs. (3.79) and (3.80)] for τ_d is usually some effective value during the whole switching process. In a more detailed analysis, Dixon and Joyce (1979) showed that the nature of τ_n changes with the level of prebias. For switching with prebias close to the threshold, i.e., $I_o \to I_{th}$, the measured quantity is actually differential lifetime, τ'_n, written as

$$\frac{dR}{dn} = \frac{d}{dn}(Bn^2) = 2Bn = \frac{1}{\tau'_n} \quad (3.82)$$

where R is the recombination rate written here in only terms of the bimolecular recombination.

In a parallel-data-link application, in addition to the delay of the solitary laser diode, one also has to take into account the skew, i.e., the differences in the delay among the laser transmitter array elements. This is often referred to as the *jitter* in the optical signal. Both the delay and the jitter will lead to additional power penalties in the data-transmission system (Shen, 1989).

For small turn-on delays required for high-speed data links, aside from designing for low-threshold-current laser structures, one has to also ensure that the carrier lifetime is not degraded and kept small. In Sect. 3.2, where the dynamic properties of semiconductor lasers were derived, we showed that the carrier lifetime τ_n is increased to $\chi\tau_n$ [where $\chi = 1+(\tau_r/\tau_e)$] by carrier transport. This trend has been observed experimentally (Odagawa et al., 1993; Nobuhara et al., 1993; Fukushima et al., 1993b). They observed that for lasers in the InP system, as the bandgap energy in the SCH was increased, the measured carrier lifetime decreased. This has been related to the favorable increase in the thermionic emission time τ_e into the SCH, which reduces the value of χ as close as possible to unity. Laser turn-on delays as small as 60 ps have been reported at room temperature for InGaAs/InGaAsP tensile-strained multiple quantum well lasers with 1.1-μm bandgap SCH (Nobuhara et al., 1993).

3.4.3 Mode locking

The following subsection describes the principles of mode locking and surveys the different structures that have been demonstrated. In each case, the important parameters are (1) short pulsewidth, (2) minimum chirping for a minimum time-bandwidth product or linear chirping for a large compression ratio, (3) high energy per pulse, (4) minimum phase noise and timing jitter, (5) minimum amplitude noise, and (6) simple integrated structures. The parameters that limit each of these characteristics will be discussed in each of the following subsections.

Principles of mode locking

To achieve short optical pulses from a semiconductor laser, the different longitudinal modes of the laser cavity must have a relatively fixed phase relationship between them such that at periodic intervals all the modes are approximately in phase. This is the origin of the term *mode locking* and is illustrated in Fig. 3.39(a, b). In the ideal case of perfect mode locking, all the modes are exactly in phase, and for the case of Gaussian-shaped pulses, the product of pulse width and spectral width is 0.44.

However, in practice, not all the modes are locked in phase, resulting in "poor" mode locking. Also, because of chirping and amplitude and phase noise, the time bandwidth product is often far above the minima of approximately 0.5. Amplitude noise degrades the pulse quality, however, as shown in Fig. 3.39(c); even with random amplitudes, the pulse shapes are reasonable if all the modes are in phase. However, the conjugate case of random phases with equal amplitudes results in completely unusable output [Fig. 3.39(d)].

A set of definitions is needed to distinguish mode locking from simple noisy output [Fig. 3.39(a) versus Fig. 3.39(d)], and one such set of definitions is

Pulsewidth to repetition period: $\tau/T < 0.25$
Time bandwidth product: $\tau\Delta f < 1$
Timing jitter to pulsewidth: $\tau_j/t < 1$
Power in primary pulse: $\int I\,dt < 2I_p\tau$

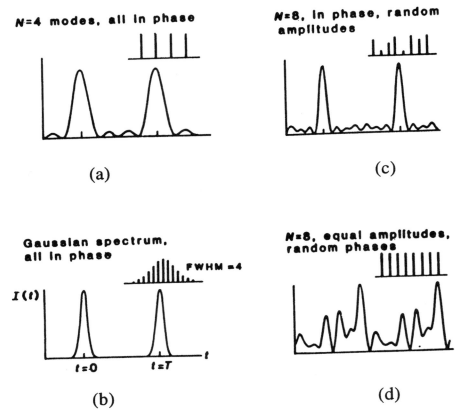

Figure 3.39: Examples of different intensity patterns versus time from several modes with different relative amplitudes and phases: (a) all in phase; (b) Gaussian spectrum, all in phase; (c) all in phase with random amplitudes; (d) equal amplitudes with random phases (after Siegman, 1986). This illustrates how important phase locking is and how much less important amplitude variation is.

where τ is the FWHM of the pulse intensity, T is the repetition period, Δf is the spectral FWHM, τ_j is the rms timing jitter over some appropriate range, I is the pulse intensity, and I_p is the peak intensity. These definitions and the ranges of possible pulse characteristics are shown in Fig. 3.40. The best pulse characteristics fall in the lower left portion of both plots, i.e., minimum pulsewidth, minimum time bandwidth product, and minimum timing jitter. A pulsewidth-to-period ratio of 0.5 corresponds

3.4 Large-signal modulation

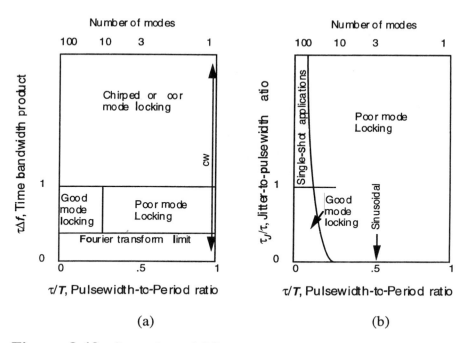

Figure 3.40: Comparison of different regions of mode locking in terms of (a) pulsewidth and (b) timing jitter. In each diagram, the lower left corner represents high-quality mode locking.

to sinusoidal output. Very good pulsed outputs ($0 < \tau/T \ll 0.5$) or very good gaps in the optical output ($0.5 \ll \tau/T < 1$) are the extreme edges of the plot, and the region in the middle corresponds to poor mode locking. In the plot in Fig. 3.40(a), the region with $\tau\Delta f > 1$ may correspond to highly chirped output, which could be compressed if the chirp is linear, or may correspond to many unlocked modes without a stable phase relationship. In the plot in Fig. 3.40(b), the region where the jitter is larger than the pulse width is not generally useful in repetitive applications except as a source for pump/probe measurements. The final definition in the preceding equation examines the issues of pulse shape. One common problem with mode-locked pulses is long tails that may contain more energy than the original pulse. Another common problem is multiple pulses in the output. A third problem is small extinction ratios. A good mode-locked pulse should have extinction ratios of 100 or more, but ratios of 0.1 have been reported as mode locking. All three of these examples would fail the criterion given above.

The maximum pulsewidth is limited by the spectral width of the gain medium. With a gain width of 50 nm, semiconductor lasers should be capable of very short pulsewidths, under 100 fs. As we shall see below, other effects such as self phase modulation are presently a more severe limit than the spectral width of the gain medium.

Mode-locking techniques can be classified as active, passive, or hybrid (Fig. 3.41). Active mode locking employs modulation of the gain to shape the pulse and works extremely well in semiconductor lasers because the gain can be easily modulated with very high extinction ratios at rates

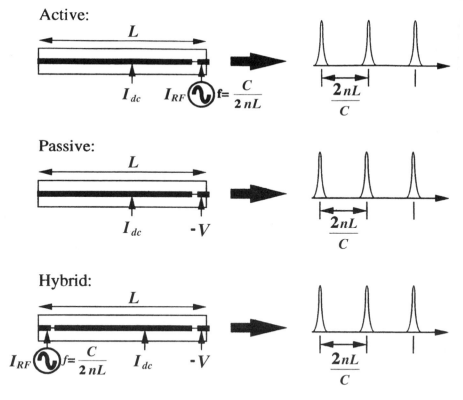

Figure 3.41: Comparison of different mode-locking techniques for a monolithic semiconductor laser: (a) active; (b) passive; and (c) hybrid.

3.4 Large-signal modulation

from kilohertz to tens of gigahertz, as described in preceding sections. Passive mode locking utilizes saturation of an absorber and has the advantage of causing pulse width reduction even for very short pulses. Passive mode locking works extremely well in semiconductors because an electric field can be placed across a *pn* junction waveguide, and the absorption, absorption recovery time, and absorption length can be adjusted easily over a wide range. Figure 3.42 shows the recovery of a reverse-biased gain region. The absorption can recover in under 10 ps, and there are subpicosecond absorption features that can be utilized for mode locking as well. A variation of this technique is to have two pulses in the cavity collide in the saturable absorber (colliding-pulse mode locking) to reduce

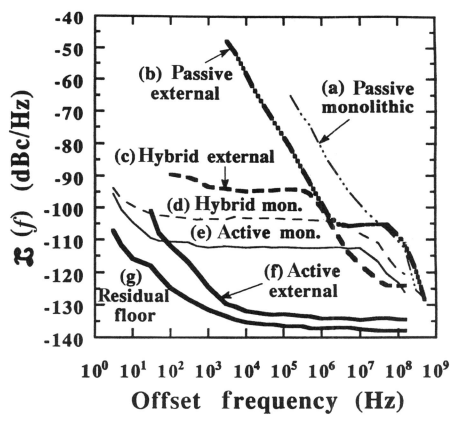

Figure 3.42: Comparison of phase noise for several types of mode locking.

the saturation energy (Chen et al., 1991). Derickson et al. (1992a) has shown, however, that the best results are obtained when a pulse collides with itself off a mirror. The problem with multiple-pulse mode locking is that as the power is increased, the pulses do not stay exactly the same in shape, energy, or timing and the collision is not as effective as when the pulse collides with itself. Another type of mode locking is Kerr lens mode locking, where nonlinear focusing effects are combined with spatial apertures. This technique works well in Ti:sapphire lasers and also has been demonstrated in semiconductor lasers. It should work well with the high-power structures (VCSELs and arrays) described below.

The third mode-locking technique of hybrid mode locking involves combining active and passive mode elements, as shown in Fig. 3.41. A structure like this could be as simple as a multielectrode laser (Morton et al., 1990) or could combine passive waveguides and DBR tuning elements (Hansen et al., 1992). As we shall see below, hybrid mode-locked lasers combine the short, single-pulse capabilities of passive mode-locked lasers with the small timing jitter and stable operation of active mode-locked lasers.

These three classes of structures could be demonstrated in monolithic cavities as in Fig. 3.41 or in external cavities. Thus there are a minimum of six such structures, and their characteristics are compared in Table 3.3. Derickson et al. (1992b) fabricated all the structures from the same wafer with the same process technology, as shown in the mask layout in Fig. 3.41. However, in a few cases, other results are substituted when other wafers produced much better results for a particular structure. It can be seen that there are wide variations in characteristics depending on the structure. The subsequent subsections of this chapter describe these differences and some of the limits to achieving minimum pulsewidth, maximum energy, and minimum timing jitter.

Pulsewidth, energy, and spectral width

The process of short pulse generation in a laser is the result of competition between the mechanisms that tend to shorten the optical pulsewidth on each pass with the mechanisms that tend to widen the pulsewidth. The pulse-shortening mechanisms are saturable absorption and active gain modulation. The pulse-widening mechanisms are the gain response of the lasing medium, saturable gain, self phase modulation due to saturable gain, saturable absorption and gain modulation, dispersion in the cavity,

Cavity Type	Modulation Technique	Pulse Width (pcs)	Spectral Width (GHz)	Time-Bandwidth Product	Pulse Energy (pJ)	Repetition Rate (GHz)	Wavelength (μm)	Active Region	Absolute rms Timing Jitter 150 Hz–50 MHz	Residual rms Timing Jitter 150 Hz–50 MHz	Relative Intensity Noise at 100 MHz
Ext.	Active two-seg.	1.4	342	0.48	0.28	3	1.3	Bulk	240 fs	65 fs	< −126 dB/Hz
Ext.	Passive two-seg.	2.5	720	1.8	0.7	5	0.84	4 QW	12.2 ps (1.5 kHz–50 MHz)	612.2 ps (1.5 kHz–50 MHz)	−103 dB/Hz
Ext.	Hybrid two-seg.	2.5	1000	2.5	0.8	5	0.84	4 QW	1060 fs	980 fs	−105 dB/Hz
Ext.	Hybrid three-seg.	1.9	900	1.71	0.18	6	0.83	Bulk			
Mon.	Active two-seg.	13	330	4.3	0.19	5.5	0.84	4 QW	600 fs	530 fs	−122 dB/Hz
Mon.	Hybrid three-seg.	6.5	540	3.5	0.18	5.5	0.84	4 QW			
Mon.	Passive two-seg.	10	400	4.0	0.25	5.5	0.84	4 QW			
Mon.	Passive two-seg.	5.5	550	3.0	0.53	11	0.84	4 QW	12.5 ps (150 kHz–50 MHz)	12.5 ps (150 kHz–50 MHz)	−116 dB/Hz
Mon.	Hybrid three-seg.	2.2	500	1.1	0.03	21	1.58	4 QW	1200 fs	1130 fs	−109 dB/Hz
Mon.	Passive two-seg.	1.3	600	.78	0.02	41	1.58	4 QW			
Mon.	Q-switch two-seg.	15	2400	36	4	1	0.825	Bulk			
Mon.	Gain-switch two-seg.	13	4000	52	3.4	1	0.822	Bulk			

Table 3.3: Comparison of mode-locking characteristics from different types of mode-locked lasers (Derickson et al., 1992b).

other wavelength-limiting elements such as antireflection coatings, birefringent tuning filters, and spontaneous emission noise.

For low pulse energies, the semiconductor medium is a linear amplifier and does not significantly change the pulse shape or spectrum. However, once the pulse energy becomes comparable with the saturation energy of the material, then significant carrier changes occur during the pulse. This gain saturation causes pulse shape changes because the gain is lower during the later portion of the pulse. It causes spectral changes because the index is significantly changing during the pulse. The saturation energy of a semiconductor medium is given by

$$E_s = \frac{h\nu\sigma}{\Gamma g} \qquad (3.83)$$

where σ/Γ is the effective cross-section area of the optical mode.

Using typical values for InGaAsP, the saturation energy for a 1-μm-wide bulk gain region is around 1 to 2 pJ. The shape and spectral changes that occur in optical amplifiers have been well described using the formalism developed for picosecond pulses by Agrawal and Olsson (1989) and extended for femtosecond pulses by Hong et al. (1994). Derickson et al. (1992b) has extended this analysis to mode-locked lasers.

This limit on achievable pulse energies is illustrated in Table 3.3, which shows the pulse energies that have been achieved by different mode-locking groups. To increase the pulse energy, we must increase the saturation energy of the gain and absorber regions. One way to increase the saturation energy is to lower the confinement factor with the use of quantum well regions. Another way is to increase the lateral width of the semiconductor region. Because of filamenting and other nonlinear processes, it is usually necessary to control the lateral beam profile. One way to do this is to use coupled arrays of lasers. One particularly successful way is to use resonant optical waveguide (ROW) lasers, where the array elements are locked in phase. With hybrid mode locking, Mar et al. (1993) achieved pulse energies of 30 pJ with a 6-ps pulsewidth, and small timing jitter has been generated. A final approach has been to amplify the mode-locked pulses with a semiconductor amplifier (Delfyett et al., 1990). The ultimate approach that minimizes spurious spontaneous emission but has the largest saturation energy is to match the saturation energy of the amplifier to the energy of the pulse at each point during amplification using a tapered amplifier, as reported by Mar et al. (1995).

Timing jitter

An important characteristic of any repetitive pulsed output is the pulse-to-pulse jitter. It is rarely useful to have pulsewidths smaller than the timing jitter. As shown earlier in Table 3.3, a wide range of timing jitter has been measured for different structures, and quite often the timing jitter of passively mode-locked lasers is substantially larger than the pulsewidth. The timing jitter is related to the phase noise of the photodetected output by the relation

$$\sigma_{rms} = \frac{1}{2\pi n f_{mod}} \sqrt{2\int_{f_{low}}^{f_{high}} \mathscr{L}(f)\,df} \qquad (3.84)$$

where σ_{rms} is the root mean square timing jitter, n is the harmonic number at which the measurement data is taken, f_{mod} is the fundamental repetition rate frequency, f_{low} is the lower offset frequency from the repetition rate harmonic carrier frequency, f_{high} is the upper offset frequency from the carrier and $\mathscr{L}(f)$ is the single sideband phase noise relative to the carrier and normalized to a 1 Hz bandwidth.

The phase noise can be measured by connecting the photodetected output of the pulsed laser into a spectrum analyzer. This is the absolute phase noise of the laser output. In most cases, though, the important parameter is the relative phase noise of the laser output relative to a microwave oscillator. This is the *residual phase noise*. This can be measured by mixing the photodetector output with the master oscillator and taking the downconverted output in a spectrum analyzer. Figure 3.43 shows the measured phase noise for several types of mode locking.

The jitter is driven by spontaneous emission noise in the laser. A spontaneous emission event causes timing jitter through three processes: (1) index of refraction change causing direct phase modulation, (2) carrier density noise causing gain fluctuations and pulse-position fluctuations, and (3) photon number fluctuations converting to phase fluctuations.

The timing jitter can be reduced by comparing the output to a source oscillator and feeding the error signal (with proper sign) to the gain or absorber segments. The advantage of semiconductors for mode locking is that their high-speed modulation capability can be used to advantage without adding additional elements in the cavity. Helkey and Bowers (1994) have reduced the timing jitter in passively mode-locked lasers by two orders of magnitude in this way.

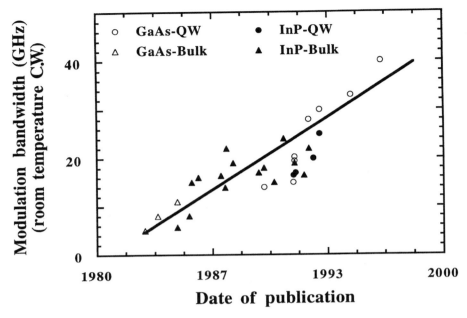

Figure 3.43: The trend in the C.W. −3-dB modulation bandwidth at room temperature reported in the GaAs and the InP systems for the past 10 years. The modulation bandwidth has roughly doubled once every 4 years for the 10 years.

3.5 Conclusions and outlook

The bandwidth requirements in a modern communications system are on the increase. As a result, the quest for devices operating at ever-increasing speeds has been in the forefront of semiconductor laser research for a number of years now and will continue to be so for years to come. Figure 3.43 shows the −3-dB room-temperature CW small-signal modulation bandwidths that have been reported in semiconductor lasers as a function of the publication date for the past decade. There has been a steady increase in the modulation bandwidths from less than 5 GHz in 1982 to 40 GHz in present-day devices. There are also a number of other trends that also can seen from this plot. The early high-speed devices had bulk GaAs active areas, and impressive progress was made in the technologically more important InP material system. Gradually, the strained and

3.5. Conclusions and outlook

unstrained quantum well lasers began to dominate in both these material systems.

A number of issues were discussed with respect to designing and fabricating lasers with large modulation bandwidths. There is an obvious need for good-quality optical material with low internal loss to fabricate lasers with low threshold currents and large output powers. This ensures that the threshold carrier density is low (resulting in a large differential gain) and that the devices can be biased easily at many times the threshold current without the detrimental thermal effects. A large differential gain is always the fundamental factor at the heart of any high-speed laser design. A combination of completely doping or modulation doping the active region with acceptors, wavelength detuning, and the use of quantum wells or even lower-dimensional active areas can address this problem. Adequate carrier confinement is essential to suppress carrier overflow under bandfilling in lower-dimensional active regions. Unconfined carriers do not contribute to gain and lead to degradation of the static differential gain. Another major issue is carrier transport in the optical and carrier confinement regions of the quantum well lasers. This not only causes a parasitic-like roll-off in the amplitude-modulation response but also degrades the dynamic differential gain in high-speed laser devices.

There is also a real need for novel ideas to beat the limits to the speed of operation set by device parasitics as well as device size. The biggest problem with present-day high-speed laser diodes is the device and package parasitics. Typically, this limits the frequency of operation of these diodes to the 20- to 30-GHz region. New fabrication techniques and device structures are required to beat this limit. Good microwave packaging is as much an art as it is a science, and more work is needed in this field to make it more of a science. The most successful present-day laser diodes have cleaved facets for a long operating lifetime and a large power output at low operating currents. There is a cavity-length limit to which these devices may be usefully cleaved. Good cleaves require the cavity length to be larger than the substrate thickness. Substrate thickness of the order of 50 μm or so becomes difficult to handle, and this places the lower limit of 100 μm or so on the length of cleaved cavities. This will be a problem when the wavelength of the microwave signal becomes the same order of magnitude as the cavity length at very high frequencies.

Acknowledgments

Our high-speed laser effort at the University of California, Santa Barbara, has benefited much from collaboration with Prof. Larry Coldren, Dr. Dennis Derickson, Dr. Toru Fukushima, Dr. Randall Geels, Dr. Roger Helkey, Masayuki Ishikawa, Dr. Takuya Ishikawa, Dr. Alan Mar, Dr. Richard Mirin, Thomas Reynolds, Dr. Daniel Tauber, Dr. Gary Wang, and Dr. Ralph Spickerman.

The financial support for our work presented here was provided by DARPA, the Rome Laboratories (Hanscom AFB), and the Office of Naval Technology block program in Electro-Optics Technology.

References

Adams, A. R. (1986). *Electron. Lett.* **22**, 249.

Agrawal, G. P. (1987). *IEEE J. Quantum Electron.* **23**, 860.

Agrawal, G. P. (1988). *J. Appl. Phys.* **63**, 1232.

Agrawal, G. P., and Dutta, N. K. (1986). *Long Wavelength Semiconductor Lasers*, Chap. 6, Van Nostrand Reinhold, New York.

Agrawal, G. P., and Olsson, N. A. (1989) *J. Quantum Electron.*, **25**(11), 2297

Aoki, M., Uomi, K., Tsuchiya, T., Suzuki, M., and Chinone, N. (1990). *Electron. Lett.* **26**, 1841.

Arakawa, Y., and Yariv, A. (1986). *IEEE J. Quantum Electron.* **22**, 1887.

Asada, M. (1989). *IEEE J. Quantum Electron.* **25**, 2019.

Asada, M., and Suematsu, Y. (1985). *IEEE J. Quantum Electron.* **21**, 434.

Asada, M., Kameyama, A., and Suematsu, Y. (1984). *IEEE J. Quantum Electron.* **20**, 745.

Asada, M., Miyamoto, Y., and Suematsu, Y. (1985). *Jpn. J. Appl. Phys.* **24**, L95.

Asada, M., Miyamoto, Y., and Suematsu, Y. (1986). *IEEE J. Quantum Electron.* **22**, 1915.

Babiker, M., and Ridley, B. K. (1986). *Superlat. Microstruct.* **2**, 287.

Bastard, G. (1988). *Wave Mechanics Applied to Semiconductor Heterostructures*, p. 14, John Wiley, New York.

Bell, J. A., Hamilton, D. A., Leep, D. A., Taylor, H. F., and Lee, Y.-H. (1991). In *Optical Technology for Microwave Applications V* (S.-K. Yao, ed.), Proc. SPIE 1476, p. 326.

Bernard, M. G. A., and Duraffourg, G. (1961). *Physica Status Solidi* **1**, 699.

Blom, P. W. M., Mols, R. F., Haverkort, J. E. M., Ley, M. R., and Wolter, J. H. (1990) *Superlattice Microstructure* **7**, 319.

Blom, P. W. M., Haverkort, J. E. M., and Wolter, J. H. (1991). *Appl. Phys. Lett.* **58**, 2767.

Blom, P. W. M., Haverkort, J. E. M., van Hall, P. J., and Wolter, J. H. (1993). *Appl. Phys. Lett.* **62**, 1490.

Blood, P., Fletcher, E. D., Woodbridge, K., Heasman, K. C., and Adams, A. R. (1989). *IEEE J. Quantum Electron.* **25**, 1459.

Bowers, J. E. (1985). *Electron. Lett.* **21**, 1195.

Bowers, J. E. (1987). *Solid State Electron.* **20**, 1.

Bowers, J. E. (1989). In *Optoelectronic Technology and Lightwave Communications Systems* (C. Lin, ed.), pp. 299–334, Van Nostrand Reinhold, New York.

Bowers, J. E., and Pollack, M. A. (1988). In *Optical Fiber Communications II* (S. E. Miller and I. P. Kaminow, eds.), pp. 509–568, Academic Press, Boston.

Bowers, J. E., Koch, T. L., Hemenway, B. R., Wilt, D. P., Bridges, T. J., and Burkhardt, E. G. (1985). *Electron. Lett.* **21**, 392.

Bowers, J. E., Hemenway, B. R., Gnauck, A. H., and Wilt, D. P., (1986). *IEEE J. Quantum Electron.* **22**, 833.

Bowers, J. E., Koren, U., Miller, B. I., Soccolich, C., and Jan, W. Y. (1987). *Electron. Lett.* **23**, 1263.

Brown, M. G., Hu, P. H. -S., Kaplan, D. R., Ota, Y., Seabury, C. W., Washington, M. A., Becker, E. E., Johnson, J. G., Koza, M., and Potopowicz, J. R. (1986). *IEEE J. Lightwave Technol.* **4**, 283.

Brum, J. A., and Bastard, G. (1986). *Phys. Rev.* **B33**, 1420.

Brum, J. A., Weil, T., Nagle, J., and Vinter, B. (1986). *Phys. Rev.* **B34**, 2381.

Casey, H. C., Jr., and Panish, M. B. (1978). *Heterostructure Lasers,* Part A: *Fundamental Principles,* Chap. 3, Academic Press, New York.

Channin, D. J. (1979). *J. Appl. Phys.* **50**, 3858.

Chen, T. R., Chen. P. C., Gee, C., and Bar-Chiam, N. (1993). *IEEE Photon. Technol. Lett.* **5**, 1.

Chen, Y. K., Wu, M. C., Tanbun-Ek, T., Logan, R. A., and Chin, M. A. (1991). *Appl. Phys. Lett.* **58**, 1253

Cheng, W. H., Buehring, K. D, Appelbaum, A., Renner, D., Chin, S., Su, C. B., Mar, A., and Bowers, J. E. (1991). *IEEE J. Quantum Electron.* **27**, 1642.

Chinn, S. R., Zory, P. S., and Reisinger, A. R. (1988). *IEEE J. Quantum Electron.* **24**, 2191.

Chinone, N., Ito, R., and Nakada, O. (1974). *IEEE J. Quantum Electron.* **10**, 81.

Corzine, S. W., and Coldren, L. A. (1991). *Appl. Phys. Lett.* **59**, 588.

Corzine, S. W., Yan, R. H., and Coldren, L. A. (1990). *Appl. Phys. Lett.* **57**, 2835.

Deimel, P. P., Cheng, J., Forrest, S. R., Hu, P. H. -S., Huntington, R. B., Miller, R. C., Potopowicz, J. R., Roccasecca, D. D., and Seabury, C. W. (1985). *IEEE J. Lightwave Technol.* **3**, 988.

Delfyett, P. J., Lee, C.-H., Alphonse, G. A., and Connolly, J. C., (1990). *Appl. Phys. Lett.* **57**, 971.

Derickson, D. J., Wasserbauer, J. G., Helkey, R. J., Mar, A., Bowers, J. E., Coblentz, D., Logan, R., and Tanbun-Ek, T., (1992a). Optical Fiber Communication Conference (OFC), paper ThB3, San Jose, CA.

Derickson, D. J., Helkey, R. J., Mar, A., Karin, J. R., Wasserbauer, J. G. and Bowers, J. E. (1992b). *J. Quantum Electron.* **28**(10), 2186.

Deveaud, B., Shah, J., Damen, T. C., Lambert, B., and Regreny, A. (1987). *Phys. Rev Lett.* **58**, 2582.

Deveaud, B., Shah, J., Damen, T. C., and Tsang, W. T. (1988). *Appl. Phys. Lett.* **52**, 1886.

Dixon, R. W., and Joyce, W. B. (1979). *J. Appl. Phys.* **50**, 4591.

Downey, P. M., Bowers, J. E., Tucker, R. S., and Agyekum, E. (1987) *J. Quantum Electron.*, **23**(6), 1039

Dutta, N. K., Olsson, N. A., and Tsang, W. T. (1984). *Appl. Phys. Lett.* **45**, 836.

Dutta, N. K., Chand, N., and Lopata, J. (1992). *Appl. Phys. Lett.* **61**, 7.

Dutta, N. K., Wang, S. J., Wynn, J. D., Lopata, J., and Logan, R. A. (1992). *Appl. Phys. Lett.* **61**, 130.

Dyment, J. C., Ripper, J. E., and Lee, T. P. (1972). *J. Appl. Phys.* **43**, 452.

Eisenstein, G., Wiesenfeld, J. M., Wegener, M., Sucha, G., Chemla, D. S., Weiss, S., Raybon, G., and Koren, U. (1991). *Appl. Phys. Lett.* **58**, 158.

Eom, J., Su, C. B., Rideout, W., Lauer, R. B., and LaCourse, J. S. (1991). *Appl. Phys. Lett.* **58**, 234.

Frankenberger, R., and Schimpe, R. (1990). *Appl. Phys. Lett.* **57**, 2520.

Frankenberger, R., and Schimpe, R. (1992). *Appl. Phys. Lett.* **60**, 2720.

Fukushima, T., Bowers, J. E., Logan, R. A., Tanbun-Ek, T., and Temkin, H. (1991a). *Appl. Phys. Lett.* **58**, 1244.

Fukushima, T., Nagarajan, R., Bowers, J. E., Logan, R. A., and Tanbun-Ek, T. (1991b). *Photon. Technol. Lett.* **3**, 691.

Fukushima, T., Nagarajan, R., Ishikawa, M., and Bowers, J. E. (1993a). *Jpn. J. Appl. Phys.* **32**(1), 89.

Fukushima, T., Kasukawa, A., Iwase, M., Namegaya, T., and Shibata, M. (1993b). *IEEE J. Quantum Electron.* **29**, 1536.

Furuya, A., Makiuchi, M., Wada, O., and Fujii, T. (1988). *IEEE J. Quantum Electron.* **24**, 2448.

Ghiti, A., and O'Reilly, E. P. (1990). *Electron. Lett.* **26**, 1978.

Gomatam, B. N., and DeFonzo, A. P. (1990). *IEEE J. Quantum Electron.* **26**, 1689.

Grabmaier, A., Hangleiter, A., Fuchs, G., Whiteaway, J. E. A., and Glew, R. W. (1991). *Appl. Phys. Lett.* **59**, 3024.

Grabmaier, A., Schöfthaler, M., Hangleiter, A., Kazmierski, C., Blez, M., and Ougazzaden, A. (1993). *Appl. Phys. Lett.* **62**, 52.

Hall, K. L., Lenz, G., Ippen, E. P., Koren, U., and Raybon, G. (1992). *Appl. Phys. Lett.* **61**, 2512.

Hansen, P. B., Raybon, G., Koren, U., Miller, B. I., Young, M. G., Chien, M., Burrus, C. A., and Alferness, R. C. (1992). *IEEE Photon. Technol. Lett.* **4**, 215

Harder, Ch., Katz, J., Margalit, S., Shacham, J., and Yariv, A. (1982). *IEEE J. Quantum Electron.* **18**, 333.

Hasegawa, H., Furukawa, M., and Yanai, H. (1971). *IEEE Trans. Microwave Theory Technol.* **19**, 869.

Helkey, R., and Bowers, J. E. (1994). In *Semiconductor Lasers: Past, Present and Future* (G. Agrawal, ed.), American Institute of Physics Press, New York.

Henry, C. H. (1982). *IEEE J. Quantum Electron.* **18**, 259.

Hermann, C., and Weisbuch, C. (1984). In *Optical Orientation* (F. Meier and B. P. Zakharchenya, eds.), Chap. 11, North-Holland, Amsterdam.

Hirayama, Y., Morinaga, M., Suzuki, N., and Nakamura, M. (1991). *Electron. Lett.* **27**, 875.

Hirayama, H., Yoshida, J., Miyake, Y., and Asada, M. (1992a). *Appl. Phys. Lett.* **61**, 2398.

Hirayama, Y., Morinaga, M., Onomura, M., Tanimura, M., Tohyama, M., Funemizu, M., Kushibe, M., Suzuki, N., and Nakamura, M. (1992b). *IEEE J. Lightwave Technol.* **10**, 1272.

Hong, M. Y., Chang, Y. H., Dienes, A., Heritage, J. P., and Delfyett, P. J., (1994). *IEEE J. Quantum Electron.* **30**, 1122.

Howard, N. R., and Johnson, G. W. (1965). *Solid State Electron.* **8**, 275.

Huang, R. T., Wolf, D., Cheng, W. H., Jiang, C. L., Agarwal, R., Renner, D., Mar, A., and Bowers, J. E. (1992). *IEEE Photon. Technol. Lett.* **4**, 293.

Ishikawa, H., Soda, H., Wakao, K., Kihara, K., Kamite, K., Kotaki, Y., Matsuda, M., Sudo, H., Yamakoshi, S., Isozumi, S, and Imai, H. (1987). *IEEE J. Lightwave Technol.* **5**, 848.

Ishikawa, M., Fukushima, T., Nagarajan, R., and Bowers, J. E. (1992a). *Appl. Phys. Lett.* **61**, 396.

Ishikawa, M., Nagarajan, R., Fukushima, T., Wasserbauer, J., and Bowers, J. E. (1992b). *IEEE J. Quantum Electron.* **28**, 2230.

Ishikawa, T., Nagarajan, R., and Bowers, J. E. (1994). *Optical and Quantum Electron.* **26**, S805.

Jones, E. D., Lyo, S. K., Fritz, I. J., Klem, J. F., Schirber, J. E., Tigges, C. P., and Drummond, T. J. (1989). *Appl. Phys. Lett.* **54**, 2227.

Kamite, K., Sudo, H., Yano, M., Ishikawa, H., and Imai, H. (1987). *IEEE J. Quantum Electron.* **23**, 1054.

Kan, S. C., Vassilovski, D., Wu, T. C., and Lau, K. Y. (1992). *Appl. Phys. Lett.* **61**, 752.

Kane, E. O. (1957). *J. Phys. Chem. Solids* **1**, 249.

Kano, F., Yamanaka, T., Yamamoto, N., Yoshikuni, Y., Mawatari, H., Tohmori, Y., Yamamoto, M., and Yokoyama, K. (1993). *IEEE J. Quantum Electron.* **29**, 1553.

Kazarinov, R. F., Henry, C. H., and Logan, R. A. (1982). *J. Appl. Phys.* **53**, 4631.

Kersting, R., Schwedler, R., Wolter, K., Leo, K., and Kurz, H. (1992). *Phys. Rev.* **B46**, 1639.

Kesler, M. P., and Ippen, E. P. (1987). *Appl. Phys. Lett.* **51**, 1765.

Kikuchi, K., and Okoshi, T. (1985). *IEEE J. Quantum Electron.* **21**, 1814.

Kitamura, M. (1989). In *Digest of the 7th International Conference on Integrated Optics and Optical Fiber Communication*, paper 19C1-5.

Kitamura, M., Yamazaki, H., Yamada, H., Takano, S., Kosuge, K., Sugiyama, Y., Yamaguchi, M., and Mito, I. (1993). *IEEE J. Quantum Electron.* **29**, 1728.

Kinoshita, J., and Matsumoto, K. (1988). *IEEE J. Quantum Electron.* **24**, 2160.

Koch, T. L., and Bowers, J. E. (1984). *Electron. Lett.* **20**, 1038.

Koch, T. L., and Linke, R. A. (1986). *Appl. Phys. Lett.* **48**, 613.

Konnerth, K., and Lanza, C. (1964). *Appl. Phys. Lett.* **4**, 120.

Kroemer, H., and Okamoto, H. (1984). *Jpn. J. Appl. Phys.* **23**, 970.

Kucharska, A. I., and Robbins, D. J. (1990). *IEEE J. Quantum Electron.* **26**, 443.

Landsberg, P. T., and Robbins, D. J. (1985). *Solid State Electron.* **28**, 137.

Lau, K. Y., and Yariv, A. (1985a). *IEEE J. Quantum Electron.* **21**, 121.

Lau, K. Y., and Yariv, A. (1985b). In *Semiconductors and Semimetals: Lightwave Communications Technology* (R. K. Willardson and A. C. Beer, eds.), Vol. 22B (W. T. Tsang, vol. ed.), pp. 69–152, Academic Press, Orlando.

Lealman, I. F, Cooper, D. M., Perrin, S. D., and Harlow, M. J. (1992). *Electron. Lett.* **28**, 1032.

Leo, K., Shah, J., Gordon, J. P., Damen, T. C., Miller, D. A. B., Tu, C. W., and Cunningham, J. E. (1990). *Phys. Rev.* **B42**, 7065.

Lester, L. F., Offsey, S. D., Ridley, B. K., Schaff, W. J., Foreman, B. A., and Eastman, L. F. (1991b). *Appl. Phys. Lett.* **59**, 1162.

Lester, L. F., Schaff, W. J., Offsey, S. D., and Eastman, L. F. (1991a). *IEEE Photon. Technol. Lett.* **3**, 403.

Liou, J. -C., and Lau, K. M. (1993). *IEEE Trans. Microwave Theory Technol.* **41**, 824.

Lipsanen, H., Coblentz, D. L., Logan, R. A., Yadvish, R. D., Morton, P. A., and Temkin, H., (1992). *IEEE Photon. Technol. Lett.* **4**, 673.

Lu, H., McGarry, S., Li, G. P., and Makino, T. (1993). *Electron. Lett.* **29**, 1369.

Manning, J., Olshansky, R., and Su, C. B. (1983). *J. Quantum Electron.* **19**, 1525.

Mar, A., Morton, P. A., and Bowers, J. E. (1990). *Electron. Lett.* **26**, 1382.

Mar, A., Helkey, R. J., Reynolds, T., Bowers, J. E., Zmundzinski, C., Mawst, L., and Botez, D. (1993). *IEEE Photon. Technol. Lett.,* **5**, 1357.

Mar, A., Helkey, R., Zou, W. X., Young, D. B., and Bowers, J. E. (1995). *Appl. Phys. Lett.* **66**, 3559.

Mark, J., and Mørk, J. (1992). *Appl. Phys. Lett.* **61**, 2281.

Masu, K., Tokumitsu, E., Konagai, M., and Takahashi, K. (1983). *J. Appl. Phys.* **54**, 5785.

Miller, B. I., Koren, U., Young, M. G., and Chien M. D. (1991). *Appl. Phys. Lett.* **58**, 1952.

Morin, S., Deveaud, B., Clerot, F., Fujiwara, K., and Mitsunaga, K. (1991). *IEEE J. Quantum Electron.* **27**, 1669.

Morton, P. A. and Bowers, J. E. (1990). *Appl. Phys. Lett.,* **56**(2), 111.

Morton, P. A., Logan, R. A., Tanbun-Ek, T., Sciortino, P. F., Jr., Sergent, A. M., Montgomery, R. K., and Lee, B. T. (1992). *Electron. Lett.* **28**, 2156.

Morton, P. A., Tanbun-Ek, T., Logan, R. A., Sciortino, P. F., Jr., Sergent, A. M., and Wecht, K. W. (1993). *Electron. Lett.* **29** 1429.

Morton P. A., Tanbun-Ek, T., Logan, R. A., Chand, N., Wecht, K. W., Sergent, A. M., and Sciortino, P. F., Jr. (1994). *Electron. Lett.* **30**, 2044.

Nagarajan, R. (1994). *Optical and Quantum Electron.* **26** (7). This a special issue of the journal devoted exclusively to carrier transport effects in quantum well lasers.

Nagarajan, R., and Bowers, J. E. (1993). *IEEE J. Quantum Electron.* **29**, 1601.

Nagarajan, R., Kamiya, T., and Kurobe, A. (1989). *IEEE J. Quantum Electron.* **25**, 1161.

Nagarajan, R., Fukushima, T., Bowers, J. E., Geels, R. S., and Coldren, L. A. (1991a). *Appl. Phys. Lett.* **58**, 2326.

Nagarajan, R., Fukushima, T., Bowers, J. E., Geels, R. S., and Coldren, L. A. (1991b). *Electron. Lett.* **27**, 1058.

Nagarajan, R., Fukushima, T., Corzine, S. W., and Bowers, J. E. (1991c). *Appl. Phys. Lett.* **59**, 1835.

Nagarajan, R., Fukushima, T., Ishikawa, M., Bowers, J. E., Geels, R. S., and Coldren, L. A. (1992a). *IEEE Photon. Technol. Lett.* **4**, 121.

Nagarajan, R., Ishikawa, M., and Bowers, J. E. (1992b). *Electron. Lett.* **28**, 846.

Nagarajan, R., Mirin, R. P., Reynolds, T. E., and Bowers, J. E. (1992c). *IEEE Photon. Technol. Lett.* **4**, 832.

Nagarajan, R., Ishikawa, M., Fukushima, T., Geels, R. S., and Bowers, J. E. (1992d). *IEEE J. Quantum Electron.* **28**, 1990.

Nagarajan, R., Mirin, R. P., Reynolds, T. E., and Bowers, J. E. (1993). *Electron. Lett.* **29**, 1688.

Nagle, J., Hersee, S., Krakowski, M., Weil, T., and Weisbuch, C. (1986). *Appl. Phys. Lett.* **49**, 1325.

Nishimoto, H., Yamaguchi, M., Mito, I., and Kobayashi, K. (1987). *IEEE J. Lightwave Technol.* **5**, 1399.

Nishimura, Y., and Nishimura, Y. (1973). *IEEE J. Quantum Electron.* **9**, 1011.

Nobuhara, H., Nakajima, K., Tanaka, K., Odagawa, T., Fujii, T., and Wakao, K. (1993). *Electron. Lett.* **29**, 138.

Odagawa, T., Nakajima, K., Tanaka, K., Inoue, T., Okazaki, N., and Wakao, K. (1993). *IEEE J. Quantum Electron.* **29**, 1682.

Offsey, S. D., Schaff, W. J., Tasker, P. J., and Eastman, L. F. (1990). *IEEE Photon. Technol. Lett.* **2**, 9.

Ohtoshi, T., and Yamanishi, M. (1991). *IEEE J. Quantum Electron.* **27**, 46.

Okai, M., Tsuji, S., and Chinone, N. (1989). *IEEE J. Quantum Electron.* **25**, 1314.

Okoshi, T., and Kikuchi, K. (1988). *Coherent Optical Fiber Communications*, p. 67, KTK Scientific, Tokyo.

Olshansky, R., Hill, P., Lanzisera, V., and Powazinik, W. (1987). *IEEE J. Quantum Electron.* **23**, 1410.

Polland, H. -J., Leo, K., Rother, K., Ploog, K., Feldman, J., Peter, G., and Göbel, E. O. (1988). *Phys. Rev.* **B38**, 7635.

Preisel, M., Mork, J., and Haug, H. (1994). *Phys. Rev.* **B49**, 14478.

Ralston, J. D., Weisser, S., Esquivias, I., Larkins, E. C., Rosenzweig, J., Tasker, P. J., and Fleissner, J. (1993). *IEEE J. Quantum Electron.* **29**, 1648.

Ralston J. D., Weisser S., Eisele K., Sah R. E., Larkins, E. C., Rosenzweig, J., Fleissner, J., and Bender, K. (1994). *IEEE Photon. Technol. Lett.* **6**, 1076.

Ramo, S., Whinnery, J., and Van Duzer, T. (1984). *Fields and Waves in Communication Electronics*, 2d ed., p. 152, John Wiley, New York.

Rideout, W., Yu, B., LaCourse, J., York, P. K., Beernink, K. J., and Coleman, J. J. (1990). *Appl. Phys. Lett.* **56**, 706.

Rideout, W., Sharfin, W. F., Koteles, E. S., Vassell, M. O., and Elman, B. (1991). *IEEE Photon. Technol. Lett.* **3**, 784.

Ridley, B. K. (1990). *J. Appl. Phys.*, **68**, 4667.

Saxena, A. K. (1981). *Phys. Rev.* **B24**, 3295.

Schneider, H., and Klitzing, K. v. (1988). *Phys. Rev.* **B38**, 6160.

Seltzer, C. P., Perrin, S. D., Tatham, M. C., and Cooper, D. M. (1991). *Electron. Lett.* **27**, 1269.

Shank, S. M., Varriano, J. A., and Wicks, G. W. (1992). *Appl. Phys. Lett.* **61**, 2851.

Sharfin, W. F., Schlafer, J., Rideout, W., Elman, B., Lauer, R. B., LaCourse, J., and Crawford, F. D. (1991). *IEEE Photon. Technol. Lett.* **3**, 193.

Shen, T.-M. (1989). *IEEE J. Lightwave Technol.* **7**, 1394.

Shen, T. C., Gao, G. B., and Morkoç, H. (1992). *J. Vacuum Sci. Technol.* **B10**, 2113.

Shichijo, M., Kolbas, R. M., Holonyak, N., Jr., Dupuis, R. D., and Dapkus, P. D. (1978). *Solid State Commun.* **27**, 1029.

Shimizu, J., Yamada, H., Murata, S., Tomita, A., Kitamura, M., and Suzuki, A. (1991). *Photon. Technol. Lett.* **3**, 773.

Shimoyama, K., Katoh, M., Suzuki, Y., Satoh, T., Inoue, Y., Nagao, S., and Gotoh, H. (1988). *Jpn. J. Appl. Phys.* **27**, L2417.

Soda, H., Kotaki, Y., Sudo, H., Ishikawa, H., Yamakoshi, S., and Imai, H. (1987). *IEEE J. Quantum Electron.* **23**, 804.

Streifer, W., Burnham, R. D., and Scifres, D. R. (1983). *Optics Lett.* **8**, 283.

Su, C. B. (1988). *Electron. Lett.* **24**, 371.

Su, C. B., and Lanzisera, V. (1984). *Appl. Phys. Lett.* **45**, 1302.

Su, C. B., and Lanzisera, V. (1986). *IEEE J. Quantum Electron.* **22**, 1568.

Suemune, I. (1991a). *J. Quantum Electron.* **27**, 1149.

Suemune, I. (1991b). *Phys. Rev.* **B43**, 14099.

Suemune, I., Coldren, L. A., Yamanishi, M., and Kan, Y. (1988). *Appl. Phys. Lett.* **53**, 1378.

Sugano, M., Sudo, H., Soda, H., Kusunoki, T., and Ishikawa H. (1990). *Electron. Lett.* **26**, 95.

Sze, S. M. (1981a). *Physics of Semiconductor Devices,* 2d ed., p. 159, John Wiley, New York.

Sze, S. M. (1981b). *Physics of Semiconductor Devices,* 2d ed., p. 255. John Wiley, New York.

Tauber, D. A., Spickerman, R., Nagarajan, R., Reynolds, T., Holmes, A. L., Jr., and Bowers, J. E. (1994). *Appl. Phys. Lett.* **64**, 1610.

Takahashi, T., and Arakawa, Y. (1989). *Electron. Lett.* **25**, 169.

Takahashi, T., and Arakawa, Y. (1991). *IEEE J. Quantum Electron.* **27**, 1824.

Tang, J. Y., Hess, K., Holonyak, N., Jr., Coleman, J. J., and Dapkus, P. D. (1982). *J. Appl. Phys.,* **53**, 6043.

Tessler, N., and Eisenstein, G. (1993a). *Appl. Phys. Lett.* **62**, 10.

Tessler, N., and Eisenstein, G. (1993b). *IEEE J. Quantum Electron.* **29**, 1586.

Tessler, N., and Eisenstein, G. (1994). *Optical and Quantum Electron.* **26**, S767.

Tessler, N., Nagar, R., Abraham, D., Eisenstein, G., Koren, U., and Raybon, G. (1992a). *Appl. Phys. Lett.* **60**, 665.

Tessler, N., Nagar, R., and Eisenstein, G. (1992b). *IEEE J. Quantum Electron.* **28**, 2242.

Tomita, A., and Suzuki, A. (1991). *IEEE J. Quantum Electron.* **27**, 1630.

Tsang, D. Z., and Liau, Z. L. (1987). *IEEE J. Lightwave Technol.* **5**, 300.

Tucker, R. S. (1985). *IEEE J. Lightwave Technol.* **3**, 1180.

Tucker, R. S., Wiesenfeld, J. M., Downey, P. M., and Bowers, J. E. (1986). *Appl. Phys. Lett.* **48**, 1707.

Uomi, K., and Chinone, N. (1989). *Jpn. J. Appl. Phys.* **28**, L1424.

Uomi, K., Nakano, H., and Chinone N. (1989). *Electron. Lett.* **25**, 668.

Uomi, K., Mishima, T., and Chinone N. (1990). *Jpn. J. Appl. Phys.* **29**, 88.

Uomi, K., Tsuchiya, T., Aoki, M., and Chinone, N. (1991). *Appl. Phys. Lett.* **58**, 675.

Vahala, K. J., and Zah, C. E. (1988). *Appl. Phys. Lett.* **52**, 1945.

Wang, G., Nagarajan, R., Tauber, D., and Bowers, J. E. (1993). *IEEE Photon. Technol. Lett.* **5**.

Weiss, S., Wiesenfeld, J. M., Chemla, D. S., Raybon, G., Sucha, G., Wegener, M., Eisenstein, G., Burrus, C. A., Dentai, A. G., Koren, U., Miller, B. I., Temkin, H., Logan, R. A., and Tanbun-Ek, T. (1992). *Appl. Phys. Lett.* **60**, 9.

Weisser, S., Ralston, J. D., Larkins, E. C., Esquivias, I., Tasker, P. J., Fleissner, J., and Rosenzweig, J. (1992). *Electron. Lett.* **28**, 2141.

Weisser, S., Larkins, E. C., Czotscher K., Benz W., Daleiden, J., Esquivias, I., Fleissner, J., Ralston, J. D., Romero, B., Sah, R. E., Schönfelder, and Rosenzweig, J. (1996). *IEEE Photon. Technol. Lett.* **8**, 608.

Westbrook, L. D., Fletcher, N. C., Cooper, D. M., Stevenson, M., and Spurdens, P. C. (1989). *Electron. Lett.* **25**, 1183.

Willatzen, M., Uskov, A., Mørk, J., Olessen, H., Tromborg, B., and Jauho, A.-P. (1991). *IEEE Photon. Technol. Lett.* **3**, 606.

Wright, A. P., Garrett, B., Thompson, G. H. B., and Whiteaway, J. E. A. (1992). *Electron. Lett.* **28**, 1911.

Wu, T. C., Kan, S. C., Vassilovski, D., Lau, K. Y., Zah, C. E., Pathak, B., and Lee, T. P. (1992). *Appl. Phys. Lett.* **60**, 1794.

Yablonovitch, E., and Kane, E. O. (1986). *IEEE J. Lightwave Technol.* **4**, 504; errata, *IEEE J. Lightwave Technol.* **4**, 961.

Yamada, M. (1983). *IEEE J. Quantum Electron.* **19**, 1365.

Yamada, M., and Suematsu, Y. (1979). *IEEE J. Quantum Electron.* **15**, 743.

Yamamoto, T., Watanabe, T., Ide, S., Tanaka, K., Nobuhara, H., and Wakao, K. (1994). *IEEE Photon. Technol. Lett.* **6**, 1165.

Yamanaka, T., Yoshikuni, Y., Liu, W., Yokoyama, K., and Seki, S. (1992). *IEEE Photon. Technol. Lett.* **4**, 1318.

Yamanishi, Y., and Lee, Y. (1987). *IEEE J. Quantum Electron.* **23**, 367.

Yamanishi, M., and Suemune, I. (1984). *Jpn. J. Appl. Phys.* **23**, L35.

Yamazaki, H., Yamaguchi, M., Kitamura, M., and Mito, I. (1993). *IEEE Photon. Technol. Lett.* **5**, 396.

Yan, R. H., Corzine, S. W., Coldren, L. A., and Suemune, I. (1990). *IEEE J. Quantum Electron.* **26**, 213.

Yariv, A. (1989). *Quantum Electronics,* 3d ed., p. 176. John Wiley, New York.

Zah, C. E., Bhat, R., Menocal, S. G., Favire, F., Andreadakis, N. C., Koza, M. A., Caneau, C., Schwarz, S. A., Lo, Y., and Lee T. P. (1990). *IEEE Photon. Technol. Lett.* **2**, 231.

Zarem, H., Vahala, K., and Yariv, A. (1989). *IEEE J. Quantum Electron.* **25**, 705.

Zee, B. (1978). *IEEE J. Quantum Electron.* **14**, 727.

Zhao, B., Chen, T. R., and Yariv, A. (1991). *Electron. Lett.* **27**, 2343.

Zhao, B., Chen, T. R., Yamada, Y., Zhuang, Y. H., Kuze, N., and Yariv, A. (1992). *Appl. Phys. Lett.* **61**, 1907.

Chapter 4

Quantum Wire and Quantum Dot Lasers

Eli Kapon

Department of Physics, Swiss Federal Institute of Technology (EPFL), Lausanne, Switzerland

4.1 Introduction

Quantum confinement of the charge carriers in more than one dimension in quantum wire (QWR) and quantum dot (QD) heterostructures has been predicted to yield improved static and dynamic performance of semiconductor lasers as compared with quasi-two-dimensional (2D) quantum well (QW) devices (Arakawa and Sakaki, 1982; Asada et al., 1985; Arakawa and Yariv, 1986). The lateral *spatial confinement* of the excess carriers to active regions with lateral dimensions comparable with the de Broglie wavelength (typically on the order of 10 nm in III-V compound semiconductors) results in a greatly reduced volume of the active region. In addition, the narrower distributions of the density of states (DOS) achieved with reduced dimensionality lead to *spectral confinement* of the carriers near the energies of the quantum confined states. In this sense, the integration of active regions composed of QWRs and/or QDs into the bulk laser diode structure offers both the advantage of the delocalized bulk electronic states, useful in the transport of the charge carriers to the recombination region, and the enhanced oscillator strength associated with the molecular- or atomic-like localized states in the wires or dots. The spatial and spectral carrier confinements are expected to lead to reduced threshold currents (in the microampere or even the nanoampere range), reduced temperature sensitivity of the threshold current, higher modulation bandwidths, narrower spectral linewidths, and reduced chirp. The modification of the spectral gain properties also should lead to a new means for controlling the polarization of the laser beam and to new schemes for tailoring the emission wavelength. Another advantage of these laser structures lies in the wire/dot geometry of the active regions, which makes them less susceptible to dislocations originating at the interface with the device's substrate. This might be useful in lasers employing foreign substrates, e.g., III-V lasers grown on silicon. These potential advantages of QWR and QD lasers could make them most suitable for a variety of applications requiring coherent light sources with extremely low power consumption and high-speed digital modulation capability. Of particular interest in this context are integrated optoelectronic circuits involving large, dense arrays of diode lasers and integration of lasers with low-power electronics for optical interconnects and parallel signal processing.

A major challenge in this area has been the development of fabrication technologies for preparing QWR and QD heterostructures compatible with laser applications (Kapon, 1992). This has been important not only for the realization of QWR and QD lasers according to currently available models of these devices but also in order to provide much needed experimental data on the physics of these low-dimensional structures. In view of the 3D structure inherent to the QWR and QD active regions (Fig. 4.1), the main difficulty with the realization of such lasers is the fact that they are not directly compatible with *planar* growth technologies, such as conventional molecular beam epitaxy (MBE) or organometallic chemical vapor deposition (OMCVD). The conventional modes of these growth methods can provide high-quality 2D thin-film structures compatible with quasi-2D QW laser designs. The lower dimensionality of QWR and QD structures, on the other hand, necessitates *lateral patterning* of the heterostructure in order to provide the extra degrees of quantum confinement. This, in turn, requires either the addition of other fabrication steps before and/or after the growth step or resorting to completely different growth modes in order to realize the 3D structures. Moreover, the resulting patterned heterostructures need to be compatible with efficient carrier injection into the QWRs/QDs and should allow efficient interaction of the

4.1 Introduction

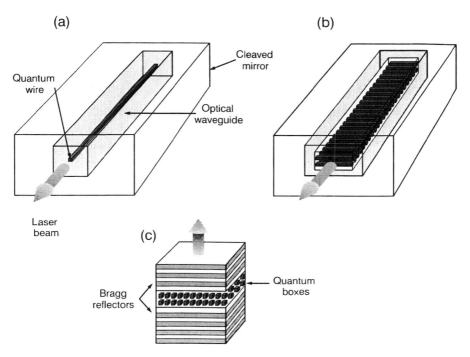

Figure 4.1: Schematic illustrations of several possible QWR and QD laser configurations: (a) longitudinal cavity QWR laser; (b) transverse cavity QWR-array laser; (c) vertical cavity QD laser.

carriers with the modes of the optical cavity. The increase in surface-to-volume ratio in a quantum-confined structure as the degree of quantum confinement is increased requires virtually perfect interfaces to suppress parasitic nonradiative recombination processes.

Besides the simulation of 1D QWR and 0D potential wells using high magnetic fields, approaches for the realization of QWR and QD lasers fall into two general categories: (1) techniques relying exclusively on post-growth processing steps utilizing lithography for defining the wire and dot structures and (2) methods employing growth-related phenomena for creating the wire and dot potential wells. As a lithography-based technique, the former approach offers a high degree of flexibility in the design of the lateral potential wells and the eventual laser configuration. However, it suffers from the fact that defects are introduced into the QWR and QD interfaces during fabrication, leading to nonradiative carrier recombination that limits the achievable optical gain. In addition, current (relevant) lithography techniques barely reach the size required for significant lateral confinement, and their absolute size fluctuations are rather large (more than a few nanometers). With the second approach, the interfaces of the wires and dots are formed in situ and are thus (potentially) defect-free. A particularly interesting class of approaches in this category relies on *self-ordering* of arrays of wire- and dot-shaped heterostructures during growth on planar or nonplanar substrates. The main disadvantage of this technique is that it limits the design flexibility of the QWR and QD devices. However, most of the recent progress in the area of 1D and 0D lasers has been achieved owing to this second approach.

This chapter reviews the principles of QWR and QD lasers and describes the recent developments in this field. Section 4.2 discusses the expected impact of the reduced dimensionality and quantum confinement on the optical properties of QWRs and QDs, the effects on semiconductor laser performance, and the special design issues applicable to these devices. Sections 4.3 and 4.4 present the state of the art of QWR and QD lasers, respectively, discussing the experimentally observed properties of wires and dots relevant to laser operation and comparing different designs, fabrication techniques, and performance. Finally, Sect. 4.5 brings the conclusions and an outlook on possible future directions.

4.2 Principles of QWR and QD lasers

The electrons and holes in QWRs and QDs are spatially confined to wire-shaped and box-shaped potential wells, whose small dimensions induce quantum confinement in one or both lateral directions. As in the case of 2D QW structures, this results in formation of discrete bound states, with electron and hole energies determined by the size and shape of the potential well. This, in turn, modifies the energy distribution of the charge carriers, with consequently an important impact on the lasing characteristics. Moreover, the modified Coulomb correlation in low-dimensional structures is expected to increase the importance of excitonic effects (Ogawa and Takagahara, 1991a, 1991b; Glutsch and Bechstedt, 1993; Rossi and Molinari, 1996) and could further alter the recombination mechanisms normally encountered in semiconductor lasers. In what follows, the main modifications in the band structure and laser properties brought about by the multidimensional quantum confinement are summarized.

4.2.1 Density of states

The reduced carrier dimensionality in ideal quantum structures leads to a dramatic redistribution of the density of energy states (DOS). Within the envelope function approximation, the carrier wave functions can be written as $\Psi(\mathbf{r}) \approx u(\mathbf{r})\psi(\mathbf{r})$, where $u(\mathbf{r})$ is the Bloch function at $\mathbf{k} = 0$ ($\mathbf{k}$ being the carrier wave vector) and $\psi(\mathbf{r})$ is the envelope function. The envelope function is a solution of the time-independent Schrödinger equation:

$$\left[-\frac{\hbar^2}{2m^*}\nabla^2 + V(\mathbf{r})\right]\psi(\mathbf{r}) = E\psi(\mathbf{r}) \tag{4.1}$$

where m^* is the effective carrier mass, $V(\mathbf{r})$ is the heterostructure potential defining the quantum structure, and E is the confinement energy (measured with respect to the bottom of the potential well V). In the simple (but instructive) case of rectangular, infinitely deep potential wells V, the confined energy levels of charge carriers confined in one, two, and three dimensions (along the z, y, and x directions, respectively) are given by

$$E_l = \frac{\hbar^2 \pi^2 l^2}{2m^* L_z^2} + \frac{\hbar^2 k_y^2}{2m^*} + \frac{\hbar^2 k_x^2}{2m^*} \quad \text{(2D)} \tag{4.2a}$$

$$E_{l,m} = \frac{\hbar^2 \pi^2 l^2}{2m^* L_z^2} + \frac{\hbar^2 \pi^2 m^2}{2m^* L_y^2} + \frac{\hbar^2 k_x^2}{2m^*} \quad \text{(1D)} \tag{4.2b}$$

$$E_{l,m,n} = \frac{\hbar^2 \pi^2 l^2}{2m^* L_z^2} + \frac{\hbar^2 \pi^2 m^2}{2m^* L_y^2} + \frac{\hbar^2 \pi^2 n^2}{2m^* L_x^2} \quad \text{(0D)} \tag{4.2c}$$

where L_x, L_y, and L_z denote the size of the confining well in each direction, $l, m, n = 1,2,3 \ldots$ are the level indices, and $k_{x,y}$ is the wavevector along the free propagation directions. Confinement in one or two directions results in formation of subbands whose edges are given by the eigen energies (4.2a) and (4.2b). Full 3D confinement gives rise to discrete 0D states with energies as shown in Eq. (4.2c).

The increasing degree of quantum confinement yields a corresponding decrease in carrier dimensionality. The density (per unit volume) of allowed energy states is then given by

$$\rho_{3D} = \frac{(2m^*/\hbar^2)^{3/2}}{2\pi^2}\sqrt{E} \quad \text{(3D)} \tag{4.3a}$$

$$\rho_{2D} = \frac{m^*}{\pi\hbar^2 L_x}\sum_l \theta(E - E_l) \quad \text{(2D)} \tag{4.3b}$$

$$\rho_{1D} = \frac{(2m^*)^{1/2}}{\pi\hbar L_x L_y}\sum_{l,m}(E - E_{l,m})^{-1/2} \quad \text{(1D)} \tag{4.3c}$$

$$\rho_{0D} = \frac{2}{L_x L_y L_z}\sum_{l,m,n}\delta(E - E_{l,m,n}) \quad \text{(0D)} \tag{4.3d}$$

where E is measured with respect to the band edge (or state) and $\theta(x)$ is the Heaviside function ($\theta = 0$ for $x < 0$; $\theta = 1$ for $x \geq 0$). Figure 4.2 shows these DOS functions for different dimensionalities in the case of ideal, perfectly uniform quantum structures. At sufficiently low carrier densities, only the ground subbands or states are populated, and the decrease in the DOS at higher energies in the QWR and QD case leads to *spectral confinement* of the charge carriers. In this ideal case, the carriers are confined to a single, low-dimensional subband or state, and the peaked energy distributions result in enhanced optical gain and differential gain, as further discussed below.

Another important outcome of lateral quantum confinement is the modification of the valence band structure, particularly due to valence band mixing (Sercel and Vahala, 1991; Bockelmann and Bastard, 1991; Vouilloz et al., 1997). The lateral confinement results in polarization anisotropy of the optical absorption and gain coefficients, usually favoring the amplification of a light beam polarized along the longer dimension of the structure. This confinement also induces mixing between the heavy-hole and the light-hole states even at the Brillouin zone center, which modifies the nature of the hole states and the details of the polarization anisotropy of the gain spectra.

The lateral quantum confinement in QWR and QD structures increases the importance of Coulomb correlations, particularly those related to the formation of excitonic states. Theoretical studies have indicated a significant increase in the exciton binding energy for strongly confined wires and dots (Degani and Hipolito, 1987; Rossi et al., 1997) as well as

4.2 Principles of QWR and QD lasers

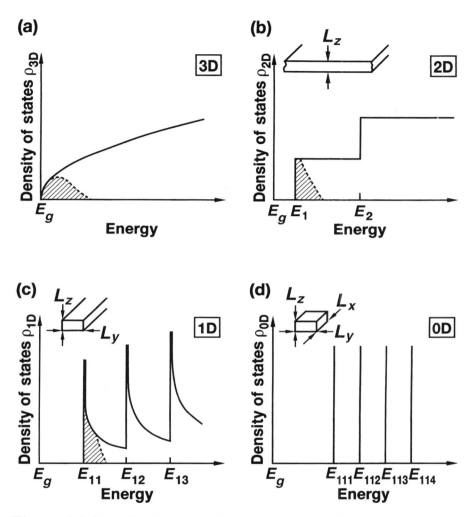

Figure 4.2: Density of states profiles as a function of dimensionality. (a) 3D; (b) 2D; (c) 1D; (d) 0D. The crossed regions denote filled states.

a reduced oscillator strength for the free carrier transitions as compared with the excitonic ones in QWRs (Ogawa and Takagahara, 1991a, 1991b; Glutsch and Bechstedt, 1993; Rossi and Molinari, 1996). Experimental studies of the luminescence spectra of QWRs at high carrier densities also demonstrated features that are consistent with the dominance of excitonic recombination at carrier densities that are higher than the corresponding ones in QWs, at least at low temperatures (Grundmann et al., 1994, 1995; Ambigapathy et al., 1997). The larger importance of excitonic recombination at carrier densities compatible with lasing thresholds would thus provide yet another means for achieving spectral carrier confinement in these low-dimensional structures.

However, to benefit from the above-mentioned spectral confinement of 1D and 0D carriers, several rather stringent requirements should be imposed on the wire/dot structure:

1. The size and composition fluctuations in the wires or dots should be small enough so that the low-dimensional states remain well resolved (Vahala, 1988; Zarem et al., 1989; Singh et al., 1992). For example, in the presence of width fluctuations δL_j, the inhomogeneous broadening for infinitely deep potential wells is given by $|\delta E_{\text{conf}}| = 2E_{\text{conf}} (\delta L_j/L_j)$, where E_{conf} is the confinement energy [the term proportional to $1/L_j^2$ in Eq. (4.2)]. Thus the relative size fluctuations must satisfy $\delta L_j/L_j \ll 1$ so that the peaked DOS distributions are preserved.

2. Thermal population of higher-order subbands (or states) should be avoided to obtain the full benefit of spectral confinement. This can be achieved with a sufficiently large subband separation for the electrons and the holes. For infinitely deep potential wells, a subband (state) separation larger than $k_B T$ is achieved for wire/dot widths L_j that satisfy

$$L_j < \sqrt{\frac{3\hbar^2 \pi^2}{2k_B m^* T}} \qquad (4.4)$$

where T denotes the carrier temperature. For room-temperature applications, this requires wire/dot widths of 10 to 20 nm at most for typical III-V semiconductors employed in diode lasers (the widths for dots should not be too small in order to ensure at least one confined state for electrons). Combined with the first requirement, this means that the effective size fluctuations should not increase a *few monolayers* over the *entire device area*.

3. The confined carriers should preferably reside in the ground states, which imposes a restriction on the quasi-Fermi levels at threshold. For example, for wires with infinitely deep potential wells near

4.2 Principles of QWR and QD lasers

$T = 0$, this implies a carrier density of $N < 10^6$ cm^{-1}. To keep the carrier density sufficiently low at threshold requires, in turn, a high internal quantum efficiency (i.e., minimizing interface and bulk defects) and high wire/dot densities, tight confinement of the optical mode, and low optical cavity losses in order to minimize the threshold gain.

4.2.2 Optical gain

The spectral carrier confinement due to narrower DOS profiles has a direct impact on the optical gain and the differential gain in lasers incorporating QWRs or QDs. The optical (material) gain function associated with transitions between free carrier states is given by

$$g(E) = \alpha_0 \int dE' \, \rho_{\text{red}}(E') |M_{cv}(E')|^2 L(E' - E)[f_c(E') - f_v(E')] \quad (4.5)$$

where α_0 is a constant, ρ_{red} is the reduced density of states, M_{cv} is the momentum matrix element, $L(E)$ is a lineshape function, and $f_{c,v}$ denote the Fermi functions for the conduction and valence bands (Zory, 1993) (see also Chap. 1). The momentum matrix element between the conduction band $|S\rangle$ and the four valence band states $|V_i\rangle = |\frac{3}{2}, \pm\frac{3}{2}\rangle, |\frac{3}{2}, \pm\frac{1}{2}\rangle$ ($\pm\frac{3}{2}$ and $\pm\frac{1}{2}$ correspond to heavy holes and light holes in bulk materials, respectively) is expressed as

$$|M_{cv}(E)|^2 = |M_b|^2 \sum_{i=1}^{4} |\langle S|\mathbf{e}\cdot\mathbf{k}|V_i\rangle \int dxdy \, \psi^*_{c\mathbf{k}}(x,y)\psi_{v\mathbf{k}}(x,y)|^2 \quad (4.6)$$

where $\mathbf{e}$ is the direction of polarization of the linearly polarized photons, and $\psi_{c,v\mathbf{k}}(x, y)$ are the carrier envelope functions.

The form of the gain function [Eq. (4.5)] illustrates the effect of the quantum confinement on the optical gain: The narrower DOS function enhances the gain (for a given carrier density) near the photon energies corresponding to transitions between the bound states in the wires or dots. In addition, the differential gain, i.e., the rate of increase in gain with increasing carrier density, is also enhanced (Arakawa and Yariv, 1986). Asada et al. (1986) have compared the gain spectra of low-dimensional structures neglecting inhomogeneous broadening effects. Figure 4.3 shows the calculated (material) gain spectra for lattice-matched bulk, 10-nm-thick QW, 10 × 10 nm^2 square cross section QWR, and 10 × 10

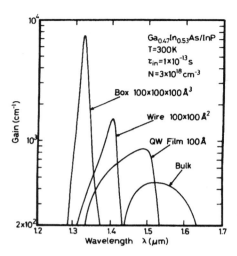

Figure 4.3: Calculated gain spectra for bulk, 10-nm-thick QW, 10×10 nm^2 QWR, and $10 \times 10 \times 10$ nm^3 QD In$_{0.53}$Ga$_{0.47}$As/InP lasers at 300 K (Asada et al., 1986, see reference).

$\times 10$ nm^3 cubic QD InGaAs/InP lasers. Simultaneously with the obvious narrowing in the gain spectra with decreasing dimensionality, the peak of the gain increases. The peak gain for the QD case is about 15 times higher than that for bulk material; a tenfold increase in the peak gain for the QDs was calculated for similar GaAs/AlGaAs structures. However, it is important to recall that both homogeneous and inhomogeneous broadening mechanisms, particularly due to size fluctuations, can smear out the peaked gain at the wire or dot transition energies (Vahala, 1988; Zarem et al., 1989).

The different symmetry of the valence band states in low-dimensional structures introduces a polarization anisotropy of the matrix elements. For example, in the case of QWRs of square or circular cross section, light polarized along the wire axis experiences a higher gain due to electron to heavy-hole-like transitions, whereas for the perpendicular polarization, transitions involving light-hole-like states dominate (the importance of each hole type generally depends on the shape of the wire cross section) (Sercel and Vahala, 1991). Similar gain anisotropy related to the valence band structure is also exhibited by elongated QDs (Willatzen et al., 1994). This offers new possibilities for controlling the polarization mode of diode lasers.

4.2 Principles of QWR and QD lasers

Introducing strain into the active QWR or QD region has been predicted to yield improved gain and differential gain characteristics (Ueno et al., 1992; Arakawa, 1992; Yamauchi et al., 1993). In compressively strained structures, the otherwise heavier effective hole mass can be reduced, leading to both larger subband (confined state) separations and a lower carrier density at transparency (see Chap. 2 for a detailed account on the effect of strain on the performance of semiconductor lasers). In addition, broadening effects also can be reduced in strained wires or dots because the lateral dimensions required for a given confinement energy are larger (Ueno et al., 1992).

It should be emphasized that Eq. (4.5) corresponds to the *material* gain, i.e., the gain experienced by the optical field if it were completely confined in the wires or dots. However, since the wavelength of the photons is much larger than the characteristic wire or dot dimensions, only a fraction Γ, the so-called optical confinement factor, is confined to the gain medium, and hence the effective gain is correspondingly reduced. This represents an inherent disadvantage of QWR and QD structures and requires laser designs that maximize the value of Γ.

4.2.3 Threshold current

Extremely low threshold currents are expected to be achieved with QWR and QD lasers due to both the small volumes of their active regions and the higher optical gain (Yariv, 1988; Miyamoto et al., 1989). The threshold gain g_{th} is related to the optical cavity loss coefficient α_{cav} via

$$\Gamma g_{th} = \alpha_{cav} = \alpha_i + \frac{1}{2L} \ln\left(\frac{1}{R_1 R_2}\right) \tag{4.7}$$

where α_i is the internal loss coefficient, L is the cavity length, and R_1 and R_2 are the mirror reflectivities. In the linear gain approximation, the variation of the gain coefficient with the carrier density N (per unit volume) is written

$$g(N) = g'(N - N_{tr}) \tag{4.8}$$

where g' is the differential gain coefficient, and N_{tr} is the carrier density at transparency. The diode current (excluding leakage components) is given by $I = eVN/\tau_c$, where V is the volume of the active region, and τ_c is the carrier lifetime. Using this relation with Eqs. (4.7) and (4.8) yields the threshold current:

$$I_{th} = \frac{eVN_{tr}}{\tau_c} + \frac{eV\alpha_{\text{cav}}}{g'\tau_c\Gamma} \equiv I_{tr} + I_{\text{cav}} \qquad (4.9)$$

In this linear gain approximation, the threshold current consists of two terms: the transparency current I_{tr} required for rendering the active region transparent and the cavity current I_{cav} needed for offsetting the cavity losses. The transparency current is reduced mainly due to the extremely small active volume, while the cavity current can be reduced because of the increase in differential gain.

Equation (4.9) illustrates the basic requirements for achieving very low threshold currents with QWR or QD lasers. First, the carrier lifetime should be sufficiently long to keep both terms small. Hence it is important to avoid nonradiative recombination centers near the active region, which might be introduced due to damaging processing steps. Second, the cavity losses should be kept low to minimize the second term. This can be achieved by ensuring low waveguide losses and low mirror losses. The latter requirement can be met using sufficiently long cavities and/or highly reflecting mirrors. However, this can be effective only by simultaneously keeping very low internal losses to allow for useful output power levels. Finally, the optical confinement factor should be sufficiently high. This calls for the use of tight optical confinement. Note that although a higher confinement factor can be achieved by increasing the active region volume (i.e., by increasing the wire or dot density), an increase in V also leads to an increase in both current terms, and thus an optimum value for the active volume should exist.

Threshold currents of ~ 1 μA were predicted for ideal QWR lasers in the limit of zero optical cavity losses (Yariv, 1988). Miyamoto et al. (1989) have calculated the threshold current densities for optimally designed InGaAsP/InP QD lasers and found values of 14 and 27 A/cm^2 assuming internal losses of 10 and 20 cm^{-1}, respectively, and neglecting size fluctuations. These values increase to 61 and 190 A/cm^2, respectively, assuming 10% width fluctuations in the lateral dot dimensions. Asryan and Suris (1996) modeled the threshold current density of QD lasers including inhomogeneous broadening effects and showed that a minimum threshold current is obtained for an optimal surface density. As an example, they calculated $J_{th} \approx 10$ A/cm^2 for lattice-matched InGaAsP/InP QD lasers with 10% size fluctuations at an optimal surface density of $\sim 6 \times 10^{10}$ dots per cm^2 (15-nm-wide dots, 10 cm^{-1} internal losses).

Besides the current components discussed earlier, one also should consider leakage currents, which can significantly increase the threshold current in real QWR or QD lasers. Different from the case of QW lasers, the configuration of laterally confined wires and dots with barriers in between them may give rise to substantial leakage currents due to carriers flowing in between the wires or dots. In addition, efficient carrier capture into the QWR or QD potential wells is necessary to minimize this leakage current. Current blocking schemes thus need to be considered to direct the diode current to the wires and dots. Quantum well or QWR regions connected to the active wires and dots also might be useful for improving the carrier capture efficiency (see Sects. 4.3 and 4.4).

The spectral confinement of the carriers due to the narrow DOS distribution is also expected to reduce the temperature sensitivity of the threshold current of QWR and QD lasers. The temperature dependence of the current density on temperature, of the form $J_{th} = J_0 \exp(T/T_0)$, was calculated for GaAs/AlGaAs QWR and QD lasers by Arakawa and Sakaki (1982). A significant increase in T_0, to 481°C for QWRs and to essentially infinity for QD lasers, has been predicted. The increase in T_0 for realistic QD structures is expected to be less dramatic because of the finite width of the DOS in that case. It should be noted, however, that the temperature dependence of the threshold current also should be affected by the temperature dependence of other mechanisms involved in the carrier transport and recombination. In structures with significant current leakage, the dependence of carrier mobility and carrier capture efficiency on temperature will result in a T_0 that does not reflect solely the effect of the DOS. The different temperature dependence of the radiative lifetime of 1D or 0D carriers (Citrin, 1992) also should be taken into account properly when assessing the effect of the DOS on the threshold current sensitivity.

4.2.4 High-speed modulation

The carrier and photon dynamics in a semiconductor laser are reflected by the so-called relaxation oscillations, whose frequency is given by (Lau et al., 1983) (see also Chap. 3)

$$f_r = \frac{1}{2\pi} \sqrt{\frac{v_{gr}\Gamma g'}{\tau_{ph}} \frac{S_0}{1 + \varepsilon S_0}} \qquad (4.10)$$

where S_0 is the steady-state photon density, v_{gr} is the photon group velocity, τ_{ph} is the cavity photon lifetime, and ε is the gain compression coefficient. This parameter gives a measure for evaluating the maximum direct modulation frequency of the device (see Chap. 3 for a detailed account on this limit). The higher differential gain in the low-dimensional structures was thus expected to lead to higher modulation speeds (Arakawa at al., 1984). For example, an approximately 50% increase in the relaxation oscillation frequency of well-designed QWR lasers with respect to QW devices has been predicted (Arakawa et al., 1986a). The further increase in the differential gain obtained by introducing compressive strain (e.g., in InGaAs/GaAs QWR structures) (Arakawa, 1992) should yield a corresponding increase of the modulation bandwidth in strained QWR lasers.

However, these predictions do not account either for nonlinear gain effects, which also can be modified by the reduced dimensionality, or for limitations set by carrier capture and relaxation effects. The latter are determined by carrier transport toward the wires or dots, followed by carrier relaxation from the edge of the potential well to its bottom. The relaxation process is mediated by emission of polar optic phonons, carrier-acoustic phonon scattering, and carrier-carrier scattering. These processes might limit the ultimate modulation speed, particularly for decreasing lateral dimensions of the confining potential. The carrier transport time can be minimized by designing active regions in which the separation of the potential wells is much smaller than the carrier diffusion length. Barrier regions with intermediate dimensionality, e.g., QW sections connected to QWR recombination regions, also may be useful in this respect. Minimizing the carrier relaxation time requires an understanding of the impact of the multidimensional confinement on the carrier scattering mechanisms. Kan et al. (1993) formulated the effect of carrier capture in low-dimensional lasers and introduced an effective capture time scaled up by the ratio of the volume of the 3D carrier confined region to that of the low-dimensional active region.

In this context, the discrete level structure in QDs has been predicted to result in a dramatically reduced relaxation efficiency, the so-called phonon bottleneck effect (Bockelmann and Bastard, 1990; Benisty et al., 1991). However, more recent theoretical studies have shown that the relaxation rates can be enhanced due to electron–hole interaction (Bockelmann, 1993), Auger recombination (Bockelmann and Egeler, 1992), and multiphonon processes (Inoshita and Sakaki, 1992).

Vurgaftman and Singh considered the tradeoff between minimizing the threshold current density and minimizing the relaxation time of the carriers in QWR laser structures (Vurgaftman and Singh, 1993, 1994). They calculated the electron relaxation time due to polar optical phonon scattering and found that it decreases from 120 to 30 ps for GaAs/AlGaAs QWRs of square cross sections ranging between 10×10 and 20×20 nm^2. The corresponding threshold current densities, however, increase from 510 to 740 A/cm^2, respectively. These rather long relaxation times are partly due to effective screening of the polar optical phonon scattering. However, electron–hole interaction was predicted to reduce the relaxation time to about 50 ps for the 10×10 nm^2 wire configuration. Much shorter capture and relaxation times (a few picoseconds or even subpicoseconds) were calculated for V-groove QWR structures by Ammann et al. (1997).

It also should be noted that the low threshold currents achievable with QWR or QD lasers make them useful for high-speed *digital* modulation applications. The time delay in the onset of lasing following current increase from below threshold is given (assuming constant carrier lifetime) by (Konnerth and Lanza, 1964)

$$t_d = \tau_c \ln\left(\frac{I}{I - I_{th}}\right) \to \frac{\tau_c I_{th}}{I} \tag{4.11}$$

where the limit is valid for $I \gg I_{th}$ (Lau et al., 1987). Since typical carrier lifetimes are in the nanosecond range, achievement of $t_d \approx 10$ ps would require $I/I_{th} \approx 100$, which demonstrates the benefit of ultralow threshold currents for such applications.

4.2.5 Spectral control

Several features of QWR and QD lasers may be used for controlling the emission spectrum. The peaked DOS profile introduces a spectral modulation of the optical gain, which results in much narrower gain profiles near each 1D or 0D transition energy. Even for an optical cavity without any mode-selection capabilities (e.g., a Fabry Perot cleaved-mirror cavity), the modes near these transition energies are selected because of their smaller gain-to-cavity-loss difference. A QWR or a QD laser structure thus may be designed to offer a better wavelength stability near the subband or state transitions.

The bandgap and refractive index modulation in the plane of the optical cavity (e.g., in an edge-emitting array configuration) inherent to QWR and QD lasers makes it possible to generate distributed feedback (DFB) effects that are useful for selecting a particular lasing mode via

Bragg scattering (Kogelnik and Shank, 1972) (see also Chap. 3 in Volume II). Whereas such DFB effects also can be obtained in more conventional DH or QW lasers by introducing a grating into the cavity, periodic QWR and QD structures also offer the possibility of periodically modulating the optical gain, in addition to the more usual modulation of the real refractive index. The combined periodic variation of the real and imaginary parts of the refractive index leads to a *complex* coupling coefficient, which breaks the degeneracy of the DFB modes found in real-index structures and gives rise to a better wavelength selectivity and stability (Kogelnik and Shank, 1972; Kapon et al., 1982).

The modification in the optical absorption and gain spectra also alters the refractive index spectra near the subband transitions. This, in turn, may lead to a significant reduction in the spectral linewidth of each cavity mode. The field spectrum linewidth of a single mode in a semiconductor laser cavity is given by (Henry, 1982)

$$\Delta\nu = \frac{R_{sp}}{4\pi S}(1 + \alpha^2) \tag{4.12}$$

The first factor in this expression is the conventional Schawlow-Townes term, where R_{sp} is the rate of spontaneous emission into the lasing mode and S is the total number of photons in this mode. The second factor is unique to semiconductor lasers and accounts for the broader linewidths usually observed due to the linewidth enhancement factor $\alpha = \delta n'/\delta n''$, where $\delta n'$ and $\delta n''$ are the changes in the real and imaginary parts of the refractive index due to the coupling of spontaneous emission photons into the lasing mode. This linewidth enhancement results from the detuning of the spectrum of $\delta n''$ with respect to that of $\delta n'$. Typical values of α in bulk lasers are 4 to 7, resulting in linewidth-power products of 50 to 100 mW·MHz (Fleming and Moradian, 1981). Lower values of α are expected with the introduction of lateral quantum confinement due to the higher differential gain (and hence higher $\delta n''$) and the smaller detuning of $\delta n'$ and $\delta n''$, particularly in the case of QD lasers. Values of α of ~1 have been calculated for QWR lasers for sufficiently high Fermi levels (Arakawa et al., 1986a). Further reduction of the linewidth enhancement factor to ~0.2 was predicted for strained QWR lasers (Arakawa, 1992). For QD lasers, a zero value of α was calculated near the gain peak, with negative values obtained for higher photon energies (Miyake and Asada, 1989; Willatzen et al., 1994). The reduction in the linewidth enhancement factor in QWR and QD lasers is also expected to reduce wavelength chirping effects during high-speed modulation (Koch and Linke, 1986). As in the case of the modulation bandwidth, a higher nonlinear gain coefficient also may reduce the effect of linewidth narrowing discussed above.

4.3 Quantum wire lasers

Approaches for realizing 1D semiconductor lasers have included immersion of conventional diode lasers in high magnetic fields, etching and regrowth of QW laser structures, cleaved-edge overgrowth of T-shaped QWRs, growth on vicinal substrates, structures incorporating strain-induced wire superlattices, and growth on nonplanar substrates, particularly using V-groove QWR structures. Figure 4.4 summarizes the structures of several QWR configurations that have been employed in lasers. Some details on the performance of these QWR lasers are given in what follows.

4.3.1 Semiconductor lasers in high magnetic fields

Early attempts to realize QWR lasers consisted of immersing diode lasers in high magnetic fields to achieve *magnetic confinement* of the carriers in two directions (in the plane normal to the axis of the field B). For magnetic fields sufficiently high that $\omega_c \gg 1/\tau_{sc}$, where $\omega_c = eB/m^*$ is the cyclotron frequency and τ_{sc} is the scattering time of the circulating carriers, application of the field results in formation of quasi-1D Landau levels with energies

$$E_l = (l - 1/2)\hbar\omega_c + \hbar^2 k_z^2/2m^* \qquad l = 1, 2, 3, \ldots \qquad (4.13)$$

and a DOS profile similar to that in Eq. (4.2c). A conventional double-heterostructure (DH) laser in a sufficiently high magnetic field then behaves as a QWR laser [see Fig. 4.4(a)]. Similarly, a QW laser in a high magnetic field should behave as a QD laser in which the confinement along the field vector is provided by the QW heterostructure and the confinement in the plane of the QW is due to the magnetic field. For electrons in GaAs, $B = 10$ T yields Landau level separation of 17 meV. Clearly, these rather high magnetic fields required for achieving magnetic confinement applicable at room temperature limit the usefulness of this approach. However, this simulation of a quasi-1D or 0D potential wells

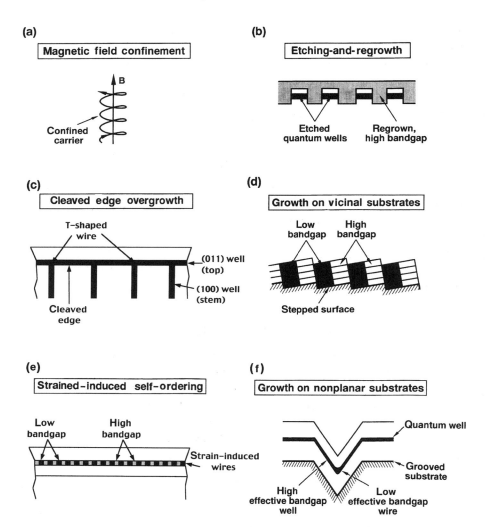

Figure 4.4: Several approaches for the realization of QWRs for laser applications. (a) immersion in high magnetic fields; (b) etching and regrowth of QW structures; (c) cleaved-edge overgrowth; (d) growth on vicinal substrates; (e) strained-induced lateral oredring; and (f) V-groove structures.

4.3 Quantum wire lasers

can yield useful insight on the impact of such confinement on the laser characteristics.

GaAs/AlGaAs DH lasers showed a reduced temperature sensitivity of the threshold current in high magnetic fields: An increase of T_0 from 155 K at $B = 0$ to 313 K at $B = 24$ T has been measured near room temperature (Arakawa and Sakaki, 1982). An increase in T_0 from 200 to 260 K for fields between 0 and 20 T also was observed for GaAs/AlGaAs QW lasers at room temperature (Berendschot et al., 1989). A relaxation oscillation frequency higher by a factor of 1.4 was measured for DH GaAs/AlGaAs with $B = 20$ T at room temperature (Arakawa et al., 1985), and a twofold reduction in the spectral linewidth was observed for DH and QW GaAs/AlGaAs lasers at low temperatures with B around 20 T (Arakawa et al., 1986b; Vahala et al., 1987).

4.3.2 QWR lasers fabricated by etching and regrowth

Early attempts to make "rigid" QWR (and QD) lasers involved the definition of the laterally confined structures on QW heterostructures using lithography and epitaxial regrowth. Starting with a QW laser structure (with or without the top cladding layers), one defines wirelike (dotlike) mesas by etching through a lithographically prepared mask. The upper part of the laser structure is then completed by regrowth in order to provide passivated wire (dot) interfaces [see Fig. 4.4(b)]. Since the effective width of the lateral potential well is (ideally) identical to that of the etched wires, the fabrication techniques involved should be capable of producing structures as narrow as a few 10 nm in width. Usually, wet chemical etching is employed to minimize interface defects. Moreover, high-density grating-like wire structures are desirable to maximize the modal gain. An optical waveguide is usually fabricated on top of the regrown structure to provide for lateral optical confinement. The preferred cavity configuration is often such that the laser beam propagates normal to the wires in order to benefit from the polarization anisotropy of the optical gain and from DFB effects obtained for properly designed array periodicities.

The first etched-and-regrown QWR-sized laser structures were fabricated in the lattice-matched InP/InGaAsP system (Cao et al., 1988, 1990). The OMCVD grown InGaAs wires were 30 nm wide, with an array pitch of 70 nm. Holographic photolithography or electron-beam (EB) lithography was used to define the mask of the wire array. The lasers operated at 77 K at 1.36-μm wavelength with threshold current densities as low as

3.8 kA/cm². Room-temperature operation was later achieved with similar structures using improved OMCVD regrowth conditions and *p*-type substrates to obtain a more efficient hole injection (Miyake et al., 1991, 1992).

Room-temperature CW operation of similar laser structures incorporating InGaAs/InP compressive- or tensile-strained wire active regions of comparable dimensions was demonstrated more recently (Miyake et al., 1993; Kudo et al., 1992, 1993). Figure 4.5 shows a schematic diagram and a scanning electron microscope (SEM) cross section of such a structure (Kudo et al., 1992). The strained active regions yield higher optical gain, which leads to a reduced threshold current density. In addition, a buried heterostructure (BH) stripe geometry, fabricated by liquid-phase epitaxial third-step regrowth, was employed to reduce the total diode current. Room-temperature CW threshold current and threshold current density

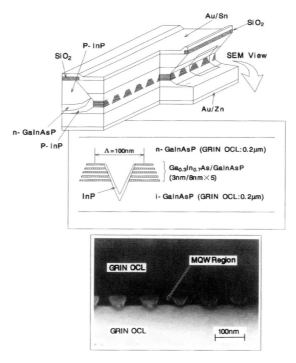

Figure 4.5: Schematic description and SEM cross section of InGaAs/InGaAsP QWR laser fabricated by etching and regrowth (Kudo et al., 1992, see reference).

4.3 Quantum wire lasers

for the tensile-strained structures were 16 mA and 816 A/cm^2 (measured in a 2-μm-wide device). The lasing wavelength was 1.46 μm. The threshold currents were about twice as large as the corresponding ones in a control tensile-strained QW laser. A study of the lasing characteristics as a function of temperature revealed that this difference is not caused by a reduced internal quantum efficiency, indicating high interface quality of the wire structures (Shin et al., 1995). The difference was rather attributed to the lower optical confinement factor (0.95% versus 1.9%) for the wire lasers. However, effects due to lateral carrier confinement were not observed in the lasing characteristics of these lasers.

Nishida et al. (1992) used EB lithography, reverse mesa wet etching, and OMCVD overgrowth to fabricate InGaAs/InP wires as narrow as 10 nm with 70-nm periodicity. The reverse mesa etching allows achieving wire widths that are smaller (by about 30 nm) than the resist width. Furthermore, facet formation during the anisotropic wet etching considerably reduces wire width fluctuations, as evidenced by atomic force microscopy (AFM) measurements. The standard deviation in the width of a given wire was thus reduced from 1.5 to 2.5 nm to only 0.7 nm (Notomi et al., 1993a). Mass transport of InP during the overgrowth steps was used to planarize the wire array and to provide higher-bandgap current-blocking InP regions between the wires, which is useful for reducing the leakage current. Increase in blue shift and polarization anisotropy of the luminescence with decreasing wire widths were interpreted as evidence for 2D quantum confinement in these structures (Notomi et al, 1993b).

Such QWRs were incorporated in diode laser structures employing 5-μm-wide ridge waveguides for lateral optical confinement with 400- or 700-μm-long optical cavities (Notomi and Tamamura, 1993). Devices with 60-nm-wide wires (single level) lased at 7 K at 1.33-μm wavelength with a threshold current of 235 mA. Lasers incorporating narrower wires (as narrow as 20 nm) did not lase and exhibited saturation in the intensity of the electroluminescence at lower energies related to transitions between the ground quantum well states. Laterally confined subbands were not observed, probably due to excessive wire-width fluctuations. However, the difficulty in achieving lasing with the narrow wires was attributed to the low optical confinement factor rather than to inhomogeneous broadening; model calculations indicated that the width fluctuations severely decrease the gain peak only for narrower wires (~10 nm).

Wang et al. have investigated the dynamics of optically pumped InGaAs/InGaAsP quasi-QWR lasers made by etching and regrowth, emitting at 1.5-μm wavelength (Wang et al., 1996a, 1996b). The DFB lasers, with a first-order grating configuration and with wire widths ranging between 100 and 140 nm, were pumped with short (3-ps) optical pulses. The time delay of the laser emission was found to decrease with decreasing wire width. Simulation of the carrier and photon dynamics indicated that this decrease is due to an increase in the differential gain for narrower wires (Wang et al., 1996a). Analysis of the decay time of the emitted pulse for 100-nm-wide wire lasers versus that for QW lasers indicated a much longer carrier capture time for the wire lasers (90 versus 56 ps) because of the smaller volume of the recombination region (Wang et al., 1996b).

4.3.3 QWR lasers made by cleaved-edge overgrowth

Cleaved-edge overgrowth (CEO) has been proposed as a technique for producing QWR structures with uniformity comparable with that of high-quality QW heterostructures (Chang et al., 1985; Goni et al., 1992). In this approach, a conventional QW heterostructure (e.g., GaAs/AlGaAs) is grown on a (100) substrate. The substrate is then cleaved in situ in an MBE growth system, and another QW heterostructure is overgrown on the (011) cleaved facet. T-shaped QWR structures are thus formed in which the carrier wavefunctions are quantum confined in two perpendicular directions [see Fig. 4.4(c)]. The (100) QW at the stem of the T-shaped QWR provides a perturbation to the planar potential distribution of the (011) QW, localizing the wavefunctions in the vicinity of the T-junction. Since the dimensions of all parts of the T-shaped QWRs are determined by epitaxial growth steps, and since atomically smooth cleaved (011) planes can be achieved, highly uniform wire structures are expected with this approach.

Both optically pumped (Wegscheider et al., 1993) and current injection (Wegscheider et al., 1994) semiconductor lasers incorporating multiple (15 to 22 periods) T-shaped $GaAs/Al_{0.35}Ga_{0.65}As$ QWRs have been demonstrated. Figure 4.6(a) illustrates schematically a T-QWR laser structure, and Fig. 4.6(b) shows a transmission electron microscope (TEM) cross-sectional image of the active region. The QWs composing the T-QWRs are 7 nm thick each and exhibit well-defined, sharp interfaces. Only the ground 1D states of the carriers are confined in these T-shaped structures. An $Al_{0.1}Ga_{0.9}As$ waveguide layer is grown on top of the wires, leading to the formation of a T-shaped 2D optical waveguide that yields a relatively high optical confinement factor in the wires (the confinement factor is estimated to be $\Gamma \approx 3 \times 10^{-3}$). In the diode laser version, a *p-n* junction

4.3 Quantum wire lasers

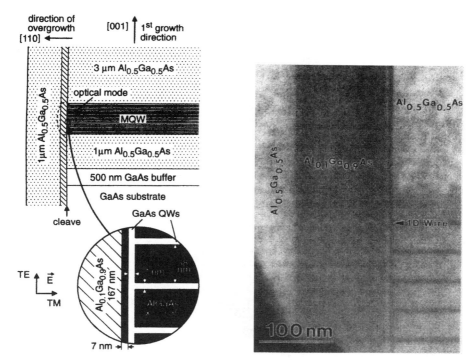

Figure 4.6: Schematic illustration (a) and TEM cross section (b) of a GaAs/AlGaAs laser incorporating T-shaped QWRs formed by cleaved-edge overgrowth (Wegscheider et al., 1993, see reference).

is formed by δ-doping the barriers of the (100) QWs with Be and by Si-doping the (011) layers starting from the waveguide layer. Laser contacts are then made on both the (100) and the CEO surfaces.

Low-temperature (liquid He) polarized photoluminescence (PL) spectra of the undoped T-QWR laser structure are shown in Fig. 4.7 (Wegscheider et al., 1993). At low pump levels, luminescence from both the QWR and the connected top- and stem-QW regions is observed, with optical gain at the QWR transition evidenced by the FP mode modulation of the spectra. Lasing is observed at the T-QWR transitions with a threshold current density of about 600 W/cm^2. The occurrence of lasing only at the QWR transition is partly due to the optical confinement by the T-shaped waveguide and partly due to the efficient transfer of the photoexcited carriers from the connected QWs into the QWRs. The fact that the QWR emission energy does not change much with increasing carrier

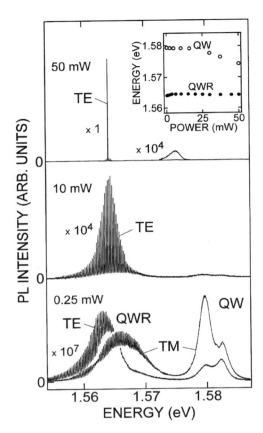

Figure 4.7: Photoluminescence and lasing spectra of a T-shaped QWR laser made by cleaved-edge overgrowth measured at several power levels (optical pumping). The inset shows the power dependence of the QW and the QWR emission energy (Wegscheider et al., 1993, see reference).

density (up to about 10^6 cm^{-1} at threshold) was interpreted as a signature of the enhanced stability of the 1D exciton gas. An enhancement of about 50% in the exciton binding energy was estimated for these structures from the difference in the energies of the QWR and the QW emission lines. The fixed QWR emission energy is in contrast to the red shift in the QW emission line, indicating a "normal" bandgap renormalization effect associated with the high-density 2D electron–hole plasma. These observations suggest that, at least at low temperatures, the confinement to a 1D structure leads to lasing dominated by excitonic effects.

4.3 Quantum wire lasers

The diode T-QWR lasers also were characterized at low temperature (4.2 K), showing CW threshold currents of 0.4 and 0.6 mA for uncoated devices with cavity lengths of 400 and 800 μm, respectively (Wegscheider et al., 1994). The emission wavelength was 786 nm. Constant energy of the QWR emission line at currents above threshold also was observed in this case and was again attributed to the dominance of excitonic effects. Room-temperature operation of T-QWR lasers has not been achieved so far, probably due to the relatively small confinement energy in the QWRs (i.e., the difference between the energy of the 1D subband edge and that of the 2D QW region), measured to be about 17 meV. However, recent studies have demonstrated that these confinement energies can be increased up to about 50 meV by optimizing the T-shaped QWR structure (Gislason et al., 1996).

4.3.4 QWR lasers grown on vicinal substrates

Growth on *vicinal substrates* makes use of the natural stepped profile of a misoriented substrate as the template for subsequent formation of wire or dot arrays (Petroff et al., 1984; Fukui and Saito, 1988) [see Fig. 4.4(d)]. As an example, a 2° misoriented (100) GaAs substrate ideally exhibits an 8-nm-pitch monolayer step profile. Preferential growth at the step edges would then yield lateral (fractional layer) superlattice (SL) structures via alternate growth of fractions of monolayers of, e.g., GaAs and AlAs. Sandwiching such lateral SLs between higher-bandgap layers results in extremely dense arrays of QWRs. However, a precise control of the growth rates (less than 1% variations) is required to obtain reproducible structures, since the fraction of monolayers that is actually grown determines the angle between the wire sidewalls and the wafer plane. In particular, if the GaAs and the AlAs depositions deviate from an exact 0.5/0.5 monolayer scheme, a *tilted*-wire SL results.

Such MBE-grown tilted-SL GaAs/AlGaAs lasers with a nominally 8-nm wire periodicity were studied by Tsuchiya et al. (1989). The tilted SL active region was sandwiched between waveguide layers to form a separate confinement QWR heterostructure. Broad area (BA) lasers operated at room temperature under pulsed conditions with threshold current densities as low as 470 A/cm^2; the lasing wavelength was 828 nm. However, lasing characteristics were not different from those of conventional QW lasers. This can be explained by the relatively weak lateral modulation of the bandgap and the considerable nonuniformities from wire to wire

(Yi et al., 1991), as observed in TEM cross sections of the structures. The weak lateral bandgap modulation results from exchange of Al and Ga atoms at the step edges, which effectively smears out the otherwise deeply modulated structure. The nonuniformities are due partly to imperfect periodicity of the step structure and partly due to growth-rate variations across the sample.

Saito et al. (1993) fabricated a fractional-layer SL GaAs/AlGaAs laser on a GaAs (100) vicinal substrate misoriented by 2° toward [01$\bar{1}$] using OMCVD. Short-period $(AlAs)_1(GaAs)_2$ superlattices were used in this case at the waveguide layers in order to avoid random step bunching and achieve a more regular stepped surface before growth of the active region. The active region consisted of a 12-nm-thick undoped $(AlAs)_{1/4}(GaAs)_{3/4}$ fractional-layer SL. Low-diffusive C and Si atoms were used for doping the carrier confinement layers in order to avoid disordering of the QWR active region. Similar lateral SL structures have shown 50% polarization anisotropy in PL spectra at 9 K (PL intensity larger for polarization along the wires), which has been attributed to 2D quantum confinement in the wires.

The OMCVD grown fractional-layer SL lasers were tested under pulsed conditions below 195 K (Saito et al., 1993). A schematic diagram and a TEM cross section of the laser are shown in Fig. 4.8. For 24-μm-wide stripes oriented along or perpendicular to the wires, the threshold currents were 35 mA (275 A/cm^2) and 30 mA (228 A/cm^2), respectively, for similar cavity lengths (about 530 μm). The lower threshold current for the perpendicular configuration was accompanied by a longer lasing wavelength (667 versus 662 nm). This red shift was attributed to the fact that lasing due to electron-heavy hole transitions occurred for the perpendicular configuration, while it was due to electron-light hole transitions for the longitudinal one.

Serpentine SL (SSL) structures, in which the growth rate is deliberately varied during growth of a tilted SL, have been developed in order to better control the lateral SL structure (Miller et al., 1992). In these structures, the growth rate is varied *intentionally* during the growth of the lateral SL layer, which results in formation of crescent-shaped wire structures laterally stacked in a dense array configuration (Fig. 4.9). The uncontrolled lateral variation in step periodicity or growth rates should then translate into variations in the positions of the wires rather than in their size and shape.

4.3 Quantum wire lasers

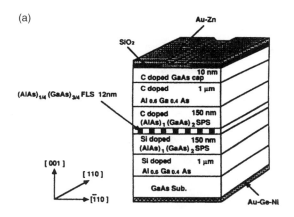

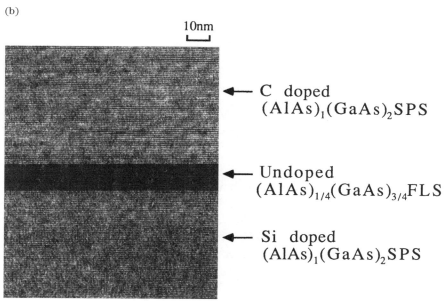

Figure 4.8: Schematic illustration (a) and TEM cross section (b) of a separate confinement, fractional superlattice GaAs/AlAs laser grown by OMCVD (Saito et al., 1993, see reference).

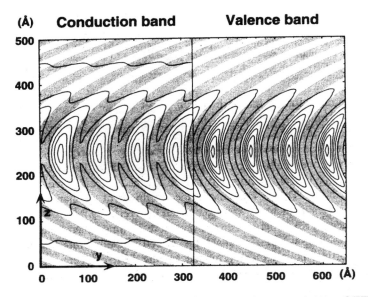

Figure 4.9: Schematic cross section of a serpentine superlattice QWR structure. The darker areas represent the higher-bandgap AlGaAs regions. Probability density contours of the calculated conduction band and valence band wavefunctions are shown (Hu et al., 1995, see reference).

Diode lasers incorporating such lateral SLs were fabricated and studied by Hu et al. (1993, 1995). The structures were grown on a vicinal (100) GaAs substrate tilted by 2° toward [011]. The recombination region consisted of a 68-nm-thick AlGaAs lateral SL layer, with an Al mole fraction varying nominally between 0 and 0.33 in the lateral direction. Low-temperature photoluminescence spectra of the structure showed polarization anisotropy, with the emission intensity higher for polarization vector oriented along the lateral SL wires than for the perpendicular orientation. The measured polarization anisotropy implies a sinusoidal lateral variation in the Al mole fraction of the lateral SL active layer, with minimum and maximum values of 0.135 and 0.195, respectively. Ridge waveguide lasers were fabricated with stripes oriented parallel or perpendicular to the wires, and lasing was observed in both types up to room temperature (CW). Threshold currents and their temperature dependence were comparable with those of conventional QW lasers. The optical gain spectra were measured at 1.4 K for both configurations. For

4.3 Quantum wire lasers

the perpendicular ridge configuration, a much higher TE gain was measured, whereas the TM mode gain was higher for the parallel configuration (Fig. 4.10). The threshold currents exhibited a corresponding behavior. These results indicate higher optical gain for polarization vectors oriented along the wire (in the perpendicular configuration) or along the larger dimension of the cross section of the wire (parallel configuration), consistent with the polarization anisotropy observed in the low-temperature PL spectra.

4.3.5 QWR lasers made by strained-induced self-ordering

Strained-induced, spontaneous formation of wirelike structures during epitaxial growth of lattice-mismatched thin layers has been used to fabricate extremely dense QWR heterostructures suitable for incorporation in diode lasers (Pearah et al., 1993; Chen et al., 1993) [see Fig. 4.4(e)]. Growth of a short-period superlattice (SPS) of $(GaP)_n/(InP)_n$ layers (where

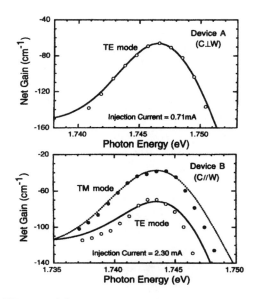

Figure 4.10: Measured low-temperature optical gain spectra of serpentine SL GaAs/AlGaAs lasers for stripes oriented perpendicular (device A) and parallel (device B) to the wires (Hu et al., 1995, see reference).

typically $n = 2$ monolayers) on a (100) GaAs substrate by gas source MBE results in lateral quasi-periodic modulation of the Ga/In composition along the [011] direction, with a typical pitch of 10 nm. Embedding such a laterally ordered layer in higher-bandgap GaInP barriers yields approximately 5-nm-wide QWRs whose thickness is directly determined by the thickness of the SPS layer. The average wire length was estimated to be approximately 300 nm. Strongly polarized luminescence, with linear polarization oriented parallel to the wire axis, was observed in both low-temperature PL (intensity ratio of up to 24) (Chen et al., 1993) and electroluminescence (Pearah et al., 1993) spectra and was interpreted as due to 2D quantum confinement in the wires.

Edge-emitting diode lasers incorporating such self-ordered wires in graded index, separate confinement heterostructure (GRIN-SCH), and step-index SCH configurations were studied by Stellini et al. (1993) and Pearah et al. (1994). The schematic and TEM cross sections of their laser structure are shown in Fig. 4.11. The wires have a cross-sectional area of approximately 5×10 nm^2 and are vertically embedded in $Al_{0.15}Ga_{0.35}In_{0.5}P$ barrier layers. Devices with BA stripe contacts aligned parallel and perpendicular to wires were tested. The perpendicular-stripe devices showed much lower thresholds and operated up to room temperature, whereas the ones with the parallel stripes lased only at low temperatures. The light versus current characteristics of the two devices at 77 K (pulsed conditions) are shown in Fig. 4.12; threshold current densities were 400 and 1500 A/cm^2 for perpendicular and parallel stripes, respectively (GRIN-SCH structures). This anisotropy was attributed to a higher optical gain for light polarized along the wire axis. The lasing wavelength was approximately 702 nm (TE mode) and 693 nm (TM mode) for laser stripes oriented perpendicular and parallel to the wires, respectively (GRIN-SCH structures, 77 K). The blue shift in lasing wavelength for the parallel stripe is due to the higher Fermi levels associated with the gain anisotropy in the wires. For the step-index SCH configuration, the emission wavelengths were 729 nm (TE) and 714 nm (TM), and the threshold current densities were 250 and 2190 A/cm^2 for the perpendicular and parallel stripe configurations, respectively.

Similar structures, emitting around 1.55-μm wavelength, were fabricated by MBE of $(GaAs)_2(InAs)_2$ SPS on (100) InP substrates (Chou et al., 1995). The active region consisted of five SPS layers, each about 10 nm thick, separated by 7.5-nm $Al_{0.24}Ga_{0.24}In_{0.52}As$ barrier layers, and was embedded in an $Al_{0.24}Ga_{0.24}In_{0.52}As$ waveguide layer bounded by

4.3 Quantum wire lasers

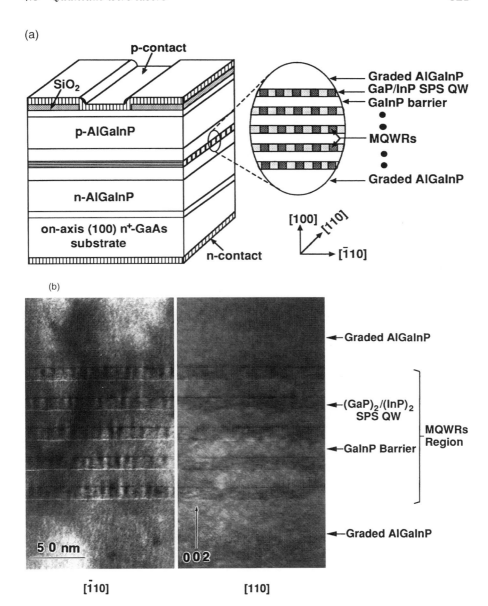

Figure 4.11: Schematic illustration (a) and TEM cross section (b) of an InGaP/AlGaInP QWR laser formed by strained-induced self-ordering. The wires are perpendicular to the [110] direction (Stellini et al., 1993, see reference).

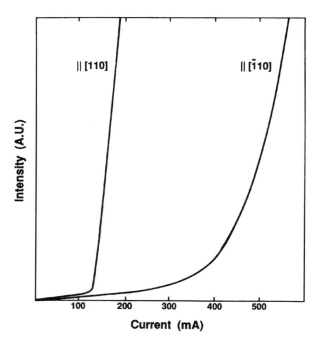

Figure 4.12: Light versus current characteristics of the QWR laser of Fig. 4.11 for injection stripes aligned along two different crystal directions (Stellini et al., 1993, see reference).

$Al_{0.48}In_{0.52}As$ cladding layers. Modulation of the Ga/In composition in the [011] direction resulted in formation of lower-bandgap, In-rich wires oriented in the [01$\bar{1}$] direction, with an average cross section of about 15 × 10 nm^2 and a lateral periodicity of about 30 nm. Top-view TEM images indicate that the wires are segmented, with an average length of 300 to 500 nm. Polarization anisotropy of the emission from these wires was observed in PL experiments, with a ratio of up to 4 between the intensity of emission polarized parallel to the wires relative to that polarized perpendicular to the wires. Broad-area diode lasers were fabricated with stripes oriented parallel and perpendicular to the wires and were operated under pulsed conditions. Much lower threshold current densities were obtained (at 77 K) for light polarized along the wires (propagation normal to the wires) as compared with light polarized perpendicular to the wires (200 A/cm^2 versus 2 kA/cm^2, emission wavelengths of 1.69 and 1.575 μm, respectively). As in the case of the shorter-wavelength GaInP/GaAs self-ordered QWR lasers, only the lasers with light polarized along the wires operated at room temperature.

4.3.6 QWR lasers grown on nonplanar substrates

The formation of buried QWR and QD structures by growth on nonplanar substrates relies on the characteristic faceting and thickness variations that occur during epitaxial growth on the patterned substrate due to the dependence of growth rate on the crystalline direction (Kapon, 1994) [see Fig. 4.4(f)]. The variation in growth rate is brought about by variations in the flux of the growth species and depends on the growth method, dimensions and orientation of the pattern, the growth species employed, and the growth conditions. As a result, a QW heterostructure grown on the nonplanar substrate exhibits lateral thickness and composition variations that are translated into lateral potential variations, partly via the 1D quantum size effect in the growth direction. The lateral potential wells thus formed can be used to quantum-confine carriers in 1D or 0D structures formed in situ (Kapon et al., 1987). Although narrow potential wells can be formed with this approach using, e.g., extremely narrow ($\sim$ 10 nm) groove patterns, *self-ordering* processes play an important role in obtaining such narrow structures without compromising their uniformity. In particular, OMCVD growth in [01$\bar{1}$]-oriented grooves on (100) GaAs substrates leads to self-ordering of crescent-shaped QWRs owing to the formation of extremely narrow, faceted V-grooves at the bottom of the etched channels (Bhat et al., 1988; Kapon et al., 1989a, 1989b, 1996). The in situ formation of the QWRs leads to high interface quality, and the self-limiting growth of the faceted surface results in wire structures whose composition, size, and shape depend mainly on the growth conditions rather than on the lithography details. In addition, tight 2D optical waveguide structures can be formed around the QWRs, which is useful for achieving relatively high optical confinement factors in the wires.

Evidence for formation of 1D subbands in GaAs/AlGaAs V-groove QWRs grown by OMCVD has been provided by low-temperature photoluminescence (PL), PL excitation, and cathodoluminescence (CL) spectroscopy (Kapon et al., 1992a; Walther et al., 1992a; Gustafsson et al., 1995; Oberli et al., 1995; Wang et al., 1995). Subband separations as large as about 50 meV have been observed in PL excitation spectra for the smallest V-groove wires obtained so far (Kapon et al., 1996). A distinct polarization anisotropy in the optical absorption, directly attributed to the 2D quantum confinement and related to valence band mixing, has been measured in such wires (Vouilloz et al., 1997), implying a corresponding anisotropy for the gain spectra. It has been observed that the energies of the 1D transitions observed in low-temperature PL and CL spectra of such wires stay constant up to relatively high carrier densities (a few times 10^6 cm^{-1}) (Grundmann et al., 1994, 1995; Gustafsson et al., 1995; Ambigapathy et al., 1997). This suggests that the recombination in these structures is dominated (at least at low temperatures) by excitonic processes and that the bandgap renormalization is smaller than in 2D systems (<10–15 meV) (Ambigapathy et al., 1997). The high interface quality of V-groove wires is evidenced by the relatively long (300–500 ps) carrier lifetimes at low temperatures. This is in contrast to the much shorter lifetimes observed, e.g., in etched wires, in which excessive nonradiative recombination at the sidewalls dramatically shortens the carrier lifetimes (Christen et al., 1992a).

A generic V-groove GaAs/AlGaAs QWR laser structure with a longitudinal cavity configuration (i.e., beam propagating *along* the wire) is depicted schematically in Fig. 4.13 (Kapon, 1993). The current is injected through a narrow ($\sim$1-μm) window in the contact defined by proton implantation. Since growth of an AlGaAs layer above a given GaAs QWR crescent results in a resharpening of the V-groove profile, recovering the self-limiting groove profile, it is possible to vertically stack several such wires of identical size and shape within the core of the optical waveguide (Christen et al., 1992b).

Separate-confinement heterostructure V-groove GaAs/AlGaAs diode lasers incorporating a single wire at the center of the V-shaped waveguide lased at room temperature under pulsed conditions at a wavelength of 800 to 830 nm and with threshold currents as low as 3.5 mA (Kapon et al., 1989a, 1989b). Peaks in the amplified spontaneous emission spectra and the envelopes of the longitudinal modes were observed and interpreted as due to the enhanced gain at the transition energies between the 1D subbands. The observed subband separation was 10 to 13 meV, in agreement with model calculations. Lasing occurred at an excited 1D subband in all these devices. Lasers with longer cavities (i.e., smaller cavity losses) or tighter optical confinement lased at a lower-energy subband. The relatively high threshold currents and the high lasing energy were due to the small optical confinement factor, which results in high quasi-Fermi levels. In addition, the relatively wide contacts lead to exces-

4.3 Quantum wire lasers

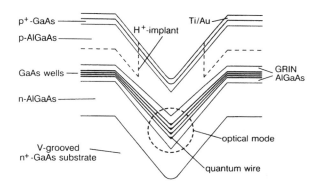

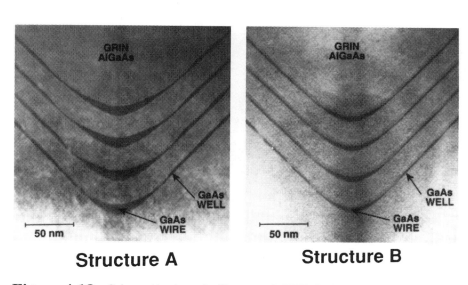

Figure 4.13: Schematic view of a V-groove 4-QWR GaAs/AlGaAs laser (upper panel) and TEM cross sections showing the cores of two lasers with different wire size (structures A and B). The devices were grown by atmospheric pressure OMCVD (Kapon et al., 1992c, see reference).

sive leakage currents. The lasers operated at the fundamental optical mode that is TE-polarized. The mode shows a circular near-field pattern, approximately 1 μm in diameter, as measured using conventional far-field (lens) optics. More recently, higher spatial resolution ($\sim$0.1 μm) scanning near-field optical microscope measurements of similar V-groove lasers showed that the fundamental mode is heart-shaped, with a width of approximately 0.5 μm (Ben-Ami et al., 1997), in agreement with an optical waveguide model of the structure (Kapon et al., 1992b).

Vertically stacked multiple-QWR V-groove lasers with improved optical confinement factors also were demonstrated and studied (Simhony et al., 1990, 1991; Kapon et al., 1992c). Figure 4.13 also shows TEM cross sections of the cores of two 4-QWR V-groove GaAs/AlGaAs QWR lasers with different wire sizes (denoted structures A and B in the figure). The dark, vertical stripes running through the centers of the wires are self-ordered AlGaAs vertical QWs (VQWs), formed due to Ga segregation at the bottom of the grooves.

The relatively slow lateral variations in the effective bandgap at the QWR crescents allow the evaluation of the 1D subband energies and wave functions in the framework of an adiabatic approximation (Kapon et al., 1989a). In this approximation, the envelope wave functions are factorized as $\psi(y,z) \approx \chi_y(z)\phi(y)$, where a slow variation with y is assumed. Substitution into Eq. (4.1) yields two coupled 1D Schrödinger equations describing the transverse and lateral wave function distributions:

$$\left[-\frac{\hbar^2}{2m^*}\frac{\partial^2}{\partial z^2} + V_y(z)\right]\chi_y(z) = E_{\text{conf}}(y)\chi_y(z) \qquad (4.14a)$$

$$\left[-\frac{\hbar^2}{2m^*}\frac{\partial^2}{\partial y^2} + E_{\text{conf}}(y)\right]\phi(y) = E\phi(y) \qquad (4.14b)$$

where $V_y(z)$ is the 1D potential distribution in the transverse z direction at each lateral position y, and $E_{\text{conf}}(y)$ is the corresponding transverse confinement energy. This adiabatic approximation yields a simple, intuitive procedure for determining the solutions for the eigen functions and eigen solutions of the quantum crescent. First, the lateral thickness variations of the QW at the crescent are evaluated (e.g., using TEM data), and the transverse potential distribution at each lateral position is employed to yield the transverse wavefunctions and the lateral variation of the confinement energy E_{conf}. The confinement energy distribution then serves as the potential well providing the lateral quantum confinement, and solution of the lateral 1D problem finally yields the lateral variation of the wavefunctions and the eigen energies of the 1D subbands. This procedure is illustrated in Fig. 4.14 for the QWR crescent structures of Fig. 4.13 (Kapon et al. 1992c). The lateral variation in confinement energy, due to the lateral tapering, results in this case in parabolic-shaped lateral potential wells with harmonic oscillator-like 1D subbands. The electron

4.3 Quantum wire lasers

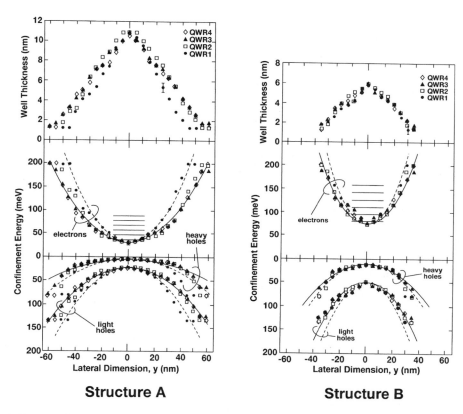

Figure 4.14: Lateral thickness variation (upper parts) and subband structure (lower parts) of the two 4-QWR structures of Fig. 4.13. Curves are parabolic fits for the lateral variations in effective confinement energies of the electrons and holes, deduced from the thickness measurements. Horizontal lines represent the calculated 1D subband energies for electrons (Kapon et al., 1992c, see reference).

subband separation is about 16 meV in the smaller wire structure and about 11 meV in the larger wire structure. Note that the 1D subband separations for heavy holes are much smaller due to their heavier mass, and therefore, the inaccuracy of this single-band model for the case of the valence band is less important for comparisons with the observed transition energies.

The gain spectra of GaAs/AlGaAs V-groove 4-QWR laser structures were evaluated by measuring the cavity finesse as a function of wavelength (Hakki and Paoli, 1975). The modal gain spectra at several diode currents are shown in Fig. 4.15(a) (Walther et al., 1992b; Kapon, 1993). The spectra were measured at room temperature under CW conditions, and the threshold current was 4 mA (high reflection coated facets). The emission is strongly TE polarized, consistent with the asymmetric shape of the wire cross section and the longitudinal cavity geometry. The gain spectra exhibit several peaks, whose energy separations agree well with the calculated energy separations between electron–hole transitions conserving the 1D subband index. Gain spectra exhibiting a similar spectral modulation were calculated for GaAs/AlGaAs V-groove QWR lasers by Citrin and Chang (1993). These gain peaks are attributed to the higher DOS at the 1D subbands. At higher diode currents, the gain at the higher-index subbands increases due to sequential filling of the higher-energy subbands. The measured values at the gain peaks are displayed in Fig. 4.15(b) as a function of the diode current. The peak (modal) gain at the energy corresponding to the transition between the ground electron and hole subbands saturates at approximately 13 cm^{-1}. Lasing cannot be observed at this transition because the cavity loss (23 cm^{-1}) exceeds the saturated gain value. Subsequent subband filling at higher diode currents leads to larger gain and finally lasing at the fourth subbands.

The 1D subband structure in these lasers is clearly observed in the amplified spontaneous emission (ASE) spectra close to their threshold. The ASE spectra of the two V-groove 4-QWR lasers of Fig. 4.13 are shown in Fig. 4.16 for several currents near threshold (Kapon et al., 1992c). The subband separations agree well with the calculated ones, thus demonstrating the scaling of the 2D confinement energy with the size of the wires. Above threshold, lasing takes place at an energy corresponding to transitions between excited 1D electron and hole subbands due to band filling. The larger-wire and the smaller-wire lasers operated under pulsed conditions at 830 and 800 nm wavelength, with threshold currents of 26 and 20 mA for (uncoated) cavity lengths of 2.5 and 1.7 mm, respectively. Lasing from lower-index subbands was observed for the structure with the larger subband separation ($l = 3$ versus $l = 5$).

Even with a multiple QWR structure, the optical confinement factor of the V-groove lasers was too low to allow lasing at room temperature from the ground 1D subbands. Modeling of the near-field distribution of the optical mode in the V-shaped 2D waveguides shows that the peak intensity of the fundamental mode is located above the geometric center of the waveguide (Kapon et al., 1992b). This suggests that a higher confine-

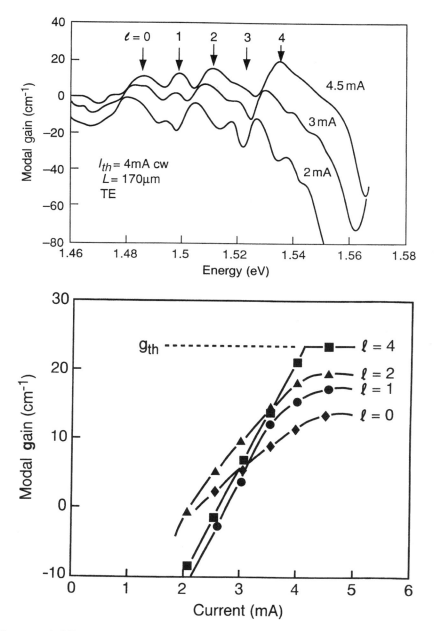

Figure 4.15: (a) Measured modal gain spectra of an asymmetric-waveguide 4-QWR V-groove GaAs/AlGaAs diode laser (room temperature, CW operation). Arrows show the calculated energies for 1D electron–heavy-hole transitions. (b) Measured peak gain at the different subbands in the spectra of (a) versus the diode current. Emission is TE polarized [parallel to the (100) substrate] (Kapon, 1993).

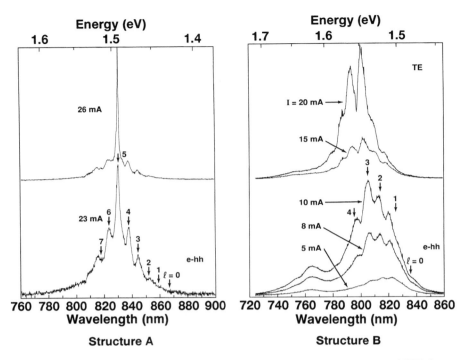

Figure 4.16: Room-temperature emission spectra of the two 4-QWR laser structures of Fig. 4.13 (pulsed operation). The arrows indicate the energies of electron–heavy-hole transitions between the calculated QWR subbands shown in Fig. 4.14 (Kapon et al., 1992c, see reference).

ment factor can be achieved with an asymmetric QWR structure in which the wires are vertically displaced above the waveguide center. In such an optimally designed 4-QWR structure, the calculated optical confinement factor increases to 1.17%, as compared with 0.63% for the symmetric structure (Kapon, 1993). Still higher confinement factors should be achievable in a waveguide with more wires. It should be noted, however, that the highest achievable confinement factor is limited due the finite barrier thickness between wires required for reestablishing the self-limiting surface profile underneath the subsequent crescent (Kapon et al., 1996).

Lower threshold currents were achieved with 3-QWR V-groove wire lasers with high-reflection facet coatings for reducing the optical cavity losses (Simhony et al., 1991). Figure 4.17 shows the emission spectra of such a device below and above threshold. Threshold is attained at room

4.3 Quantum wire lasers

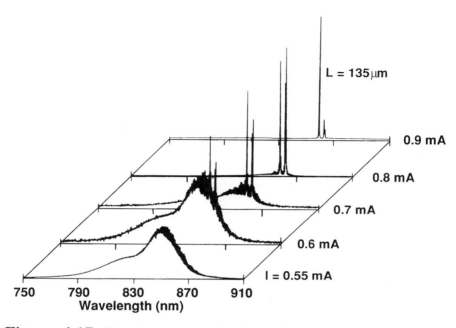

Figure 4.17: Emission spectra of a high-reflection-coated 3-QWR GaAs/AlGaAs V-groove laser. Threshold current is 0.6 mA (room temperature, pulsed operation) (Simhony et al., 1991, see reference).

temperature at 0.6 mA under pulsed conditions. Spectral modulation due to the 1D subbands is also evident in the spectra.

The effective carrier capture into the V-groove QWRs via the surrounding GaAs and AlGaAs QWs has been evidenced by the temperature dependence of their PL spectra (Walther et al., 1992a). At low temperatures ($T < 80$ K), the PL spectra are dominated by luminescence from the QW regions. However, at higher temperatures, the carriers are transported toward the QWRs more efficiently due to their higher mobility. Subsequent capture into the wires results in spectra dominated by the QWR emission line at $T > 80$ K. PL excitation spectra also support this efficient transfer from the QWs into the QWRs. This efficient transfer is particularly evident in wires made on submicron-pitch corrugations, where the pitch is smaller than the carrier diffusion length. For an array pitch of 0.25 μm, spectra dominated by QWR lines were measured even at lower temperatures in V-groove structures (Christen et al., 1992a). The capture time is limited by the diffusion in the barrier regions. Time-resolved PL studies showed capture times as short as 25 ps, corresponding (in a large part) to diffusion in approximately 100-nm-long VQW sections in 0.25-μm-pitch V-grooved QWR structures (Haacke et al., 1996). It appears that transfer into the wires via the vertical AlGaAs wells is more efficient than that due to transport in the sidewall GaAs QWs, probably due to the necking on both sides of the wires that results in effective potential barriers.

Lasing from the ground 1D states in V-groove QWR structures has not been achieved so far at room temperature because of gain saturation. Lasing from lower-index subbands, however, is possible at lower temperatures. Figure 4.18 shows the temperature dependence of the lasing energy for a single-QWR diode laser with a longitudinal cavity configuration (Chiriotti, 1996). Also depicted are the temperature dependence of the GaAs band edge as well as the variation of the calculated subband transitions. One can see that the lasing subband index decreases at lower temperatures, finally reaching the ground state at temperatures below approximately 70 K. This is due to the higher material gain at low temperatures,

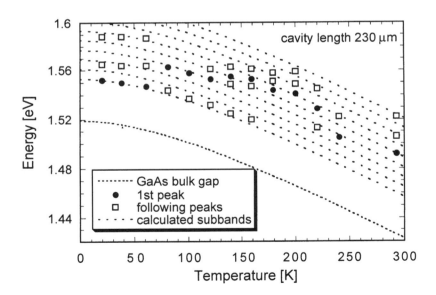

Figure 4.18: Temperature dependence of the lasing energy of a single-QWR, GaAs/AlGaAs V-groove laser. The variations with temperature of the GaAs band edge and of the 1D subbands are also indicated (Chiriotti, 1996).

4.3 Quantum wire lasers

which increases the available gain at a given carrier density. The nonmonotonous energy variation around 200 K may be due to the temperature dependence of the carrier capture efficiency, discussed earlier.

Even lower threshold currents were obtained with MBE-grown strained InGaAs/GaAs/AlGaAs V-groove QWR lasers (Tiwari, 1995; Tiwari and Woodall, 1994; Tiwari et al., 1992, 1994). Figure 4.19 displays the cross section of such a device. In this case, the [01$\bar{1}$] grooves were etched through a silicon nitride mask, which was then left on the sample during

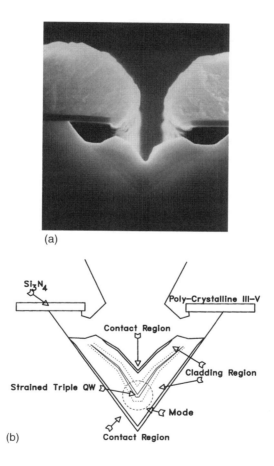

Figure 4.19: SEM (a) and schematic (b) cross sections of a V-groove InGaAs/ GaAs strained QWR laser grown by MBE (Tiwari et al., 1994, see reference).

growth. This results in growth of high-resistivity (106 Ω·cm) polycrystalline material on both sides of the laser stripe, which yields efficient current confinement. The core of the laser consisted of three strained InGaAs (20% nominal In mole fraction) crescent-shaped wires embedded in GaAs barrier layers and GRIN $Al_xGa_{1-x}As$ optical waveguide layers. Cross-sectional TEM data indicate crescent thicknesses of 10 to 11.5 nm, and modeling of the 2D confined electron subbands yields level separation of 3 to 4 meV. The estimated optical confinement factor for these lasers is about 1.6%.

The MBE-grown V-groove QWR lasers exhibited submilliampere CW room-temperature threshold currents. Threshold currents were as low as 188 μA for 1-mm-long cavities *without* mirror coatings (Fig. 4.20), the internal efficiency was 83%, and the internal optical loss coefficient was 4.2 cm^{-1}. Threshold currents as low as approximately 120 μA were obtained for shorter cavities (Tiwari, 1995). An important factor in obtaining such low threshold currents is the low leakage currents in these structures. The low internal losses led to external differential efficiency of about 0.5 μW/μA and a maximum output power of 50 μW (per facet). Further reduction in threshold current appears feasible due to the rela-

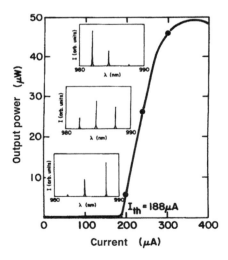

Figure 4.20: Light-current characteristic of the InGaAs/GaAs QWR laser of Fig. 4.19 (room-temperature, CW operation). Inset shows the emission spectra at several currents. (Tiwari et al., 1994, see reference).

4.3 Quantum wire lasers

tively high mirror losses in these devices. Near threshold, the lasing wavelength was about 990 nm, blue shifting at higher currents via hopping between adjacent QWR subbands (see inset in Fig. 4.20). A single longitudinal mode is obtained at each QWR subband, and the measured subband energy separation, 4 to 5 meV, agrees with the calculated value. As shown in Fig. 4.21, these QWR lasers exhibited a T_0 value of 260 K near room temperature, significantly larger than that of a ridge waveguide QW laser grown side by side ($T_0 = 206$ K). This improvement in T_0 might be limited in this case by the oscillation at several 1D subbands.

Tiwari et al. (1992) also reported the modulation characteristics of the MBE-grown V-groove QWR lasers (Fig. 4.22). The maximum 3-dB modulation bandwidth of these lasers was about 12 GHz at a diode current of less than 600 μA (Tiwari, 1995). This relatively low bandwidth was obtained despite the fact that the measured differential gain (1.5×10^{-13} cm^2) was two orders of magnitude higher than that in conventional strained QW laser. The low bandwidth was attributed to the high gain compression coefficient ($\varepsilon = 1.79 \times 10^{-16}$ cm^3), almost 10 times higher

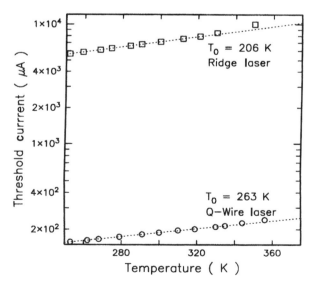

Figure 4.21: Temperature dependence of threshold current of the V-groove QWR laser of Fig. 4.19. Corresponding data for a 1.5-μm-wide ridge-waveguide QW laser grown side by side are also shown (Tiwari et al., 1994, see reference).

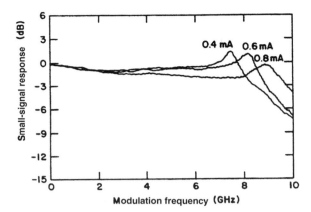

Figure 4.22: Small-signal response of the V-groove QWR laser grown by MBE as a function of modulation frequency (Tiwari et al., 1992, see reference).

than that of a strained QW laser. The increase in the gain compression was attributed to spectral hole burning and/or carrier heating.

Whereas the extremely small active volume in a single-QWR laser is beneficial for reducing the threshold current, it also leads to a limited output power. The output power of V-groove QWR diode lasers was increased by employing longitudinal cavity laser array configurations (Qian et al., 1995). The V-groove GaAs/AlGaAs structure consisted of 19 lasers on 6-μm spacings, each containing four vertically stacked GaAs QWR crescents. The structure was proton implanted to provide current confinement to each V-groove active region; all lasers were electrically pumped in parallel. Maximum peak (pulsed) output power of 115 mW was achieved with a threshold current of 0.7 A at room temperature. However, the external efficiency was rather low (maximum power at 3 A), probably due to current leakage between the V-grooves.

The QWR spectral gain modulation due to the 1D DOS profile leads to output wavelength selectivity, whereby only a few Fabry-Perot modes (and often a single one) near the gain peak oscillate (see Fig. 4.16). Further wavelength selectivity can be achieved using transverse QWR array configurations utilizing complex-coupling DFB effects. Such DFB QWR laser structures were realized by OMCVD growth on substrates patterned with submicron pitch periodic corrugations (Fig. 4.23) (Walther et al., 1993). The InGaAs wires were 14 to 17 nm thick and 70 to 80 nm wide, with a pitch of 0.25 μm. Gain-guided lasers were fabricated by proton implantation of stripes oriented parallel or perpendicular to the wires. The lasers lased up to room temperature under pulsed conditions. For the perpendicular configuration, the output wavelength was nearly constant around 920 nm for $T < 150$ K due to second-order Bragg reflection from the gain- and index-modulated grating. A complex coupling coefficient $\kappa = 2.5e^{i0.25\pi}$ to $4e^{i0.15\pi}$ was calculated for these structures, suggesting a significant gain coupling effect. The threshold current for the perpendicular configuration was found to be higher than that for the parallel one, probably because of excess cavity losses due to first-order Bragg diffraction. The threshold current density for the parallel-stripe configuration was 66 A/cm^2 at 80 K, rising sharply for $T > 200$ K to a value of 1.9 kA/cm^2 at room temperature. This sharp increase was attributed to inefficient carrier capture from the GaAs waveguide layer into the InGaAs wires; note that in these structures there is no AlGaAs VQW region, which might otherwise help in carrier capture. In the perpendicular-stripe configuration, the carrier accumulation in the GaAs waveguide resulted in lasing or ASE from the GaAs layers at room temperature. High-reflection-coated lasers with a perpendicular-stripe configuration lased at room temperature with a threshold current of 100 mA (cavity length of 360 μm), showing spectral features attributed to 1D subband transitions (Kapon, 1993).

Arakawa et al. (1992) fabricated vertical cavity surface-emitting lasers incorporating GaAs/AlAs distributed Bragg mirrors and InGaAs V-groove QWRs and tested them under optical pumping. The wires were fabricated on top of the bottom Bragg mirror using selective OMCVD growth on a submicron-pitch SiO$_2$ mask. A 4-lambda cavity was used in order to ensure complete planarization of the upper Bragg mirror. The structures were optically pumped at 77 K under CW conditions. The measured luminescence spectra, centered at 883 nm, showed a much narrower spectral width (3 nm), as compared with the spectral width of the wires without the cavity (16 nm), demonstrating the filtering effect of the cavity. The nonlinear dependence of the output intensity as a function of the pumping power was interpreted as evidence for lasing from the wires at the cavity mode.

The small cross-sectional area of a QWR may also be advantageous for minimizing the adverse effects of misfit dislocations arising during growth on mismatched substrates, e.g., in the case of III-V compound lasers grown on Si substrates. Hasegawa et al. (1994) studied V-groove GaAs/AlGaAs QWR lasers grown on GaAs/Si substrates by atmospheric-pressure OMCVD. The V-grooves were etched in an n-type GaAs layer

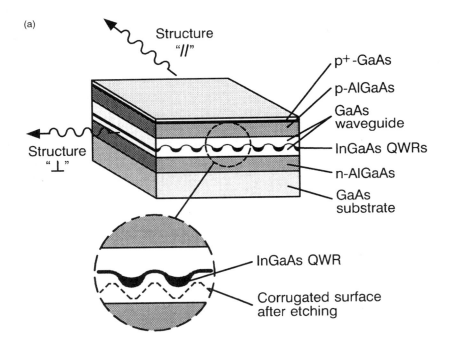
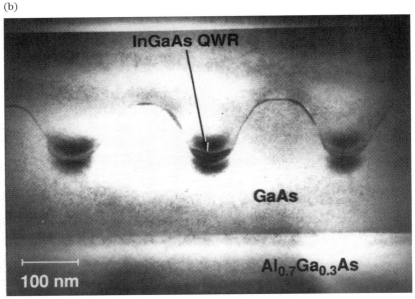

Figure 4.23: Schematic diagram (a) and TEM cross section (b) of a V-groove QWR laser incorporating strained InGaAs/GaAs QWRs grown on a submicron-pitch gratings (Walther et al., 1993).

grown on a Si substrate. Their structure consisted of three vertically stacked GaAs/AlGaAs wires at the center of a V-shaped waveguide. Room-temperature operation with threshold currents as low as 9.8 mA under pulsed conditions and 16 to 20 mA under CW conditions were demonstrated at 859-nm output wavelength. The threshold currents were significantly lower than for QW lasers grown on planar Si substrates (24–64 mA, CW). This was attributed to the small size of the active region, which effectively reduces the number of dislocations in the lasing region.

The characteristics of different QWR lasers demonstrated so far are summarized in Table 4.1 on the following pages.

4.4 Quantum dot lasers

Efficient QD lasers rely on high interface quality even more than QWR devices because of the larger surface-to-volume ratio for dots. The further reduction in optical confinement factor requires the realization of extremely high-density arrays of dots in order to exploit the expected increase in material gain. As in the case of QWR lasers, earlier attempts to realize QD lasers consisted of etching and regrowth of QW laser heterostructures [Fig. 4.24(a)]. This approach has had limited success due to the nature of the processed and regrown interfaces as well as because of the limits on dot density set by current lithography techniques. More recently, the in situ formation of high-density QD structures using various self-organized growth schemes has been demonstrated and employed in the fabrication and studies of QD laser structures [e.g., Fig. 4.24(b)].

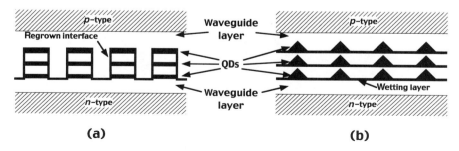

Figure 4.24: Schematic diagrams of QD structures for laser applications fabricated by (a) etching and regrowth and (b) Stranski-Krastanow self-organized growth.

Fabrication Method/Structure	Material System	Growth Method	Highest Operation Temperature	Lasing Wavelength	Threshold Current/ Current Density	Reference
Etching and regrowth/BA	InGaAs/InP (lattice matched)	OMCVD	77 K	1.36 μm	3.8 kA/cm^2	Cao et al., 1990
Etching and regrowth/BH	InGaAs/GaAs (strained)	OMCVD	RT (CW)	1.46 μm	16 mA/816 A/cm^2	Kudo et al., 1993
Etching and regrowth/ ridge waveguide	InGaAs/GaAs (lattice matched)	Gas source MBE	7 K	1.33 μm	235 mA	Notomi and Tamamura, 1993
Cleaved edge overgrowth	GaAs/AlGaAs	MBE	4 K (CW)	786 nm	0.4 mA	Wegscheider et al., 1994
Lateral superlattice on vicinal substrates/BA	GaAs/AlGaAs	MBE	RT	828 nm	470 A/cm^2	Tsuchiya et al., 1989
Lateral superlattice on vicinal substrates/BA	GaAs/AlGaAs	OMCVD	195 K	667 nm (10 K)	275 A/cm^2 (10 K)	Saito et al., 1993
Serpentine super- lattice/ridge waveguide	GaAs/AlGaAs	MBE	RT	710 nm (1.4 K)	20 mA (RT, CW)	Hu et al., 1995
Strained-induced self-ordering/BA	GaInP/GaAs	Gas source MBE	RT	729 nm (77 K)	250 A/cm^2 (77 K)	Pearah et al., 1994

4.4 Quantum dot lasers

Strained-induced self-ordering/BA	InGaAs/InP	MBE	RT	1.69 μm (77 K)	200 A/cm^2 (77 K)	Chou et al., 1995
V-groove/single QWR	GaAs/AlGaAs	OMCVD	RT (CW)	800–840 nm	4 mA	Kapon et al., 1989
V-groove/multiple QWR	GaAs/AlGaAs	OMCVD	RT	860 nm	0.6 mA (HR coated)	Simhony et al., 1991
V-groove/stripe geometry DFB	InGaAs/GaAs	OMCVD	RT	990 nm	100 mA (HR coated)	Walther et al., 1993; Kapon 1993
V-groove/multiple QWR	InGaAs/GaAs	MBE	RT (CW)	990 nm	120 μA	Tiwari et al., 1995
V-groove/VCSEL	InGaAs/GaAs	OMCVD	77 K	883 nm	Optically pumped	Arakawa et al., 1994
V-groove/multiple QWR	GaAs/AlGaAs on GaAs/Si	OMCVD	RT (CW)	859 nm	16 mA	Hasegawa et al., 1994

Note: All parameters are for pulsed operation and for edge-emitting cavity configurations unless otherwise indicated.

Table 4.1: Summary of characteristics of different quantum wire laser structures.

4.4.1 QD lasers fabricated by etching and regrowth

Electroluminescence from QD-like structures prepared by holographic photolithography, wet chemical etching, and liquid-phase epitaxy (LPE) regrowth was first reported by Miyamoto et al. (1984). The QW base structures were realized in the InGaAsP/InP lattice-matched system using OMCVD, and the pitch of the square array of dots was 204 nm. Electroluminescence from the dots (at ~1.4-μm wavelength) was observed at 77 K under pulsed conditions. However, lasing threshold could not be attained, probably due to excessive nonradiative recombination at the regrown interfaces.

Tensile-strained, buried-heterostructure $In_{0.33}Ga_{0.67}As$/InGaAsP/InP QD lasers fabricated using electron-beam lithography, wet etching, and OMCVD growth and regrowth were reported by Hirayama et al. (1994) (Fig. 4.25). Such tensile-strained structures allow the achievement of 1.5-μm wavelength emission with a QW layer that is thicker than for corresponding compressively strained structures, thus increasing the optical confinement factor. The dots were 30 nm in diameter, with a 70-nm center-to-center spacing on a square lattice; the optical confinement factor was estimated to be approximately 0.3% (a single dot layer). Lasing from the QD region was observed under pulsed conditions at 77 K at 1.26-μm wavelength, with $I_{th} = 1.1$ A and $J_{th} = 7.6$ kA/cm^2 ($L = 2080$ μm, uncoated facets). The relatively large blue shift of the lasing wavelength relative to the InGaAs band edge (0.1 μm) was attributed to large band filling in the dots due to the small optical confinement factor and dot size fluctuations.

4.4.2 QD lasers made by self-organized growth

Nötzel et al. (1994a, 1994b) have demonstrated the self-formation of disklike InGaAs/AlGaAs structures during OMCVD on (311)B GaAs substrates. The disks are formed during a growth interruption following the deposition of a strained InGaAs layer, presumably due to the lower barrier height for adatom migration as compared with (100) substrates. The self-forming disks are as small as 30 nm in diameter and 5 nm in height and are placed on a quasi-regular hexagonal lattice with approximately 200 nm pitch. Efficient luminescence due to excitonic transitions has been observed from such disk structures (Nötzel et al., 1994b).

Ridge waveguide diode lasers incorporating such quantum disks were fabricated by vertically stacking two 6-nm-thick $In_{0.25}Ga_{0.75}As$ layers be-

4.4 Quantum dot lasers

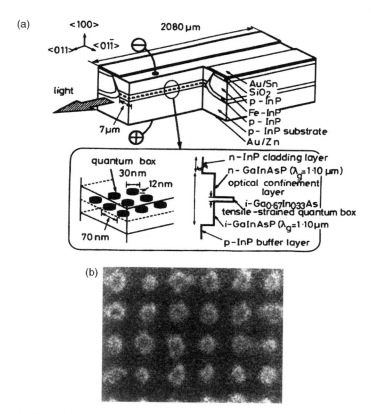

Figure 4.25: Schematic diagram (a) and SEM top view of a 70 nm-pitch GaInAs/GAInAsP/InP tensile-strained Quantum Box Laser (just after wet-chemical etching. (Hirayama et al., 1994, see reference).

tween $Al_{0.15}Ga_{0.85}As$ barrier layers in a SCH configuration (Temmyo et al., 1995). An SEM top view of the disk structure is shown in Fig. 4.26. The diameter of each InGaAs disk is approximately 60 nm, and the surface coverage density is approximately 6%. Room-temperature CW operation of approximately 900-μm-long devices with approximately 2-μm-wide ridges was achieved at a threshold current of 23 mA. The lasing wavelength was approximately 940 nm, blue shifted with respect to that of a control double-QW InGaAs/AlGaAs laser grown side by side.

Another approach for preparing QD structures whose interfaces are formed in situ is the application of the strain-driven Stranski-Krastanow

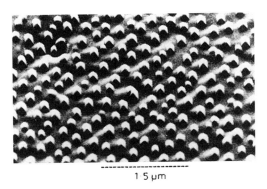

Figure 4.26: Top-view SEM image of disklike InGaAs/AlGaAs heterostructures self-formed during OMCVD on (311)B GaAs substrates (Temmyo et al., 1995, see reference).

(SK) growth mode during MBE or OMCVD (Goldstein et al., 1985). In this case, growth of a sufficiently thick, highly strained layer gives rise to formation of flat islands lying on a 2D layer, the so-called wetting layer. The resulting dots are pyramidal or lenslike in shape, typically 10 to 50 nm in diameter, and can be prepared in densities as high as 10^{10} to 10^{11} cm^{-2}. An atomic force microscope (AFM) top view of such an InGaAs/GaAs dot structure grown by MBE on a (100) GaAs substrate is shown in Fig. 4.27 (Leonard et al., 1993). Defect-free dot structures with high

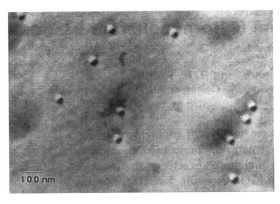

Figure 4.27: Top-view atomic force microscope image of Stanski-Krastanow InGaAs/GaAs dots (Leonard et al., 1993, see reference).

luminescence efficiency thus result, with typical low-temperature luminescence linewidths of 30 to 50 meV. The rather broad linewidths are due to inhomogeneous broadening brought about by the relatively wide size distributions of the dots, typically 10% to 20% in width. In fact, luminescence spectra of a small number of dots (or possibly a single QD) obtained using mesa-etched dot structures (Marzin et al., 1994) or by performing low-voltage CL spectroscopy (Grundmann et al., 1995) show extremely narrow (submillielectron-volt) peaks attributed to recombination of excitons in an individual QD potential well. In addition to the efficient luminescence, the fast rise time of the PL signal following a short pulse excitation (10 ps) supports the absence of a phonon bottleneck in such dots (Marzin et al., 1994; Wang et al., 1994; Kurtenbach et al., 1995). Almost constant carrier lifetime as a function of temperature was measured at low temperatures for InP/InGaP QDs (Kurtenbach et al., 1995), consistent with the expected behavior of 3D confined structures (Citrin, 1993).

Kirstaedter et al. (1994) reported lasing from electrically pumped SK $In_{0.5}Ga_{0.5}As$/GaAs QD structures embedded in a GaAs/AlGaAs SCH diode laser configuration grown by MBE. Nominal thicknesses of the (single) $In_{0.5}Ga_{0.5}As$ active layer ranged between 1 nm (where the onset of dot formation is observed) to 1.8 nm (where the dots are well developed), and dot diameters ranged between 10 and 30 nm. At low temperatures (below 150 K), lasing was observed at 1-μm wavelength, more than 100 meV lower than the emission energies of a uniform 1-nm-thick $In_{0.5}Ga_{0.5}As$ QW layer, indicating lasing from the dot structures rather than from the wetting layer. Threshold current densities were as low as 120 A/cm^2 at 77 K but increased sharply at higher temperatures (as low as 950 A/cm^2 at room temperature). A high value of T_0 (330–350 K) was measured at low temperatures and was interpreted as a signature of lasing due to transitions in a 0D structure. The sharp increase in threshold current above approximately 150 K may be due to thermal excitation of carriers into the GaAs waveguide layer.

Similar MBE grown InGaAs/GaAs QD laser structures with larger dots (~50 nm in diameter, 13 nm in height, 2–3 × 10^{10} cm^{-2} density) lased at room temperature under pulsed conditions (Mirin et al., 1996). Broad area lasers of 800-μm cavity lengths had a threshold current density of approximately 500 A/cm^2 at a wavelength of 1020 nm. Comparison with QW lasers incorporating essentially the InGaAs wetting layer only, for which the lasing wavelength was 940 nm, indicates that lasing in the

QD lasers resulted from recombination at excited 0D states because of state filling. The temperature dependence of the emission wavelength of the QD lasers was found to be much weaker than that of the QW lasers and was explained as due to a lower degree of state filling at lower temperatures. In another report, room-temperature lasing in $In_{0.4}Ga_{0.6}As/$ GaAs SK QD lasers incorporating dots of 12-nm diameter at 1028-nm wavelength was attributed to e1-hh2 QD transitions (Kamath et al., 1996).

Moritz et al. (1996) reported room-temperature lasing in optically pumped, visible-wavelength $InP/Ga_{0.52}In_{0.48}P$ SK QDs sandwiched between $Al_{0.51}In_{0.49}P$ waveguide cladding layers (grown on a GaAs substrate). The InP dots were 20 to 30 nm in diameter and 5 to 10 nm in height, with a density of 3×10^{10} cm^{-2}. Lasing from both the wetting layer (at 670 nm) and the QDs (at 705 nm) was observed. Measurements of the optical gain in these structures show a higher gain at the wetting layer, although the lasing spectra were dominated by emission from the QDs. This was attributed to the fact that the carriers in the dots were not in thermal equilibrium with those in the wetting layer, possibly due to necking in the wetting layer thickness next to the dots.

Self-formed InGaAs/GaAs dot structures also were realized by a strained-induced atomic layer epitaxy (ALE) technique, in which several short cycles of $(InAs)_1$-$(GaAs)_1$ growth sequences are obtained via a fast switching of the flow of organometallic precursors (Mukai et al., 1994). The resulting $In_{0.5}Ga_{0.5}As$ dots are typically 20 to 30 nm in diameter and approximately 10 nm in height (the latter determined by the total grown thickness of InGaAs), with up to 10% surface coverage. They were vertically cladded by GaAs and laterally delimited by $In_{0.1}Ga_{0.9}As$. The emission wavelength of such dots extends over 1.1 to 1.5 μm. Peaks separated by 50 to 80 meV were observed in low-temperature PL and electroluminescence spectra of such dots and were attributed to recombination at the 0D states (Mukai et al., 1996). Analysis of the luminescence lineshape indicated intersubband relaxation times on the order of 10 to 100 ps and was interpreted as due to phonon bottleneck effects.

Edge-emitting broad-area diode lasers incorporating a single layer of such InGaAs/GaAs dots (20-nm diameter, 10-nm height, ~10% surface coverage) in a separate confinement heterostructure were studied by Shoji et al. (1995). Lasing was obtained under pulsed conditions at 80 K at a threshold current density of 815 A/cm^2. The lasing wavelength was 911 nm, consistent with transitions between excited QD states. A low value of the diamagnetic shift in the lasing energy (2.5 meV for magnetic fields up 13 T) as compared with InGaAs QW lasers emitting at similar wavelengths (5 meV) supported the interpretation of lasing from 3D confined states.

The results just described indicate that the optical gain due to transitions between the dot ground states is not sufficient for attaining the threshold in a typical edge-emitting laser cavity. This is due to the small optical confinement factor in a single-layer QD structure, in which a typical area coverage is approximately 10%. In an attempt to increase the optical gain and reduce the carrier density at threshold, vertically stacked arrays of SK self-organized QDs have been employed. It has been demonstrated that stacking of several SK InAs/GaAs dot layers with sufficiently thin GaAs barriers results in alignment of the QDs in a vertical array configuration (Xie et al., 1995; Solomon et al., 1996). The vertical alignment results from dips in the stress fields that develop directly above each dot in the first layer, thereby inducing preferential In migration and island growth at that location in subsequent layers. Vertical alignment of more than 95% of the dots has been achieved with GaAs separation layers less than approximately 50 nm thick (Xie et al., 1995).

Edge-emitting diode lasers incorporating five layers of vertically stacked InAs/GaAs SK QDs (with an area density of $\sim 6 \times 10^{10}$ cm^{-2}) were grown using MBE by Xie et al. (1996). Figure 4.28 shows a cross section of the laser structure. The dots were vertically separated by approximately 10 nm and were therefore electronically uncoupled. The devices lased at 79 K under pulsed conditions, at a wavelength of 987 nm, and with a threshold current density of 310 A/cm^2. In comparison, the PL spectra showed emission from the ground states of the dots at 1052 nm and from the wetting layer at 863 nm.

Diode lasers incorporating *electronically coupled,* vertically stacked SK InAs/GaAs dots were reported by Ustinov et al. (1997). The dots were approximately 20 nm in diameter and approximately 5 nm in height, with a density of approximately 10^{11} cm^{-2} per layer. Edge-emitting structures with up to 10 layers of dots separated by 1 to 2 nm of GaAs were investigated. Pulsed operation was achieved up to room temperature, with the lasing wavelength increasing from approximately 950 nm for a single dot layer to 1050 nm for three layers. The multilayer structures also resulted in an improved temperature stability of the threshold current, yielding $T_0 = 430$ K up to 175 K for the three-layer devices. However, the threshold currents increased rapidly above that point ($J_{th} = 40$ A/cm^2 at 80 K and 660 A/cm^2 at 290 K). This was probably due to carrier reevaporation from

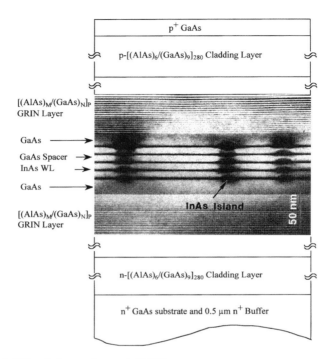

Figure 4.28: Schematic and TEM cross sections of a vertically stacked Stranski-Krastanow QD InAs/GaAs laser (Xie et al., 1996, see reference).

the dots to the GaAs waveguide layer at the high carrier densities near laser threshold. Further improvement was obtained by inserting the InAs dots in thin (10-nm) GaAs QWs, yielding $T_0 = 350$ K up to room temperature, where J_{th} was 500 A/cm^2. Employing AlGaAs as dot barriers resulted in threshold current densities as low as 63 A/cm^2 and internal efficiencies as high as 70%. Diode lasers incorporating 10 QD layers with HR (99%) coated back mirrors operated CW at room temperature.

Room-temperature operation of three-layer InAs/GaAs SK QD lasers (20-nm diameter, 5-nm height, 20-nm GaAs barriers, ~10% area coverage) also was reported by Shoji et al. (1996). Figure 4.29 shows the room-temperature electroluminescence spectra of such a laser with uncoated facets. Emission from several QD subbands is observed, with a blue shift due to state filling at higher currents. At room temperature, 900-μm-long lasers with as-cleaved mirrors lased at the first excited subband transition (~1040 nm wavelength), and lasing abruptly shifted to longer wavelengths (~1070 nm) corresponding to ground-state lasing as the tempera-

4.4 Quantum dot lasers

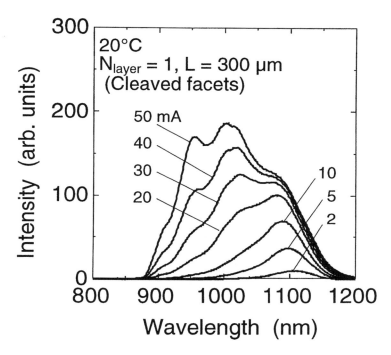

Figure 4.29: Room-temperature electroluminescence spectra of a three-layer, vertically stacked Stranski-Krastanow InAs/GaAs QD edge-emitting laser (Shoji et al., 1996, see reference).

ture was reduced to approximately 260 K (Fig. 4.30). In high-reflection-coated devices of the same length (95%, both facets), lasing from the ground state persisted up to room temperature (1105 nm wavelength). Room-temperature CW operation also was demonstrated with the HR-coated devices ($J_{th} \approx 1$ kA/cm^2 per dot layer).

A subgroup of QDs with a narrow size distribution can be effectively selected out of the broad size distribution of the SK dots using a wavelength-selective optical cavity. Saito et al. (1996) have demonstrated such a QD laser by integrating a multistack of In$_{0.5}$Ga$_{0.5}$As/Al$_{0.25}$Ga$_{0.75}$As SK dots in a VCSEL structure consisting of two GaAs/AlAs Bragg mirrors and an AlGaAs λ-spacer. The active region consisted of 10 layers of 28-nm-diameter and 3-nm-height dots with a density of approximately 2×10^{10} cm^{-2}. Single VCSEL devices with an area of 25×25 μm^2 were operated CW at room temperature with a threshold current of 32 mA,

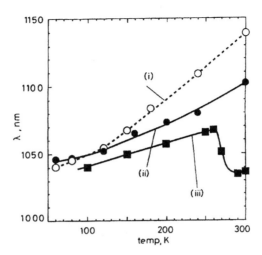

Figure 4.30: Temperature dependence of PL (i) and lasing wavelengths [(ii), HR coated; (iii), as cleaved] of a three-layer, vertically stacked SK InAs/GaAs QD edge-emitting laser (900-μm cavity length) (Shoji et al., 1996, see reference).

showing a single longitudinal mode output at 960-nm wavelength. This wavelength was intermediate between the PL wavelengths of the QD ground state and the wetting layer emission (1000 and 920 nm, respectively), suggesting that lasing occurred due to transitions between excited states of the dots.

One of the open questions regarding the SK-based QD lasers is related to the mechanism of optical gain in these devices. The very large, inhomogeneous broadening of the luminescence spectra—the measured homogeneous line widths of single dots are approximately 10^3 narrower than the inhomogeneous widths—implies that a correspondingly small fraction of the dots supports the oscillation of a given cavity mode. This suggests that the material gain needed for attaining laser threshold is approximately 10^3 larger than in a QW laser with a *similar* optical confinement factor Γ. However, gain calculations (see, e.g., Fig. 4.3) predict a much smaller gain enhancement due to the (free carrier) quantum confinement. Further work is thus needed for elucidating the lasing mechanisms in these interesting structures.

Table 4.2 summarizes the characteristis of the different QD lasers just described.

4.4 Quantum dot lasers

Fabrication Method/ Structure	Material System	Growth Method	Highest Operation Temperature	Lasing Wavelength	Threshold Current/ Current Density	Reference
Etching and regrowth/BH	InGaAs/InP	OMCVD	77 K	1.26 µm	1.1 A/7.6 kA/cm^2	Hirayama et al., 1994
Self-ordering on (311)B substrates/ridge waveguide	InGaAs/GaAs	OMCVD	RT (CW)	940 nm	23 mA	Temmyo et al., 1995
SK self ordering/BA	InGaAs/GaAs	MBE	RT	1 µm	120 A/cm^2 (77 K)	Kirstaedter et al., 1994
SK self-ordering/BA	InGaAs/GaAs	MBE	RT	1020 nm	500 A/cm^2	Mirin et al., 1996
SK self-ordering/BA	InGaAs/GaAs	MBE	RT	1028 nm	650 A/cm^2	Kamath et al., 1996
SK self-ordering	InP/GaInP/GaAs	MBE	RT	705 nm	Optically pumped	Moritz et al., 1996
Strained-induced self-ordering/BA	InGaAs/GaAs	ALE	80 K	911 nm	815 A/cm^2	Shoji et al., 1995
SK self-ordering/vertically stacked, BA	InAs/GaAs	MBE	79 K	987 nm	310 A/cm^2	Xie et al., 1996
SK self-ordering/vertically stacked, BA	InAs/GaAs	MBE	RT	950–1050 nm	63 A/cm^2	Ustinov et al., 1997
SK self-ordering/vertically stacked, ridge waveguide	InAs/GaAs	MBE	RT (CW)	1105 nm	3 kA/cm^2	Shoji et al., 1996
SK self-ordering/vertically stacked, VCSEL	InGaAs/GaAs	MBE	RT (CW)	960 nm	32 mA	Saito et al., 1996

Note: All parameters are for pulsed operation and for edge-emitting cavity configurations unless otherwise indicated.

Table 4.2: Summary of characteristics of different quantum dot laser structures.

4.5 Conclusions and outlook

Considerable progress has been achieved in recent years in the development of concepts and technologies relevant to low-dimensional 1D and 0D lasers. The traditional, lithography-based techniques for preparing QWRs and QDs using etching and regrowth have much improved, allowing the fabrication of wires and dots with reasonable interface quality and dimensions approaching those needed for efficient low-dimensional confinement. Most of the recent progress has been achieved, though, owing to the introduction of novel growth techniques in which the wires and dots are formed in situ. Most notably, a new class of growth methods utilizing self-ordering of wires and dots has emerged, permitting the demonstration and investigation of lasers incorporating wires and dots whose sizes and densities are beyond the reach of current (relevant) lithography techniques.

However, the performance of QWR and QD diode lasers has not reached yet what has been predicted by early theoretical models of these devices. These early models emphasized the role of the narrower DOS distribution in improving the static and dynamic laser properties. Currently achievable devices, on the other hand, still suffer from excessive inhomogeneous broadening due to size fluctuations. In addition, there are other important features of low-dimensional structures that have not been fully accounted for by the early models and which might be equally important in determining the lasing features. Among these are the mechanisms of carrier transport, capture, and relaxation and the interplay between excitonic and free carrier effects in the optical gain spectra, all of which are affected by the lower dimensionality. These features have only recently been addressed experimentally owing to the availability of high-quality QWR and QD structures suitable for optical studies, and the results of these investigations might contribute to establishing new, more realistic models of 1D and 0D lasers.

The realization of QWR and QD lasers with features approaching those of an ideal system requires significant improvements in the uniformity of the wires or dots, as well as efficient schemes for current confinement to the recombination regions. Size distributions of not more than a few percent and negligible leakage currents will allow exploitation of the full benefits of the narrower density of states and realization of threshold currents comparable with the radiative currents predicted by models of the ideal systems. On the other hand, other features of QWR and QD structures may prove advantageous over those of QW lasers even in the presence of significant inhomogeneous broadening. These include the small lateral size of the wires and dots, which should be effective in reducing the adverse effects of misfit dislocations in lasers grown on foreign substrates, particularly III–V lasers grown on Si substrates. Selecting a narrow subset of the size distribution using cavity-filtering effects, e.g., in a Bragg mirror microcavity, also might be an effective way to reduce the relevant size distribution. However, in this case, the extremely small effective volumes of the active region also will necessitate a corresponding reduction in the optical cavity losses. Such microcavity lasers incorporating QWRs and QDs are also of interest because they provide the ultimate control of the carrier and photon features via quantum carrier and photon confinement. For example, a laser consisting of a single QD embedded in a three-dimensional photonic bandgap (Yablonovitch, 1989) structure will offer the possibility of independently engineering the carrier and photon states via the design of the semiconductor bandgap and refractive index distributions.

Future progress in this area will thus depend on both further refinement of the relevant technology and on the understanding of the physics pertinent to low dimensional lasers.

Acknowledgments

I wish to thank Benoit Deveaud for critically reading the manuscript and Yann Ducommun for his help in preparing the figures.

References

Ammann, C., Dupertuis, M. A., Bockelmann, U., and Deveaud, B. (1997). *Phys. Rev.* **B55**, 2420.

Ambigapathy, R., Bar-Joseph, I., Oberli, D. Y., Haacke, S., Brasil, M. J., Reinhardt, F., Kapon, E., and Deveaud, B. (1997). *Phys. Rev. Lett.* **78**, 3579.

Arai, S., Kudo, K., Hirayama, H., Miyake, Y., Miyamoto, Y., Tamura, S., and Suematsu, Y. (1993). *Optoelectronics Dev. Technol.* **8**, 461.

Arakawa, Y. (1992). *IEICE Trans. Fundamentals* **E75-A**, 20.

Arakawa, Y., and Sakaki, H. (1982). *Appl. Phys. Lett.* **80**, 939.

Arakawa, Y., and Yariv, A. (1986). *IEEE J. Quantum Electron.* **QE-22**, 1887.

Arakawa, T., Nishioka, M., Nagamune, Y., and Arakawa, Y. (1994). *Appl. Phys. Lett.* **64**, 2200.

Arakawa, Y., Vahala, K., and Yariv, A. (1984). *Appl. Phys. Lett.* **45**, 950.

Arakawa, Y., Vahala, K., Yariv, A., and Lau, K. Y. (1985). *Appl. Phys. Lett.* **47**, 1142.

Arakawa, Y., Vahala, K., and Yariv, A. (1986a). *Surf. Sci.* **174**, 155.

Arakawa, Y., Vahala, K., Yariv, A., and Lau, K. Y. (1986b). *Appl. Phys. Lett.* **48**, 384.

Asada, M., Miyamoto, T., and Suematsu, Y. (1985). *Jpn. J. Appl. Phys.* **24**, L95.

Asada, M., Miyamoto, Y., and Suematsu, Y. (1986). *IEEE J. Quantum Electron.* **QE-22**, 1915.

Asryan, L. V., and Suris, R. A. (1996). *Semicond. Sci. Technol.* **11**, 554.

Ben-Ami, U., Nagar, R., Ben-Ami, N., Lewis, A., Eisenstein, G., Orenstein, M., Kapon, E., Reinhardt, F., Ils, P., and Gustafsson, A. (1997). *Conference on Lasers and Electro-Optics (CLEO '97) Technical Digest*, paper, CThA6, p. 316.

Benisty, H., Sotomayor-Torres, C. M., and Weisbuch, C. (1991). *Phys. Rev.* **B44**, 10945

Berendschot, T. T. J. M., Reinen, H. A. J. M., Bluyssen, H. J. A., Harder, Ch., and Meier, H. P. (1989). *Appl. Phys. Lett.* **54**, 1827.

Bhat, R., Kapon, E., Hwang, D. M., Koza, M. A., and Yun, C. P. (1988). *J. Crystal Growth* **93**, 850.

Bockelmann, U. (1993). *Phys. Rev.* **B48**, 17637.

Bockelmann, U., and Bastard, G. (1990). *Phys. Rev.* **B42**, 8947.

Bockelmann U., and Bastard, G. (1991). *Europhys. Lett.* **15**, 215.

Bockelmann, U., and Egeler, T. (1992). *Phys. Rev.* **B46**, 15574.

Cao, M., Daste, P., Miyamoto, Y., Miyake, Y., Nogiwa, S., Arai, S., Furuya, K., and Suematsu, Y. (1988). *Electron. Lett.* **24**, 824.

Cao, M., Miyake, Y., Tamura, S., Hirayama H., Arai, Suematsu, Y., and Miyamoto, Y. (1990). *Trans. IEICE* **73**, 63.

Chang, Y. C., Chang, L. L., and Esaki, L. (1985). *Appl. Phys. Lett.* **47**, 1324.

Chen, A. C., Moy, A. M., Pearah, P. J., Hsieh, K. C., and Cheng, K. Y. (1993). *Appl. Phys. Lett.* **62**, 1359.

Chiriotti, N. (1996). Diploma thesis, EPFL, Switzerland (unpublished).

Chou, S. T., Cheng, K. Y., Chou, L. J., and Hsieh, K. C. (1995). *Appl. Phys. Lett.* **66**, 2220.

Christen, J., Grundmann, M., Kapon, E., Colas, E., Hwang, D. M., and Bimberg, D. (1992a). *Appl. Phys. Lett.* **61**, 67.

Christen, J., Kapon, E., Colas, E., Hwang, D. M., Schiavone, L. M., Grundmann, M., and Bimberg, D. (1992b). *Surf. Sci.* **267**, 257.

Citrin, D. S. (1992). *Phys. Rev. Lett.* **69**, 3393.

Citrin, D. S. (1993). *Superlat. Microstruct.* **13**, 303.

Citrin, D. S., and Chang, Y.-C. (1993). *J. Quantum Electron.* **29**, 97.

Degani, M. H., and Hipolito, O. (1987) *Phys. Rev.* **B35**, 9345.

Fleming, M. W., and Mooradian, A. (1981). *Appl. Phys. Lett.* **38**, 511.

Fukui, T., and Saito, H. (1988). *J. Vac. Sci. Technol.* **B6**, 1373.

Gislason, H., Langbein, W., and Hvam, J. M. (1996). *Appl. Phys. Lett.* **69**, 3248.

Glutsch, S., and Bechstedt, F. (1993). *Phys. Rev.* **B47**, 4315.

Goldstein, L., Glas, F., Marzin, J. Y., Charasse, M. N., and Le Roux, G., (1985). *Appl. Phys. Lett.* **47**, 1099.

Goni, A. R., Pfeiffer, L. N., West, K. W., Pinczuk, A., Baranger, H. U., and Stormer, H. L. (1992). *Appl. Phys. Lett.* **61**, 1956.

Grundmann, M., Christen, J., Joschko, M., Stier, O., Bimberg, D., and Kapon, E. (1994). *Semicond. Sci. Technol.* **9**, 1939.

Grundmann, M., Christen, J., Bimberg, D., and Kapon, E. (1995). *J. Nonlinear Opt. Phys. Mater.* **4**, 99.

Grundmann, M. Christen, J., Ledentsov, N. N., Bohrer, J., Bimberg, D., Ruvimov, S. S., Werner, P., Richter, U., Gosele, U., Heydenreich, J., Ustinov, V. M., Egorov, A. Yu., Zhukov, A. E., Kop'ev, P. S., and Alferov, Zh. I. (1995). *Phys. Rev. Lett.* **74**, 4043.

Gustafsson, A., Reinhardt, F., Biasiol, G., and Kapon, E. (1995). *Appl. Phys. Lett.* **67**, 3673.

Haacke, S., Hartig, M., Oberli, D. Y., Deveaud, B., Kapon, E., Marti, U., and Reinhart, F. K. (1996). *Solid State Electron.* **40**, 299.

Hakki, B. W., and Paoli, T. (1975). *J. Appl. Phys.* **46**, 1299.

Hasegawa, Y., Egawa, T., Jimbo, T., and Umeno, M. (1994). *J. Crystal Growth* **145**, 728.

Henry, C. H. (1982). *J. Quantum Electron.* **QE-18**, 259.

Hirayama, H., Matsunaga, K., Asada, M., and Suematsu, Y. (1994). *Electron. Lett.* **30**, 142.

Hu, S.-Y., Miller, M. C., Young, D. B., Yi, J. C., Leonard, D., Gossard, A. C., Petroff, P. M., Coldren, L. A., and Dagli, N. (1993). *Appl. Phys. Lett.* **63**, 2015.

Hu, S.-Y., Yi, J. C., Miller, M. C., Leonard, D., Young, D. B., Gossard, A. C., Dagli, N., Petroff, P. M., and Coldren, L. A. (1995). *IEEE J. Quantum Electron.* **31**, 1380.

Inoshita, T., and Sakaki, H. (1992). *Phys. Rev.* **B46**, 7260.

Kamath, K., Bhattacharya, P., Sosnowski, T., Norris, T., and Phillips, J. (1996). *Electron. Lett.* **32**, 1374.

Kan, S. C., Vassilovski, D., Wu, T. C., and Lau, K. Y. (1993). *Appl. Phys. Lett.* **62**, 2307.

Kapon, E. (1992). *Proc. IEEE* **80**, 398.

Kapon, E. (1993). *Optoelectronics Dev. Technolo.* **8**, 429.

Kapon, E. (1994). Chapter 4 In *Semiconductors and Semimetals*, Vol. **40**, pp. 259–336. (A.C. Gossard, ed.), Academic, New York.

Kapon, E., Hardy, A.,and Katzir, A. (1982). *IEEE J. Quantum Electron.* **QE-18**, 66.

Kapon, E., Tamargo, M. C., and Hwang, D. M. (1987). *Appl. Phys. Lett.* **50**, 347.

Kapon, E., Hwang, D. M., and Bhat, R. (1989a). *Phys. Rev. Lett.* **63**, 430.

Kapon, E., Simhony, S., Bhat, R., and Hwang, D. M. (1989b). *Appl. Phys. Lett.* **55**, 2715.

Kapon, E., Kash, K., Clausen, E. M., Jr., Hwang, D. M., and Colas, E. (1992a). *Appl. Phys. Lett.* **60**, 477.

Kapon, E., Walther, M., Christen, J., Grundmann, M., Caneau, C., Hwang, D. M., Colas, E., Bhat, R., Song, G. H., and Bimberg, D. (1992b). *Superlat. Micorstruct.* **12**, 491.

Kapon, E., Hwang, D. M., Walther, M., Bhat, R., and Stoffel, N. G. (1992c). *Surf. Sci.* **267**, 593.

Kapon, E., Biasiol, G., Hwang, D. M., Colas, E., and Walther, M. (1996) *Solid State Electron.* **40**, 815.

Kirstaedter, N., Ledentsov, N. N., Grundmann, M., Bimberg, D., Ustinov, V. M., Ruvimov, S. S., Maximov, M. V., Kop'ev, P. S., Alferov, Zh. I., Richter, U., Werner, P., Gösele, U., and Heydenreich, J. (1994). *Electron. Lett.* **30**, 1416.

Koch, T. L., and Linke, R. A. (1986). *Appl. Phys. Lett.* **48**, 613.

Kogelnik, H., and Shank, C. V. (1972). *J. Appl. Phys.* **43**, 2327.

Konnerth, K., and Lanza, C. (1964). *Appl. Phys. Lett.* **4**, 120.

Kudo, K., Miyake, Y., Hirayama, H., Tamura, S., Arai, S., and Suematsu, Y. (1992). *IEEE Photon. Technol. Lett.* **4**, 1089.

Kudo, K., Nagashima, Y., Tamura, S., Arai, S., Huang, Y., and Suematsu, Y. (1993). *IEEE Photon. Technol. Lett.* **5**, 864.

Kurtenbach, A., Ruhle, W. W., and Eberl, K. (1995). *Solid State Commun.* **96**, 265.

Leonard, D., Krishnamurthy, M., Reaves, C. M., Denbaars, S. P., and Petroff, P. M. (1993). *Appl. Phys. Lett.* **63**, 3203.

Lau, K. Y., Bar-Chaim, N., Ury, I., Harder, C., and Yariv, A. (1983). *Appl. Phys. Lett.* **43**, 1.

Lau, K. Y., Bar-Chaim, N., Derry, P. L., and Yariv, A. (1987). *Appl. Phys. Lett.* **51**, 69.

Marzin, J. Y., Gerard, J. M., Izrael, A., Barrier, D., and Bastard, G. (1994). *Phys. Rev. Lett.* **73**, 716.

Miller, M. S., Weman, H., Pryor, C. E., Krishnamurthy, M., Petroff, P. M., Kroemer, H., and Merz, J. L. (1992). *Phys. Rev. Lett.* **68**, 3464.

Mirin, R., Gossard, A., and Bowers, J. (1996). *Electron. Lett.* **32**, 1732.

Miyake, Y., and Asada, M. (1989). *Jpn. J. Appl. Phys.* **28**, 1280.

Miyake, Y., Hirayama, H., Arai, S., Miyamoto, Y., and Suematsu, Y. (1991). *IEEE Photon. Technol. Lett.* **3**, 191.

Miyake, Y., Hirayama, H., Shim, J. I., Arai, S., and Miyamoto, Y. (1992). *IEEE Photon. Technol. Lett.* **4**, 964.

Miyake, Y., Hirayama, H., Kudo, K., Tamura, S., Arai, S., Asada, M., Miyamoto, Y., and Suematsu, Y. (1993). *IEEE J. Quantum Electron.* **29**, 2123.

Miyamoto, Y., Cao, M., Shingai, Y., Furuya, K., Suematsu, Y., Ravikumar, K. G., and Arai, S. (1984). *Jpn. J. Appl. Phys.* **26**, L225.

Miyamoto, Y., Miyake, Y., Asada, M., and Suematsu, Y. (1989). *J. Quantum Electron.* **QE-25**, 2001.

Moritz, A., Wirth, R., Hangleiter, A., Kurtenbach, A., and Eberl, K. (1996). *Appl. Phys. Lett.* **69**, 212.

Mukai, K., Ohtsuka, N., Sugawara, M., and Yamazaki, S. (1994). *Jpn. J. Appl. Phys.* **33**, L1710.

Mukai, K., Ohtsuka, N., Shoji, H., and Sugawara, M. (1996). *Appl. Phys. Lett.* **68**, 3013.

Nishida, T., Notomi, M., Iga, R., and Tamamura, T. (1992). *Jpn. J. Appl. Phys.* **31**, 4508.

Notomi, M., and Tamamura, T. (1993). *Optoelectronics Dev. Technol.* **8**, 563.

Notomi, M., Nakao M., and Tamamura, T. (1993a). *Appl. Phys. Lett.* **62**, 2350.

Notomi, M., Okamoto, M., Iwamura, H., and Tamamura, T. (1993b). *Appl. Phys. Lett.* **62**, 1094.

Notzel, R., Temmyo, J., and Tamamura, T. (1994a). *Nature* **369**, 131.

Notzel, R., Temmyo, J., Kamada, H., Furuta, T., and Tamamura, T. (1994b). *Appl. Phys. Lett.* **65**, 457.

Oberli, D. Y., Vouilloz, F., Dupertuis, M.-A., Fall, C., and Kapon, E. (1995). *Il Nuovo Cimento* **D17**, 1641.

Ogawa, T., and Takagahara, T. (1991a). *Phys. Rev.* **B43**, 14325.

Ogawa, T., and Takagahara, T. (1991b). *Phys. Rev.* **B44**, 8138.

Pearah, P. J., Stellini, E. M., Chen, A. C., Moy, A. M., Hsieh, K. C., and Cheng, K. Y (1993). *Appl. Phys. Lett.* **62**, 729.

Pearah, P. J., Chen, A. C., Moy, A. M., Hsieh, K. C., and Cheng, K. Y. (1994). *IEEE J. Quantum Electron.* **QE-30**, 608.

Petroff, P., Gossard, A. C., and Wiegmann, W. (1984). *Appl. Phys. Lett.* **45**, 620.

Qian, Y., Xu, Z. T., Zhang, J. M., Chen, L. H., Wang, Q. M., Zheng, L. X., and Hu, X. W. (1995). *Electron. Lett.* **31**, 102.

Rossi, F., and Molinari, E. (1996). *Phys. Rev. Lett.* **76**, 3642.

Rossi, F., Goldoni, G., and Molinari, E. (1997). *Phys. Rev. Lett.* **78**, 3527.

Saito, H., Uwai, K., and Kobayashi, N. (1993). *Jpn. J. Appl. Phys.* **32**, 4440.

Saito, H., Nishi, K., Ogura, I., Sugou, S., and Sugimoto, Y. (1996). *Appl. Phys. Lett.* **69**, 3140.

Sercel, P. C., and Vahala, K. (1991). *Phys. Rev.* **B44**, 5681.

Shin, K.-C., Arai, S., Nagashima, Y., Kudo, K., and Tamura, S. (1995). *IEEE Photon. Technol. Lett.* **7**, 345.

Shoji, H., Mukai, K., Ohtsuka, N., Sugawara, M., Uchida, T., and Ishikawa, H. (1995). *IEEE Photon. Technol. Lett.* **7**, 1385.

Shoji, H., Nakata, Y., Mukai, K., Sugiyama, Y., Sugawara, M., Yokoyama, N., and Ishikawa, H. (1996). *Electron. Lett.* **32**, 2023.

Simhony, S., Kapon, E., Colas, E., Bhat, R., Stoffel, N. G., and Hwang, D. M. (1990). *IEEE Photon. Technol. Lett.* **5**, 305.

Simhony, S., Kapon, E., Colas, E., Hwang, D. M., Stoffel, N. G., and Worland, P. (1991). *Appl. Phys. Lett.* **59**, 2225.

Singh, J., Arakawa, Y., and Bhattacharya, P. (1992). *IEEE Photon. Technol. Lett.* **4**, 835.

Solomon, G. S., Trezza, J. A., Marshall, A. F., and Harris, J. S., Jr. (1996). *Phys. Rev. Lett.* **76**, 952.

Stellini, E. M., Cheng, K. Y., Pearah, P. J., Chen, A. C., Moy, A. M., and Hsieh, K. C. (1993). *Appl. Phys. Lett.* **62**, 458.

Temmyo, J., Kuramochi, E., Sugo, M., Nishiya, T., Notzel, R., and Tamamura, T. (1995). *Electron. Lett.* **31**, 209.

Tiwari, S. (1995). In *Proceedings of the NATO Advanced Research Workshop on Low Dimensional Structures Prepared by Epitaxial Growth or Regrowth on Patterned Substrates, Ringberg in Rottach Egern, Germany,* February 20–24, 1995 (K. Eberl, P. M. Petroff, and P. Demeester eds.), Kluwer Academic, Dordrecht, pp. 335–344.

Tiwari, S., and Woodall, J. M. (1994). *Appl. Phys. Lett.* **64**, 2211.

Tiwari, S., Petit, G. D., Milkove, K. R., Davis, R. J., Woodall, J. M., and Legoues, F. (1992). In *Proceedings of the International Electron Device Meeting, San Francisco, California, IEDM '92 Digest,* paper 34.2.1, p. 859.

Tiwari, S., Petit, G. D., Milkove, K. R., Legoues, F., Davis, R. J., and Woodall, J. M. (1994). *Appl. Phys. Lett.* **64**, 3536.

Tsuchiya, M., Petroff, P. M., and Coldren, L. A., (1989). In *47th Annual Device Research Conference, MIT, Cambridge, MA,* June 19–21, paper IVA-1.

Ueno, S., Miyake, Y., and Asada, M. (1992). *Jpn. J. Appl. Phys.* **31**, 286.

Ustinov, V. M., Egorov, A. Yu., Kovsh, A. R., Zhukov, A. E., Maximov, M. V., Tsatsul'nikov, A. F., Gordeev, N. Yu., Zaitsev, S. V., Shernyakov, Yu. M., Bert, N. A., Kop'ev, P. S., Alferov, Zh. I., Ledentsov, N. N., Bohrer, J., Bimberg, D., Kosogov, A. O., Werner, P., and Gosele, U. (1997). *J. Crystal Growth* **175/176**, 689.

Vahala, K. (1988). *J. Quantum Electron. Lett.* **QE-24**, 523.

Vahala, K., Arakawa, Y., and Yariv, A. (1987). *Appl. Phys. Lett.* **50**, 365.

Vouilloz, F., Oberli, D., Dupertuis, M.-A, Gustafsson, A., Reinhardt, F., and Kapon, E. (1997) *Phys. Rev. Lett.* **78**, 1580.

Vurgaftman, I., and Singh, J. (1993), *Appl. Phys. Lett.* **63**, 2024.

Vurgaftman, I., and Singh, J. (1994), *IEEE J. Quantum Electron.* **30**, 2012.

Walther, M., Kapon, E., Christen, J., Hwang, D. M., and Bhat, R. (1992a). *Appl. Phys. Lett.* **60**, 521.

Walther, M., Kapon, E., Scherer, A., Hwang, D. M., Song, G. H., Caneau, C., Bhat, R., and Wilkens, B. (1992b). In *Proceedings of the 18th European Conference on Optical Communication, Berlin, Germany, September 1992, ECOC '92 Conference Digest,* Vol. 1, paper Mo B3.2, p. 97.

Walther, M., Kapon, E., Caneau, C., Hwang, D. M., and Schiavone, L. M. (1993). *Appl. Phys. Lett.* **62**, 2170.

Wang, G., Fafard, S., Leonard, D., Bowers, J. E., Merz, J. L., and Petroff, P. M. (1994). *Appl. Phys. Lett.* **64**, 2815.

Wang, J., Griesinger, U. A., Adler, F., Schweizer, H., Harle, V., and Scholz, F. (1996a). *Appl. Phys. Lett.* **69**, 287.

Wang, J., Griesinger, U. A., and Schweizer, H. (1996b). *Appl. Phys. Lett.* **69**, 1585.

Wang, X.-L., Ogura, M., and Matsuhata, H. (1995). *Appl. Phys. Lett.* **66**, 1506.

Wegscheider, W., Pfeiffer, L., Dignam, M. M., West, K. W., McCall, S. L., and Hull, R. (1993). *Phys. Rev. Lett.* **71**, 4071.

Wegscheider, W., Pfeiffer, L., West, K. and Leibenguth, R. E. (1994). *Appl. Phys. Lett.* **65**, 2510.

Willatzen, M., Tanaka, T., Arakawa, Y.,and Singh, J. (1994). *J. Quantum Electron.* **30**, 640.

Xie, Q., Madhukar, A., Chen, P., and Kobayashi, N. P. (1995). *Phys. Rev. Lett.* **75**, 2542.

Xie, Q., Kalburge, A., Chen, P., and Madhukar, A. (1996). *IEEE Photon. Technol. Lett.* **8**, 965.

Yablonovitch, E. (1987). *Phys. Rev. Lett.* **58**, 2059.

Yamauchi, T., Takahashi, T., Schulman, J. N., and Arakawa, Y. (1993). *IEEE J. Quantum Electron.* **29**, 2109.

Yariv., A. (1988). *Appl. Phys. Lett.* **53**, 1033.

Yi, J. C., Dagli, N., and Coldren, L. A., (1991). *Appl. Phys. Lett.* **59**, 3015.

Zarem, H., Vahala, K., and Yariv, A. (1989). *J. Quantum Electron.* **QE-25**, 705.

Zory, P. S., ed. (1993). *Quantum Well Lasers,* Academic, Boston.

Chapter 5

Quantum Optics Effects in Semiconductor Lasers

Y. Yamamoto,
Edward L. Ginzton Laboratory, Stanford University, Stanford, CA
and NTT Basic Research Laboratories, Tokyo, Japan

S. Inoue
NTT Basic Research Laboratories, Tokyo, Japan

G. Björk,
Department of Microwave Engineering, Royal Institute of Technology, Stockholm, Sweden

H. Heitmann
NTT Basic Research Laboratories, Tokyo, Japan

F. Matinaga
NTT Basic Research Laboratories, Tokyo, Japan

5.1 Introduction

Quantum statistical properties of laser light have been studied extensively for the last 30 years both theoretically and experimentally. A laser operating at far above threshold generates a coherent state of light. Various experimental facts support such theoretical prediction. However, recent studies on a semiconductor laser revealed that a semiconductor laser generates a number-phase squeezed state rather than a squeezed state. This is due primarily to the fact that a semiconductor laser is pumped by a shot-noise-free electric current. If such a squeezed state becomes phase coherent with an independent local laser oscillator, the squeezed light can be detected by optical homodyne detectors and used for various interferometric measurements. A phase-coherent squeezed state or squeezed vacuum state can be generated by injection locking the squeezed slave laser with an external master laser. This chapter reviews the theoretical and experimental aspects of squeezing of such an injection-locked semiconductor laser.

Spontaneous emission of an atom also has been studied for many years both theoretically and experimentally. A quantum electrodynamic theory of spontaneous emission was well established by various experimental results. Spontaneous emission is not an immutable property of an atom but a consequence of atom-vacuum field coupling. The decay rate, energy, and radiation pattern of spontaneous emission can be altered by a cavity wall. Since spontaneous emission is a major source of energy loss, speed limitation, and noise of a semiconductor laser, such capability of control of spontaneous emission is expected to improve the performance of a semiconductor laser. In this chapter we also review the theoretical and experimental aspects of control of spontaneous emission in various semiconductor microcavities.

5.2 Squeezing in semiconductor lasers

5.2.1 Brief review of squeezed states

The Heisenberg uncertainty principle for two conjugate observables $\hat{O}_1$ and $\hat{O}_2$ is the direct consequence of the commutation relation between them. It can be derived by the Schwartz inequality[1]:

$$[\hat{O}_1,\hat{O}_2] = i\hat{O}_3 \rightarrow \langle \Delta \hat{O}_1^2 \rangle \langle \Delta \hat{O}_2^2 \rangle \geq \tfrac{1}{4}|\langle[\hat{O}_1,\hat{O}_2]\rangle|^2 = \tfrac{1}{4}|\langle\hat{O}_3\rangle|^2 \quad (5.1)$$

where $\Delta\hat{O}_i = \hat{O}_i - \langle\hat{O}_i\rangle$, $= 1, 2$. A state that exactly satisfies the equality in Eq. (5.1) is called a *minimum-uncertainty state* and is defined mathematically as an eigenstate of the operator[2]

$$\hat{O}(r) = e^r \hat{O}_1 + ie^{-r} \hat{O}_2 \quad (5.2)$$

5.2. Squeezing in semiconductor lasers

where r is a squeezing parameter. The squeezing parameter determines the distribution of quantum noise between $\hat{O}_1$ and $\hat{O}_2$:

$$\begin{aligned} \langle \Delta \hat{O}_1^2 \rangle &= \langle \Delta \hat{O}_1^2 \rangle_{r=0}\, e^{-2r} \\ \langle \Delta \hat{O}_2^2 \rangle &= \langle \Delta \hat{O}_2^2 \rangle_{r=0}\, e^{2r} \end{aligned} \right\} \text{ (minimum-uncertainty product)} \quad (5.3)$$

When two quadrature-phase components $\hat{a}_1 = \frac{1}{2}(\hat{a} + \hat{a}^\dagger)$ and $\hat{a}_2 = (\frac{1}{2i})(\hat{a} - \hat{a}^\dagger)$ are chosen for two conjugate observables $\hat{O}_1$ and $\hat{O}_2$, then $\hat{O}_3$ is a c number ($= \frac{1}{2}$). In the special case of $r = 0$, $\hat{O}(r)$ is the photon annihilation operator $\hat{a}$, and its eigenstate is a coherent state.[3] Two quadrature components of the coherent state share the same amount of quantum noise, namely, $\langle \Delta \hat{a}_1^2 \rangle = \langle \Delta \hat{a}_2^2 \rangle = \frac{1}{4}$. The electromagnetic field generated by a classic oscillating current is in a coherent state.[4] Its quadrature-component quantum noise is referred to as the *standard quantum limit (SQL)*.

In the case of $r \neq 0$, the eigenstate of $\hat{O}(r)$ is called a *quadrature amplitude squeezed state*,[5-7] which features reduced quantum noise in one quadrature and enhanced quantum noise in the other quadrature. The quadrature amplitude squeezed state $|\beta, r\rangle$ can be generated by unitary evolution from a coherent state:

$$|\beta, r\rangle = \hat{S}(r)|\beta\rangle = \exp[\tfrac{1}{2}r(\hat{a}^2 - \hat{a}^{\dagger 2})]\,|\beta\rangle \quad (5.4)$$

This unitary evolution is realized by the second- or third-order nonlinear processes. The quadrature amplitude squeezed state is expanded in terms of photon number states:

$$\begin{aligned} |\beta, r\rangle = \exp\left[-\frac{1}{2}|\beta|^2 + \frac{1}{2}\tanh(r)\beta^2 \right] \\ \times \sum_n \frac{\left[\frac{1}{2}\tanh(r)\right]^{n/2}}{[n!\cosh(r)]^{1/2}} H_n\left[\frac{\beta}{\sqrt{\sinh(2r)}}\right]|n\rangle \end{aligned} \quad (5.5)$$

where H_n is the nth order Hermite polynomial. In order to ultimately reduce the noise in one of the quadratures to zero, the electromagnetic field mode must have an infinite photon number. This is so because the enhanced quadrature noise component will require part of the total photon number[7]

$$\langle \hat{n} \rangle_{SS} = \langle \hat{a}_1 \rangle^2 + \langle \hat{a}_2 \rangle^2 + \sinh^2(r) \underset{(r \to \infty)}{\longrightarrow} \infty \quad (5.6)$$

That is, there is a tradeoff relation between quantum-noise reduction and the required photon number.

A number-phase squeezed state of the electromagnetic field is the other kind of minimum-uncertainty state. The observables $\hat{O}_1$ and $\hat{O}_2$ correspond to the photon number and sine operators, and $\hat{O}_3$ is then the cosine operator defined by

$$\hat{O}_1 = \hat{n} \equiv \hat{a}^\dagger \hat{a}, \qquad \hat{O}_2 = \hat{S} \equiv \frac{1}{2i}[(\hat{n}+1)^{-1/2}\hat{a} - \hat{a}^\dagger(\hat{n}+1)^{-1/2}] \qquad (5.7)$$

$$\hat{O}_3 = \hat{C} \equiv \frac{1}{2}[(\hat{n}+1)^{-1/2}\hat{a} + \hat{a}^\dagger(\hat{n}+1)^{-1/2}] \qquad (5.8)$$

The normalizable number-phase minimum-uncertainty state can be constructed mathematically when $\langle \hat{S} \rangle = 0$.[34] However, the average photon number $\langle \hat{n} \rangle$ and the ratio of photon number noise to sine operator noise $\langle \Delta \hat{n}^2 \rangle / \langle \Delta \hat{S}^2 \rangle$ cannot be chosen independently but rather should satisfy the constraint

$$I_{-1-\langle \hat{n} \rangle}\left[\left(\frac{\langle \Delta \hat{n}^2 \rangle}{\langle \Delta \hat{S}^2 \rangle}\right)^{1/2}\right] = 0 \qquad (5.9)$$

where $I_\mu(x)$ is a modified Bessel function of the first kind of order μ. From the constraint in Eq. (5.9), $\langle \hat{n} \rangle$ must be chosen as $\langle \hat{n} \rangle \in [2k, 2k+1]$ ($k = 0, 1, 2, 3, \ldots$). The number-phase minimum-uncertainty state $|v, r\rangle$ is expanded in terms of photon-number states:

$$|v, r\rangle = c \sum_n I_{n-\langle \hat{n} \rangle}(e^{2r})|\hat{n}\rangle \qquad (5.10)$$

where $c = [\sum_n I^2_{n-\langle \hat{n} \rangle}(e^{2r})]^{-1/2}$ is a normalization factor. When r is equal to $-\frac{1}{2}\ln(2\langle \hat{n} \rangle)$ and the average photon number is much greater than unity, the photon-number noise and the sine-operator noise (which corresponds to the classic phase noise) are reduced to $\langle (\Delta \hat{n})^2 \rangle = \langle \hat{n} \rangle$ and $\langle (\Delta \hat{S})^2 \rangle = [\langle (\hat{C})^2 \rangle/(4\langle n \rangle)] \simeq 1/(4\langle \hat{n} \rangle)$. This state is close to a coherent state, although a coherent state does not satisfy a number-phase minimum-uncertainty product in an exact mathematical sense. Therefore, the number-phase squeezed state cannot be generated by unitary evolution from a coherent state. This number and sine-operator noise is also referred to as the SQL. A number-phase squeezed state with $r > -\frac{1}{2}\ln(2\langle \hat{n} \rangle)$ features smaller photon-number noise and larger sine-operator noise than the SQL, while the minimum-uncertainty relationship $\langle (\Delta \hat{n})^2 \rangle \langle (\Delta \hat{S})^2 \rangle = \frac{1}{4}\langle \hat{C} \rangle^2$ is preserved. The photon-number noise ultimately can be reduced to zero (photon-num-

5.2. Squeezing in semiconductor lasers

ber state) without requiring an inifinite photon number because the enhanced phase noise does not require any photons at all.

The difference between the preceding photon states is described by the quasi-probability density (Q representation)

$$Q(\alpha) \equiv \langle \alpha | \hat{\rho} | \alpha \rangle = |\langle \alpha | \psi \rangle|^2 \qquad (5.11)$$

where $|\psi\rangle$ is a state vector, and $|\alpha\rangle$ is a coherent state with the complex eigenvalue α, and the real and imaginary parts of α corresponds to $\langle \hat{a}_1 \rangle$ and $\langle \hat{a}_2 \rangle$, respectively. Neither a squeezed state nor a number-phase squeezed state possesses a positive-definite diagonal $P(\alpha)$ function for the density operator, which implies that these states cannot be realized by a classic mixture of coherent states. The term *nonclassic photon states* has been coined because of this fact. Figure 5.1 compares the quasi-probability densities $Q(\alpha)$ of various quantum states of light.

An unmistakable mark of quadrature amplitude squeezed-state generation is the observation of quadrature phase squeezing. That is, a homo-

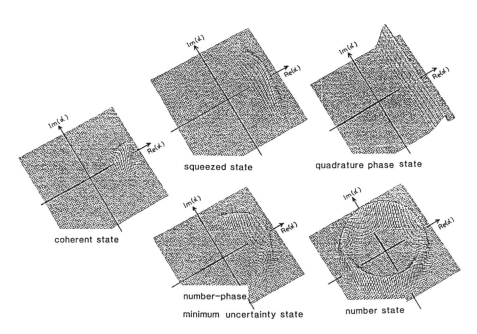

Figure 5.1: Quasi-probability densities $Q(\alpha)$ of various quantum states of light.

dyne-detector output current will feature a reduced noise level below that for vacuum fields.[35–38] On the other hand, an unmistakable mark of number-phase squeezed-state generation is the observation of amplitude squeezing. That is, a photon-counting detector output will feature a noise level below the shot-noise level and thus sub-Poissonian photon statistics.

5.2.2 Theory of squeezed-state generation in semiconductor lasers

Operator Langevin equations

In order to analyze the quantum noise properties of a semiconductor laser, we employ the operator Langevin equations. The operator Langevin equation for the cavity internal field $\hat{A}(t)$ is given by

$$\frac{d\hat{A}(t)}{dt} = -\frac{1}{2}\left[\frac{\omega}{Q} + j2(\omega_0 - \omega) - \frac{\omega}{\mu^2}(\tilde{\chi}_i - j\tilde{\chi}_r)\right]\hat{A}(t) + \hat{G}(t) + \hat{g}(t) + \hat{f}(t) \quad (5.12)$$

Here ω/Q is the photon decay rate, which is decomposed into the internal loss contribution ω/Q_0 and the output coupling contribution ω/Q_e as

$$\frac{\omega}{Q} = \frac{\omega}{Q_0} + \frac{\omega}{Q_e} \quad (5.13)$$

ω_0 and ω are the empty cavity resonance frequency and the actual oscillation frequency. μ is the nonresonant refractive index. $\tilde{\chi} = \tilde{\chi}_r + j\tilde{\chi}_i$ is the complex susceptibility operator. The real part $\tilde{\chi}_r$ represents an anomalous index dispersion, and the imaginary part $\tilde{\chi}_i$ indicates the stimulated emission gain

$$\frac{\omega}{\mu^2}\tilde{\chi}_i = \tilde{E}_{cv} - \tilde{E}_{vc} \quad (5.14)$$

where $\tilde{E}_{cv}$ and $\tilde{E}_{vc}$ are the operators of the stimulated emission and absorption rates, respectively. The noise operators $\tilde{G}(t)$, $\hat{g}(t)$, and $\hat{f}(t)$ are associated with the random processes of stimulated emission gain, internal loss, and output coupling (mirror) loss.

The operator Langevin equation for the total excited electron number operator $\tilde{N}_c(t)$ is given by

$$\frac{d\tilde{N}_c(t)}{dt} = p - \frac{\tilde{N}_c(t)}{\tau_{sp}} - (\tilde{E}_{cv} - \tilde{E}_{vc})\hat{n}(t) - \tilde{E}_{cv} + \tilde{\Gamma}_p(t) + \tilde{\Gamma}_{sp}(t) + \tilde{\Gamma}(t) \quad (5.15)$$

where p is the pumping rate and τ_{sp} is the spontaneous emission lifetime of the electrons. $\hat{n}(t) \equiv \hat{A}^\dagger(t)\hat{A}(t)$ is the photon-number operator. The fourth term, $\tilde{E}_{cv}$, is the spontaneous emission rate into the lasing mode, while $\tilde{N}_c/\tau_{sp}$ is the spontaneous emission rate into all nonlasing modes. The noise operators $\tilde{\Gamma}_p(t)$, $\tilde{\Gamma}_{sp}(t)$, and $\tilde{\Gamma}(t)$ are associated with the pump noise, spontaneous emission noise, and stimulated emission and absorption noise, including the spontaneous emission into a lasing mode.

Correlation functions of noise operators

The correlation functions of noise operators can be determined by the fluctuation-dissipation theorem. When the heat baths exhibit broad-frequency spectra and therefore the dissipation processes can be considered as Markovian, the correlation functions of the two Hermitian (real and imaginary parts) operators for each noise operator in Eq. (5.12) are [39-41]

$$\langle \tilde{G}_r(t)\tilde{G}_r(u)\rangle = \langle \tilde{G}_i(t)\tilde{G}_i(u)\rangle = \tfrac{1}{4}(\langle \tilde{E}_{cv}\rangle + \langle \tilde{E}_{vc}\rangle)\,\delta(t-u) \quad (5.16)$$

$$\langle \hat{g}_r(t)\hat{g}_r(u)\rangle = \langle \hat{g}_i(t)\hat{g}_i(u)\rangle = \frac{1}{2}\frac{\omega}{Q_0}\left(n_{th} + \frac{1}{2}\right)\delta(t-u) \simeq \frac{1}{4}\frac{\omega}{Q_0}\delta(t-u) \quad (5.17)$$

and

$$\langle \hat{f}_r(t)\hat{f}_r(u)\rangle = \langle \hat{f}_i(t)\hat{f}_i(u)\rangle = \frac{1}{2}\frac{\omega}{Q_e}\left(n_{th} + \frac{1}{2}\right)\delta(t-u) \simeq \frac{1}{4Q_e}\omega\,\delta(t-u) \quad (5.18)$$

where n_{th} is the number of thermally generated photons, which is approximately equal to zero at room temperature and optical frequency:

$$n_{th} = \frac{1}{e^{\hbar\omega/k_B T} - 1} \simeq 0 \quad (5.19)$$

The spontaneous population-inversion decay process is a stimulated-emission process induced by the zero-point fluctuations of all continuous modes in an active medium except for the lasing mode. Since these zero-point fluctuations have only a very short memory and the decay process is considered to be Markovian, the correlation function for $\tilde{\Gamma}_{sp}$ is given by [39, 41]

$$\langle \tilde{\Gamma}_{sp}(t)\tilde{\Gamma}_{sp}(u)\rangle = \frac{\langle \tilde{N}_c\rangle}{\tau_{sp}}\delta(t-u) \quad (5.20)$$

In a conventional quantum-mechanical laser theory, the pumping process is treated as the reverse process of the spontaneous emission.[39]

This treatment implicitly assumes incoherent light pumping. Incoherent light has a broad spectrum of frequency and a very short memory time. The correlation function for $\tilde{\Gamma}_p$ is then given by[39, 41]

$$\langle \tilde{\Gamma}_p(t)\tilde{\Gamma}_p(u) \rangle = p\, \delta(t-u) \qquad (5.21)$$

which holds even when the pumping light is coherent laser radiation. This is so because the pumping process, i.e., the photoelectron emission process is a self-exciting Poisson point process.

The noise operators $\tilde{G}$ and $\tilde{\Gamma}$ originate from the dipole-moment fluctuation operator. These noise operators enter into the Langevin equations through the adiabatic elimination of the dipole-moment operator. The correlation function for $\tilde{\Gamma}$ is

$$\langle \tilde{\Gamma}(t)\tilde{\Gamma}(u) \rangle = \delta(t-u)\,(\langle \tilde{E}_{cv} \rangle + \langle \tilde{E}_{vc} \rangle)\,\langle \hat{n} \rangle \qquad (5.22)$$

Since the two operators $\tilde{G}$ and $\tilde{\Gamma}$ come from the same origin, they have the correlation

$$\langle \tilde{G}_r(t)\tilde{\Gamma}(u) \rangle = -\frac{1}{2}\delta(t-u)\,(\langle \tilde{E}_{cv} \rangle + \langle \tilde{E}_{vc} \rangle)\,\langle \hat{A} \rangle \qquad (5.23)$$

and

$$\langle \tilde{G}_i(t)\tilde{\Gamma}(u) \rangle = 0 \qquad (5.24)$$

Noise power spectra of the cavity internal field

Let us consider a laser oscillator pumped at well above threshold. For this region, the operator Langevin equations [Eqs. (5.12) and (5.15)] can be solved by the quasi-linearization procedure. We expand the operators into c-number mean values and fluctuating operators according to

$$\hat{A}(t) = [A_0 + \Delta \hat{A}(t)]\exp[-j\Delta\hat{\phi}(t)] \qquad (5.25)$$

$$\tilde{N}_c(t) = N_{c0} + \Delta \tilde{N}_c(t) \qquad (5.26)$$

$$\tilde{n}(t) = \hat{A}^\dagger(t)\hat{A}(t) = [A_0 + \Delta\hat{A}(t)]^2 \simeq A_0^2 + 2A_0\Delta\hat{A}(t) \qquad (5.27)$$

$$\tilde{\chi}_i \simeq \langle \tilde{\chi}_i \rangle + \frac{d\langle \tilde{\chi}_i \rangle}{dN_{c0}}\Delta\tilde{N}_c \qquad (5.28)$$

$$\tilde{\chi}_r \simeq \langle \tilde{\chi}_r \rangle + \frac{d\langle \tilde{\chi}_r \rangle}{dN_{c0}}\Delta\tilde{N}_c \qquad (5.29)$$

Here A_0 and N_{c0} are the average field amplitude and excited electron number (c-numbers). $\Delta\hat{A}$, $\Delta\hat{\phi}$ and $\Delta\tilde{N}_c$ are the Hermitian amplitude, phase,

5.2. Squeezing in semiconductor lasers

and excited electron fluctuating operators. Although the Hermitian phase operator does not exist in a strict quantum-mechanical sense, it is known that Eq. (5.25) is still a reasonable approximation when the photon number A_0^2 is much larger than unity.[16] We also introduce the notations

$$\frac{1}{\tau_{st}} = \frac{\omega}{\mu^2} A_0^2 \frac{d\langle \tilde{\chi}_i \rangle}{dN_{c0}} \tag{5.30}$$

$$n_{sp} = \frac{\langle \tilde{E}_{cv} \rangle}{\langle \tilde{E}_{cv} \rangle - \langle \tilde{E}_{vc} \rangle} \tag{5.31}$$

$$\alpha = \frac{d\langle \tilde{\chi}_r \rangle / dN_{c0}}{d\langle \tilde{\chi}_i \rangle / dN_{c0}} \tag{5.32}$$

and

$$R = \frac{p}{p_{th}} - 1 \tag{5.33}$$

where τ_{st}, n_{sp}, α, R and p_{th} are the electron lifetime due to stimulated emission, the population inversion factor, the linewidth enhancement factor, the normalized pumping rate, and the threshold pumping rate.

Substituting Eqs. (5.25) to (5.29) into Eqs. (5.12) and (5.15), and neglecting the terms in the order of Δ^2, the linearized equations for $\Delta\hat{A}$, $\Delta\hat{\phi}$ and $\Delta\tilde{N}_c$, which are the fluctuating real and imaginary part of Eq. (5.12) and the fluctuating part of Eq. (5.15), are obtained as

$$\frac{d\Delta\hat{A}(t)}{dt} = \frac{1}{2A_0\tau_{st}} \Delta\tilde{N}_c(t) + \hat{H}_r(t) \tag{5.34}$$

$$\frac{d\Delta\hat{\phi}(t)}{dt} = \frac{\alpha}{2A_0^2\tau_{st}} \Delta\tilde{N}_c(t) - \frac{1}{A_0} \hat{H}_i(t) \tag{5.35}$$

and

$$\frac{d\Delta\tilde{N}_c(t)}{dt} = -\left(\frac{1}{\tau_{sp}} + \frac{1}{\tau_{st}}\right) \Delta\tilde{N}_c(t) - 2\frac{\omega}{Q} A_0 \Delta\hat{A}(t) + \tilde{\Gamma}_p + \tilde{\Gamma}_{sp} + \tilde{\Gamma} \tag{5.36}$$

where $\hat{H}_r(t)$ and $\hat{H}_i(t)$ are the Hermitian quadrature operators defined by

$$\hat{H}_r = \frac{1}{2}[(\tilde{G} + \hat{g} + \hat{f})e^{j\Delta\hat{\phi}} + e^{-j\Delta\hat{\phi}}(\tilde{G}^\dagger + \hat{g}^\dagger + \hat{f}^\dagger)] \tag{5.37}$$

and

$$\hat{H}_i = \frac{1}{2j}[(\tilde{G} + \hat{g} + \hat{f})e^{j\Delta\hat{\phi}} - e^{-j\Delta\hat{\phi}}(\tilde{G}^\dagger + \hat{g}^\dagger + \hat{f}^\dagger)] \tag{5.38}$$

The steady-state equations, which are the nonfluctuating part of Eqs. (5.12) and (5.15), are also obtained as

$$\frac{\omega}{Q} = \frac{1}{\tau_{ph}} = \frac{\omega}{\mu^2} \langle \tilde{\chi}_i \rangle = \langle \tilde{E}_{cv} \rangle - \langle \tilde{E}_{vc} \rangle \tag{5.39}$$

$$\omega = \omega_0 + \frac{\omega}{2\mu^2} \langle \tilde{\chi}_r \rangle \tag{5.40}$$

and

$$p = \frac{N_{c0}}{\tau_{sp}} + A_0^2 (\langle \tilde{E}_{cv} \rangle - \langle \tilde{E}_{vc} \rangle) + \langle \tilde{E}_{cv} \rangle \tag{5.41}$$

where τ_{ph} is the photon lifetime. Equations (5.39) and (5.40) are used to obtain Eqs. (5.34) to (5.36).

The Fourier-series analysis with a period T is used for the Fourier transform of Eqs. (5.34) to (5.36). The finite Fourier transform of $\Delta \hat{A}(t)$ is defined by

$$\Delta \hat{A}(T, \Omega) \equiv \sqrt{\frac{2}{T}} \int_{-T/2}^{T/2} \Delta \hat{A}(t) e^{-j\Omega t} \, dt \tag{5.42}$$

Using Eqs. (5.34) to (5.37), we can obtain Fourier transforms of $\Delta \hat{A}(\Omega)$ and $\Delta \hat{\phi}(\Omega)$ as

$$\Delta \hat{A}(\Omega) = \frac{-A_3[\tilde{\Gamma}_p(\Omega) + \tilde{\Gamma}_{sp}(\Omega) + \tilde{\Gamma}(\Omega)] + (A_1 - j\Omega)\hat{H}_r(\Omega)}{(A_2 A_3 + \Omega^2) + j\Omega A_1} \tag{5.43}$$

and

$$\Delta \hat{\phi}(\Omega) = \frac{j\hat{H}_i(\Omega)}{\Omega A_0} + \frac{jA_2 A_4 \hat{H}_r(\Omega)/\Omega - A_4[\tilde{\Gamma}(\Omega) + \tilde{\Gamma}_{sp}(\Omega) + \tilde{\Gamma}(\Omega)]}{(A_2 A_3 + \Omega^2) + j\Omega A_1} \tag{5.44}$$

Here the parameters A_1 to A_4 are given by

$$A_1 = -\left(\frac{1}{\tau_{sp}} + \frac{1}{\tau_{st}}\right) = \frac{1}{\tau_{sp}}(1 + n_{sp}R) \tag{5.45}$$

$$A_2 = -2A_0 \frac{\omega}{Q} \tag{5.46}$$

$$A_3 = \frac{1}{2A_0 \tau_{st}} = \frac{n_{sp}R}{2A_0 \tau_{sp}} \tag{5.47}$$

$$A_4 = \frac{\alpha}{2A_0^2 \tau_{st}} = \frac{\alpha n_{sp}R}{2A_0^2 \tau_{st}} \tag{5.48}$$

5.2. Squeezing in semiconductor lasers

The power spectrum of the operator $\Delta \hat{A}(t)$ is calculated by Wiener-Khintchin's theorem:

$$P_{\Delta \hat{A}}(\Omega) = \lim_{T \to \infty} \langle \Delta \hat{A}^\dagger(T,\Omega) \Delta \hat{A}(T,\Omega) \rangle \qquad (5.49)$$

The noise operators $\hat{H}_r(t)$ and $\hat{H}_i(t)$ are of the multiplicative noise type. However, the correlation functions of $\hat{H}_r(t)$ and $\hat{H}_i(t)$ are not affected by the phase noise $\Delta\hat{\phi}$, because it is a slowly varying function as compared with the Markovian noise operators $\tilde{G}, \hat{g},$ and $\hat{f}$. That is, the noise operators $\hat{H}_r(t)$ and $\hat{H}_i(t)$ change and lose their memories completely before $\Delta\hat{\phi}$ changes appreciably.[39] Therefore, we obtain

$$\langle \hat{H}_r(t)\hat{H}_r(u) \rangle = \langle \tilde{G}_r(t)\tilde{G}_r(u) \rangle + \langle \hat{g}_r(t)\hat{g}_r(u) \rangle + \langle \hat{f}_r(t)\hat{f}_r(u) \rangle \qquad (5.50)$$

and

$$\langle \hat{H}_i(t)\hat{H}_i(u) \rangle = \langle \tilde{G}_i(t)\tilde{G}_i(u) \rangle + \langle \hat{g}_i(t)\hat{g}_i(u) \rangle + \langle \hat{f}_i(t)\hat{f}_i(u) \rangle \qquad (5.51)$$

Similarly, we obtain the cross-correlation function between $\hat{H}_r(t)$ and $\tilde{\Gamma}(t)$ as

$$\langle \hat{H}_r(t)\tilde{\Gamma}(u) \rangle = \langle \tilde{G}_r(t)\tilde{\Gamma}(u) \rangle \qquad (5.52)$$

Finally, the resulting expressions for the power spectra of $\Delta\hat{A}$ and $\Delta\hat{\phi}$ are given by

$$P_{\Delta\hat{A}}(\Omega) = [A_3^2 \langle |\tilde{\Gamma}_p(\Omega)|^2 + |\tilde{\Gamma}_{sp}(\Omega)|^2 + |\tilde{\Gamma}(\Omega)|^2 \rangle + (\Omega^2 + A_1^2) \langle |\tilde{G}_r(\Omega)|^2 \\ + |\hat{g}_r(\Omega)|^2 + |\hat{f}_r(\Omega)|^2 \rangle - 2A_1 A_3 \langle \tilde{G}_r(\Omega)\tilde{\Gamma}(\Omega) \rangle]/[(\Omega^2 + A_2 A_3)^2 \qquad (5.53) \\ + \Omega^2 A_1^2]$$

and

$$P_{\Delta\hat{\phi}}(\Omega) = \frac{\langle |\tilde{G}_i(\Omega)|^2 + |\hat{g}_i(\Omega)|^2 + |\hat{f}_i(\Omega)|^2 \rangle}{\Omega^2 A_0^2} \\ + A_4^2 \bigg[\langle |\tilde{\Gamma}_p(\Omega)|^2 + |\tilde{\Gamma}_{sp}(\Omega)|^2 + |\tilde{\Gamma}(\Omega)|^2 \rangle \qquad (5.54) \\ + \left(\frac{A_2}{\Omega}\right)^2 \langle |\tilde{G}_r(\Omega)|^2 + |\hat{g}_r(\Omega)|^2 + |\hat{f}_r(\Omega)|^2 \rangle \bigg] \\ \times [(\Omega^2 + A_2 A_3)^2 + \Omega^2 A_1^2]^{-1}$$

The power spectra for the noise operator are calculated by the correlation functions. For instance, $\langle|\tilde{\Gamma}_p(\Omega)|^2\rangle$ is

$$\langle|\tilde{\Gamma}_p(\Omega)|^2\rangle \equiv 2\int_{-\infty}^{\infty}\langle\tilde{\Gamma}_p(\tau)\tilde{\Gamma}_p(0)\rangle e^{-j\Omega\tau}d\tau = 2p \quad (5.55)$$

As indicated in this equation, the power spectrum in this section is defined as the single-sided (unilateral) spectral density per cycle per second (cps).

Cavity internal field and external output field

The conventional quantum-mechanical laser theory treats only the noise properties of the cavity internal field because the internal field is the only "system" of interest and the external field outside the cavity is considered to be one of the "heat baths." Since we are interested in the external output field, it must be treated as another system.

Figure 5.2 shows the cavity internal field $\hat{A}$ and the external output field $\hat{r}$. The external output field $\hat{r}$ consists of the transmitted internal field and the reflected part of the incident zero-point fluctuation. They are quantum-mechanically correlated and interfere with each other. Therefore, the external output field fluctuation is different from the internal field fluctuation. The noise operator $\hat{f}$ in Eq. (5.12) represents the contribution made by the zero-point fluctuation $\hat{f}_e$ that enters into the cavity through the partially reflecting mirror, as shown in Fig. 5.2. If the incident field is normalized such that $\langle\hat{f}_e^\dagger\hat{f}_e\rangle$ represents the average photon flux (number per second), $\hat{f}$ is related to $\hat{f}_e$ by[17]

$$\hat{f} = \left(\frac{\omega}{Q_e}\right)^{1/2}\hat{f}_e \quad (5.56)$$

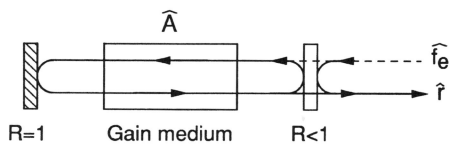

Figure 5.2: Cavity internal field and external output field.

5.2. Squeezing in semiconductor lasers

The external output field $\hat{r}$ is related to the internal field $\hat{A}$ and the reflected part of the incident zero-point fluctuation by the following equation:

$$\hat{r}(t) = -\hat{f}_e(t) + \left(\frac{\omega}{Q_e}\right)^{1/2} \hat{A}(t) \tag{5.57}$$

This relation is obtained from the argument based on time reversal[17] or from the direct analysis of waves bouncing back and fourth in a Fabry-Perot resonator.[18]

The external output field operator $\hat{r}$ is also expanded into c-number mean value and fluctuating operators according to

$$\hat{r} = (r_0 + \Delta\hat{r})e^{-j\Delta\hat{\psi}} \tag{5.58}$$

where r_0^2 is the average photon flux (number per second), and $\Delta\hat{r}$ and $\Delta\hat{\psi}$ are the Hermitian external amplitude and phase fluctuating operators. Using quasi-linearizations Eq. (5.25) for $\hat{A}$ and Eq. (5.58) for $\hat{r}$ together with Eq. (5.57), the Hermitian amplitude and phase fluctuating operators of the external output field are obtained as

$$\Delta\hat{r} = \left(\frac{\omega}{Q_e}\right)^{1/2} \Delta\hat{A} - \frac{1}{2(\omega/Q_e)^{1/2}}(\hat{f}e^{j\Delta\hat{\phi}} + e^{-j\Delta\hat{\phi}}\hat{f}^\dagger) \tag{5.59}$$

and

$$\Delta\hat{\psi} = \Delta\hat{\phi} - \frac{1}{2jr_0(\omega/Q_e)^{1/2}}(\hat{f}e^{j\Delta\hat{\phi}} - e^{-j\Delta\hat{\phi}}\hat{f}^\dagger) \tag{5.60}$$

Substituting Eqs. (5.43) and (5.44) into the Fourier transforms of Eqs. (5.59) and (5.60), we obtain

$$\Delta\hat{r}(\Omega) = -\frac{\hat{f}_1(\Omega)}{(\omega/Q_e)^{1/2}} + \left(\frac{\omega}{Q_e}\right)^{1/2} \frac{-A_3[\tilde{\Gamma}_p(\Omega) + \tilde{\Gamma}_{sp}(\Omega) + \tilde{\Gamma}(\Omega)] + (A_1 - j\Omega)\hat{H}_r(\Omega)}{(A_2 A_3 + \Omega^2) + j\Omega A_1} \tag{5.61}$$

and

$$\Delta\hat{\psi}(\Omega) = \frac{j\hat{H}_i(\Omega)}{\Omega A_0} + \frac{\hat{f}_2(\Omega)}{r_0(\omega/Q_e)^{1/2}} + \frac{jA_2 A_4 \hat{H}_r(\Omega)/\Omega - A_4[\tilde{\Gamma}(\Omega) + \tilde{\Gamma}_{sp}(\Omega) + \tilde{\Gamma}(\Omega)]}{(A_2 A_3 + \Omega^2) + j\Omega A_1} \tag{5.62}$$

where $\hat{f}_1(\Omega)$ and $\hat{f}_2(\Omega)$ are the Fourier transforms of

$$\hat{f}_1(t) = \frac{1}{2}(\hat{f}e^{j\Delta\hat{\phi}} + e^{-j\Delta\hat{\phi}}\hat{f}^\dagger) \tag{5.63}$$

and

$$\hat{f}_2(t) = \frac{1}{2j}(\hat{f}e^{j\Delta\hat{\phi}} - e^{-j\Delta\hat{\phi}}\hat{f}^\dagger) \tag{5.64}$$

Finally, the power spectra for $\Delta\hat{r}$ and $\Delta\hat{\psi}$ are obtained as

$$P_{\Delta\hat{r}}(\Omega) = \left\{ \frac{\omega}{Q_e} A_3^2 \langle |\tilde{\Gamma}_p(\Omega)|^2 + |\tilde{\Gamma}_{sp}(\Omega)|^2 + |\tilde{\Gamma}(\Omega)|^2 \rangle + \frac{\omega}{Q_e} \right.$$

$$(\Omega^2 + A_1^2)\langle |\tilde{G}_r(\Omega)|^2 + |\hat{g}_r(\Omega)|^2 \rangle + \left[\left(\frac{\omega}{Q_e} A_1 - A_2 A_3 - \Omega^2 \right)^2 \quad (5.65) \right.$$

$$\left. + \Omega^2 \left(\frac{\omega}{Q_e} + A_1 \right)^2 \right] \frac{\langle |\hat{f}_r(\Omega)|^2 \rangle}{\omega/Q_e} - 2\frac{\omega}{Q_e} A_1 A_3 \langle \tilde{\Gamma}(\Omega) \tilde{G}_r(\Omega) \rangle \right\}$$

$$[(\Omega)^2 + A_2 A_3)^2 + \Omega^2 A_1^2]^{-1}$$

and

$$P_{\Delta\hat{\psi}}(\Omega) = \frac{(\omega/Q_e)}{r_0^2} \left[\frac{1}{(\omega/Q_e)^2} + \frac{1}{\Omega^2} \right] \langle |\hat{f}_i(\Omega)|^2 \rangle$$

$$+ (\frac{\omega/Q_e}{r_0^2 \Omega^2} \langle |\tilde{G}_i(\Omega)|^2 + |\hat{g}_i(\Omega)|^2 \rangle + A_4^2 \left[\langle |\tilde{\Gamma}_p(\Omega)|^2 + |\tilde{\Gamma}_{sp}(\Omega)|^2 \right. \quad (5.66)$$

$$\left. + |\tilde{\Gamma}(\Omega)|^2 \rangle + \left(\frac{A_2}{\Omega} \right)^2 \langle |\tilde{G}_r(\Omega)|^2 + |\hat{g}_r(\Omega)|^2 + |\hat{f}_r(\Omega)|^2 \rangle \right]$$

$$\times [(\Omega^2 + A_2 A_3)^2 + \Omega^2 A_1^2]^{-1}$$

where $r_0 = \sqrt{\omega/Q_e} A_0$.

Quantum noise properties of the cavity internal field

Figure 5.3 presents characteristic examples of the normalized amplitude and phase noise spectra. The following numerical parameters are assumed for a diode laser: $\omega/Q_0 = 10^{11}$ s^{-1}, $\omega/Q_e = 4 \times 10^{11}$ s^{-1}, $\alpha = 2$, $n_{sp} = 2$, $\tau_{sp} = 2 \times 10^{-9}$ s, and $A_0^2 = 10^5 R$. When the pumping level is well above the threshold ($R \gg 1$), $1/\tau_{st}$ can be much larger than the other lifetime constants such as $1/\tau_{ph}$ and $1/\tau_{sp}$. For such a bias level, we need to consider only the fluctuation frequency below $1/\tau_{st}$, and then the following approximations hold:

$$\Omega \ll |A_1| \quad (5.67)$$

$$\frac{A_3}{A_1} \simeq -\frac{1}{2A_0} \quad (5.68)$$

$$\frac{A_2 A_3}{A_1} \simeq \frac{1}{\tau_{ph}} \quad (5.69)$$

5.2. Squeezing in semiconductor lasers

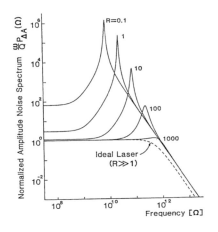
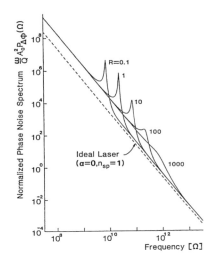

Figure 5.3: (a) Normalized amplitude-noise spectra of the cavity internal field. (b) Normalized phase-noise spectra of the cavity internal field. Numerical parameters are as follows: $\omega/Q_0 = 10^{11}$ s^{-1}, $\omega/Q_e = 4 \times 10^{11}$ s^{-1}, $\alpha = 2$, $n_{sp} = 2$, $\tau_{sp} = 2 \times 10^{-9}$ s, and $A_0^2 = 10^5 R$.

$$\frac{A_2 A_3}{A_1^2} \ll 1 \tag{5.70}$$

$$\frac{A_4}{A_1} \simeq \frac{-\alpha}{2A_0^2} \tag{5.71}$$

$$\frac{A_2 A_4}{A_1 \Omega} \simeq \frac{\alpha}{A_0 \tau_{ph} \Omega} \tag{5.72}$$

Using Eqs. (5.67) to (5.70) and power spectra for the noise operators, Eq. (5.53) is rewritten as

$$P_{\Delta\hat{A}}(\Omega) = \frac{(1/\tau_{ph}) + (N_{c0}/\tau_{sp} A_0^2)}{\Omega^2 + (1/\tau_{ph})^2} \tag{5.73}$$

where Eq. (5.41) is also used. For a high pumping level, the term $N_{c0}/\tau_{sp} A_0^2$ is negligible compared with the term $1/\tau_{ph}$, and then the amplitude-noise spectrum becomes Lorentzian:

$$P_{\Delta\hat{A}}(\Omega) = \frac{\omega/Q}{\Omega^2 + (\omega/Q)^2} \quad (R \gg 1) \tag{5.74}$$

which is shown by the dashed curve in Fig. 5.3(a). The variance of the amplitude is calculated by Parseval's theorem as

$$\langle \Delta \hat{A}^2 \rangle \equiv \int_0^\infty \frac{d\Omega}{2\pi} P_{\Delta \hat{A}}(\Omega) = \frac{1}{4} \tag{5.75}$$

The variance of the photon number is then

$$\langle \Delta \hat{n}^2 \rangle \equiv 4A_0^2 \langle \Delta \hat{A}^2 \rangle = \langle \hat{n} \rangle \tag{5.76}$$

Equation (5.76) indicates that the internal field features Poissonian photon statistics.

Half the noise power [Eq. (5.75) or Eq. (5.76)] stems from the pump fluctuation $\tilde{\Gamma}_p$ and the internal loss fluctuation $\hat{g}_r$. The remaining half is due to the incident zero-point fluctuation $\hat{f}_r$. The noise operator $\tilde{\Gamma}_{sp}$ contribution due to the spontaneous decay process is negligible as compared with the pump fluctuation at well above the threshold, since $p \gg \langle \tilde{N}_c \rangle / \tau_{sp}$. The contribution of noise operators $\tilde{G}_r$ and $\tilde{\Gamma}$ cancel each other out exactly because of their negative correlation [Eq. (5.23)].

Next, we consider the phase-noise spectrum in the high pumping level $(R \gg 1)$. Using Eqs. (5.71) and (5.72) together with power spectra for the noise operators, Eq. (5.54) is rewritten as

$$P_{\Delta\hat{\phi}}(\Omega) = \frac{(1 + \alpha^2)n_{sp}}{A_0^2 \tau_{ph} \Omega^2} \quad (R \gg 1) \tag{5.77}$$

When the linewidth enhancement factor α is equal to zero and the population inversion factor n_{sp} is equal to unity, the phase-noise spectrum is

$$P_{\Delta\hat{\phi}} = \frac{(\omega/Q)}{A_0^2 \Omega^2} \tag{5.78}$$

which is shown by the dashed curve in Fig. 5.3(b). Because of the absence of the phase-restoring force, a laser undergoes the unstationary phase diffusion process, which is responsible for the Ω^2 dependence of the noise spectral density. The phase-noise spectrum [Eq (5.54)] is contributed by the dipole noise operator $\tilde{G}_i$ as well as by the internal loss and zero-point fluctuation $\hat{g}_i$ and $\hat{f}_i$. The failure to suppress the $\tilde{G}_i$ contribution is due to the fact that the gain saturation establishes the coherence on the in-phase component of the dipole moment.

Quantum noise properties of the cavity external output field

Figure 5.4 presents characteristic examples of the amplitude and normalized phase-noise spectra for the external output field. Numerical parameters are the same as for Fig. 5.3. When the pumping level is at well above the threshold ($R \gg 1$), the amplitude noise spectrum becomes white, i.e.,

$$P_{\Delta \hat{r}}(\Omega) = \tfrac{1}{2} \quad (R \gg 1) \tag{5.79}$$

which is shown by the dashed line in Fig. 5.4(a). The power spectrum for the photon flux noise, $\Delta \hat{N} = \Delta(\hat{r}^\dagger \hat{r})$, in such a pumping level is equal to

$$P_{\Delta \hat{N}}(\Omega) \equiv 4r_0^2 \, P_{\Delta \hat{r}}(\Omega) = 2 \langle \hat{N} \rangle \tag{5.80}$$

This shot-noise-limited white-noise spectrum is what is measured by an actual photodetector placed outside the cavity.

The amplitude-noise spectrum within the cavity bandwidth, $P_{\Delta \hat{r}}(\Omega \leq \omega/Q)$, is due to the pump fluctuation $\tilde{\Gamma}_p$ and the internal-loss-induced fluctuation $\hat{g}_r$. On the other hand, the amplitude-noise spectrum above the cavity bandwidth, $P_{\Delta \hat{r}}(\Omega \geq \omega/Q)$, stems from the reflected zero-point fluctuation $\hat{f}_e$. This is shown shematically in Fig. 5.5. Also as mentioned, one-half the internal amplitude noise spectrum, $P_{\Delta \hat{A}}(\Omega)$, is contributed by the incident zero-point fluctuation $\hat{f}_r$. This part of the internal amplitude

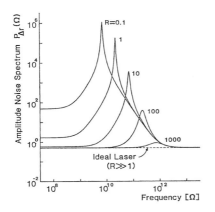
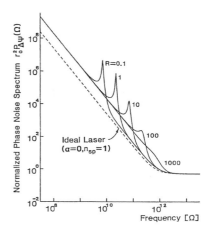

Figure 5.4: (a) Amplitude-noise spectra of the cavity external field. (b) Normalized phase-noise spectra of the cavity external field. Numerical parameters are the same as those in Fig. 5.3.

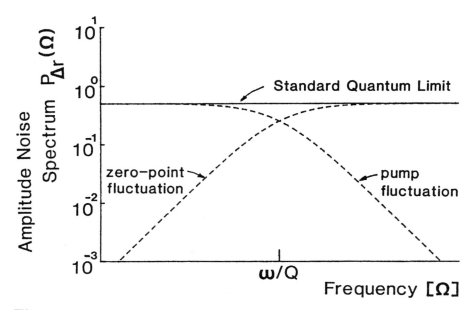

Figure 5.5: Origins for standard quantum-limited amplitude-noise spectrum of the cavity external field.

noise is completely suppressed when the internal field is coupled out of the cavity because the reflected zero-point fluctuation $\hat{f}_r$ beats against the coherent excitation r_0, and the resulting amplitude noise is the exact replica of the internal amplitude noise due to $\hat{f}_r$. Because of the minus sign in front of $\hat{f}_e$ in Eq. (5.57), the two amplitude noises are 180° out of phase and cancel each other out exactly. As a result of this destructive interference, the remaining half of the internal amplitude noise due to $\tilde{\Gamma}_p$ and $\hat{g}_r$ emerges in the external output field. In the frequency region higher than the cavity bandwidth ($\Omega \geq \omega/Q$), the internal amplitude noise is cut off as shown in Fig. 5.3(a), while the incident zero-point fluctuation is simply reflected back from the mirror. This reflected zero-point fluctuation constitutes the higher-frequency part of the white amplitude noise.

When α is equal to zero and n_{sp} is equal to unity, the phase-noise spectrum of the external output field is

$$P_{\Delta\hat{\psi}} = \frac{1}{2r_0^2} + \frac{(\omega/Q)}{A_0^2\Omega^2} = \frac{1}{2r_0^2} + \frac{(\omega/Q_e)(\omega/Q)}{r_0^2\Omega^2} \qquad (5.81)$$

which is shown by the dashed curve in Fig. 5.4(b). The external phase-noise spectrum is different from the internal phase-noise spectrum [Eq. (5.78)] by the first term of Eq. (5.81). This term is due to the reflected zero-point fluctuation. The destructive interference between the transmitted internal field and the reflected zero-point fluctuation, which suppresses the amplitude-noise spectrum within the cavity bandwidth, does not work for the phase-noise spectrum within the cavity bandwidth. This is so because the internal phase noise $\Delta\hat{\phi}$ due to the incident zero-point fluctuation, the first term of Eq. (5.60), is just in the quadrature phase with the phase noise due to the reflected zero-point fluctuation, the second term of Eq. (5.60). Therefore, the two noise powers are additive, even though they originate from the same noise source $\hat{f}_e$.

Amplitude squeezing in a pump-noise-suppressed laser

The amplitude-noise and normalized phase-noise spectra for the cavity external output field of a pump-noise-suppressed laser oscillator are calculated by Eqs. (5.65) and (5.66) with $\langle|\tilde{\Gamma}_p(\Omega)|^2\rangle = 0$. Characteristic examples of amplitude-noise spectra are shown in Fig. 5.6(a). Numerical parameters for Fig. 5.6 are the same as in Fig. 5.3 except that the internal loss is negligible as compared with the output coupling loss, i.e., $\omega/Q_0 \ll \omega/Q_e$. The external amplitude-noise spectrum at $R \gg 1$ becomes lower than the standard quantum limit within the cavity bandwidth

$$P_{\Delta\hat{r}} = \frac{1}{2}\frac{\Omega^2}{\Omega^2 + (\omega/Q)^2} \quad (5.82)$$

The infinite amplitude squeezing is obtained for the external field in the low-frequency limit $\Omega \ll \omega/Q$. The failure in suppressing the amplitude noise above the cavity bandwidth, $\Omega \geq \omega/Q$, stems from the absence of the amplitude noise in the internal field in this frequency region. The zero-point fluctuation cannot enter into the cavity and is simply reflected back from the mirror.

Although the infinite amplitude squeezing can be obtained at $\Omega \ll \omega/Q$ when $\omega/Q_0 \ll \omega/Q_e$, as shown in Fig. 5.6(a), the squeezing factor is limited when the internal loss is not negligible. This is so because the quantum-mechanical correlation between the internal amplitude noise and the reflected zero-point fluctuation becomes imperfect due to the contribution of the internal loss fluctuation $\hat{g}_r$. Figure 5.7 shows the external amplitude-noise spectrum in the low-frequency region $P_{\Delta\hat{r}}(\Omega \simeq 0)$

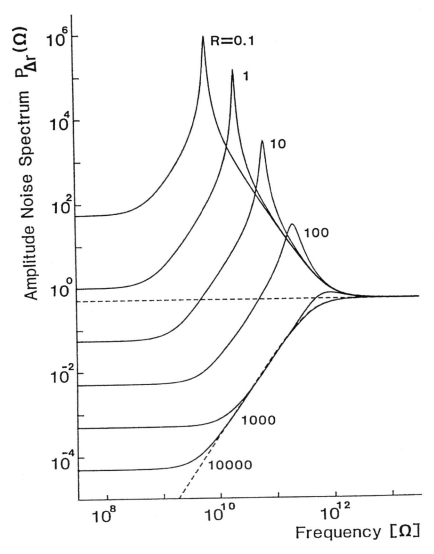

Figure 5.6: (a) External amplitude-noise spectra of a pump-noise-suppressed laser ($\omega/Q_0 = 0$). (b) External normalized phase-noise spectra of a pump-noise-suppressed laser ($\omega/Q_0 = 0$).

5.2. Squeezing in semiconductor lasers

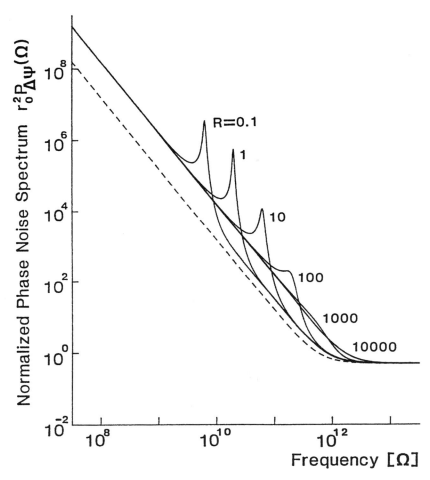

Figure 5.6: Continued

versus the normalized pumping level R as a function of the output coupling loss to internal loss ratio Q_0/Q_e. In order to obtain large-amplitude squeezing, a value of ω/Q_e that is larger than ω/Q_0 and a higher pumping level are desirable.

The characteristic examples of the normalized phase-noise spectrum are shown in Fig. 5.6(b). It is clear that the ultimate phase-noise spectrum [Eq. (5.81)] is not altered by the pump-noise suppression.

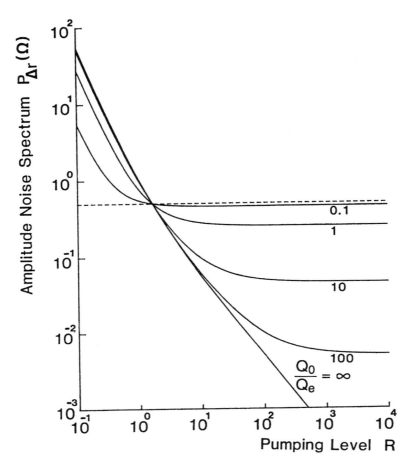

Figure 5.7: External amplitude-noise spectral density $P_{\Delta\hat{r}}$ ($\Omega \simeq 0$) at low frequency versus pumping level R as a function of the output coupling loss to internal loss ratio, Q_0/Q_e.

Uncertainty product

The product of the amplitude- and phase-noise spectra of an ideal laser satisfies the minimum-certainty product in the frequency region above the cavity bandwidth, $\Omega \geq \omega/Q$,

$$P_{\Delta\hat{r}}(\Omega)P_{\Delta\hat{\psi}}(\Omega) = \frac{1}{4r_0^2} \tag{5.83}$$

5.2. Squeezing in semiconductor lasers

However, the product of the amplitude- and phase-noise spectra in the frequency region within the cavity bandwidth, $\Omega \geq \omega/Q$, is larger than the minimum-uncertainty product because of the phase-diffusion noise. Therefore, an ordinary laser oscillator is not a quantum-limited device in this respect, even though the conditions of $R \gg 1$, $\alpha = 0$, and $n_{sp} = 1$ are all satisfied.

The product of the amplitude- and phase-noise spectra of an ideal pump-noise-suppressed laser oscillator in the frequency region below the cavity bandwidth, $\Omega \leq \omega/Q$, is

$$P_{\Delta \hat{r}}(\Omega) P_{\Delta \hat{\psi}}(\Omega) = \frac{1}{2r_0^2} \quad (5.84)$$

This is twice as large as the minimum-uncertainty product. The amplitude- and normalized phase-noise spectra of an ideal pump-noise-suppressed laser are compared with the standard quantum limit of broadband coherent states in Fig. 5.8(a,b). If the mode is defined by a bandwidth much narrower than the laser cavity bandwidth, $B \ll \omega/Q$, the amplitude and phase noise of this laser output is close to the "amplitude-squeezed state" shown in Fig. 5.8(b), even though it is not the "number-phase minimum uncertainty state." If the mode is defined by a bandwidth much broader than the laser cavity bandwidth, $B \gg \omega/Q$, the amplitude and phase noise is close to the coherent state. One great advantage of using a semiconductor laser as an amplitude-squeezed-state generator is its broad bandwidth; i.e., the value of ω/Q is in the range of 10^{12} rad/sec.

5.2.3 Experimental results

Observation of amplitude squeezing by balanced detectors with a delay line

In a usual intensity-noise measurement, two measurement steps are required: one for a laser intensity-noise measurement and the other for a shot-noise calibration with a light-emitting diode. To eliminate the error introduced by the two-step measurements and the photodetector saturation effect dependent on a beam spot size, the balanced detectors with a delay line shown in Fig. 5.9 are developed.[19] One detector output is delayed (by $\tau = 50$ nsec in our measurement), but the other detector output is not. The difference in these two outputs is produced by a differential amplifier. A coaxial cable delay line has a loss coefficient proportional to

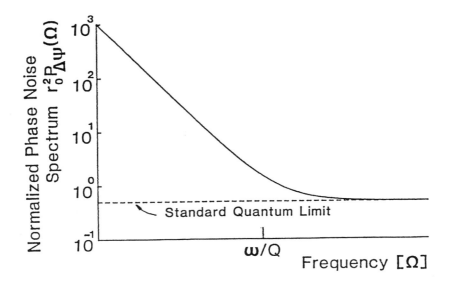

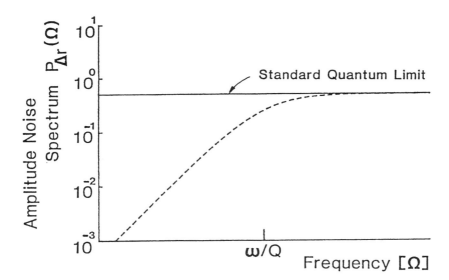

Figure 5.8: Normalized phase-noise spectrum $r_0^2 P_{\Delta\tilde{\psi}}(\Omega)$ (a) and amplitude-noise spectrum $P_{\Delta\tilde{r}}(\Omega)$ (b) of a pump-noise-suppressed laser oscillator biased at well above threshold (r_0^2 is the average photon flux and ω/Q is the cavity bandwidth).

5.2. Squeezing in semiconductor lasers

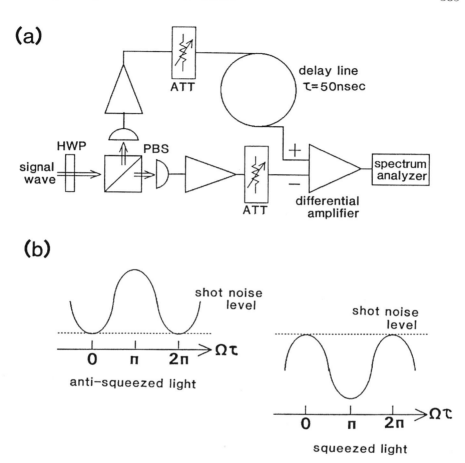

Figure 5.9: (a) Balanced direct detectors with a delay line and attenuators (ATT). (b) Current noise spectra of an amplitude-antisqueezed light and an amplitude-squeezed light.

$\sqrt{\Omega}$, where Ω is the fluctuation frequency. This frequency-dependent attenuation imposed on the delayed signal is compensated for by introducing the same attenuation to the other signal.

For a fluctuation frequency Ω_{in} satisfying inphase delay condition $\Omega_{in}\tau = 2N\pi$, where N is an integer, the differential amplifier output measures $\hat{I}_1 - \hat{I}_2$, where $\hat{I}_1$ and $\hat{I}_2$ are the two photodetector currents. The current-fluctuation spectral density is exactly equal to the shot-noise level. For a fluctuation frequency Ω_{out} satisfying out-of-phase delay condition

$\Omega_{out}\tau = (2N + 1)\pi$, the differential amplifier output measures $\hat{I}_1 + \hat{I}_2$. The quantum-mechanical theory of a balanced detector[20-22] shows that $\hat{I}_1 + \hat{I}_2$ corresponds to the quantum noise of the laser itself with polarizations parallel to the coherent excitation. Thus the detector output simultaneously displays the laser-noise level and the corresponding shot-noise level on a spectrum analyzer with a frequency period of $\Delta\Omega = 2\pi/\tau$.

When the signal wave is amplitude-antisqueezed (super-Poissonian), the current fluctuation at Ω_{in} is smaller than that at Ω_{out}, as shown in Fig. 5.9(b). On the other hand, when the signal wave is amplitude-squeezed (sub-Poissonian), the current fluctuation at Ω_{in} is larger than that at Ω_{out}, as shown in Fig. 5.9(b). This inverted modulation in the current-fluctuation spectrum is an unmistakable mark of amplitude-squeezed light. This is a single-step measurement; therefore, the ambiguity in the shot-noise level owing to photodetector saturation can be eliminated.

A short-cavity AlGaAs/GaAs transverse-junction stripe semiconductor laser with an antireflection coating on the front facet and a high-reflection coating on the rear facet was used at 77 K. The threshold current was 1 mA, and the differential quantum efficiency above the threshold was 70%. To eliminate a minute optical reflection feedback to the laser from the measurement optical elements, an optical isolator was used, and all optical elements were antireflection-coated and slanted with respect to the beam direction. The overall detection quantum efficiency, including losses from the laser collimating lens, cryostat window, gold mirror, optical isolator, half-wave plate, photodetector focusing lens, polarization beam splitter, and photodetector quantum efficiency, was 60%.

Figure 5.10 shows current noise spectra at two different bias levels. The current noise spectrum for bias level $R_P = 0.03$, which is shown by curve A, features lower noise power at Ω_{in} than at Ω_{out}. This indicates that the field is amplitude-antisqueezed (super-Poissonian). The amplifier noise level is shown by curve B. The front-end amplifier is an ac-coupled bipolar transistor (NEC 2SC3358) with a noise figure of 1.1 dB and a load resistance of 220 Ω. The current noise spectrum for bias level $R_P = 12.6$, which is shown by curve C, features higher noise power at Ω_{in} than at Ω_{out}. This indicates that the field is amplitude-squeezed (sub-Poissonian). The total dc photocurrents are 15 μA and 6.12 mA for $R_P = 0.03$ and 12.6, respectively. Curves D and E are the current noise spectra when one of the two incident signal waves for $R_P = 12.6$ is blocked. The modulation disappears, as expected, in a low-frequency region. At high frequencies,

5.2. Squeezing in semiconductor lasers

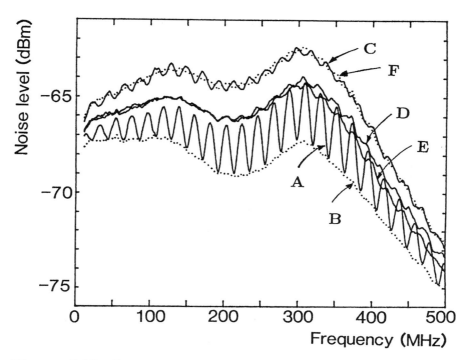

Figure 5.10: Current noise spectra for bias levels $I/I_{th} = 1.03$ (curve A) and $I/I_{th} = 13.6$ (curve C). Curve B is the amplifier thermal noise. Curves D and E are obtained when one of the two signal beams is blocked for $I/I_{th} = 13.6$. Curve F is the sum of noise curves D and E.

however, the noise power is reflected back at the differential amplifier input, and so the modulation due to the round trip in a delay line appears. Curve F is the sum of the current noise spectra indicated by curves D and E. The noise level of curve F is not equal to a 3-dB noise rise from the noise level indicated by curves D or E because of the amplifier thermal noise (curve B). Note that the current noise spectrum indicated by curve F is between the shot-noise level at Ω_{in} and the reduced noise level at Ω_{out}. This is so because the noise level of amplitude-squeezed light increases to approach the shot-noise level when the amplitude is attenuated.

Figure 5.11 shows the current noise spectra normalized by the shot-noise level for the two bias levels. The shot-noise level calibrated by the measurement of $\hat{I}_1 - \hat{I}_2$ is compared with the shot-noise level generated by the light-emitting diode with the same wavelength as the laser. The

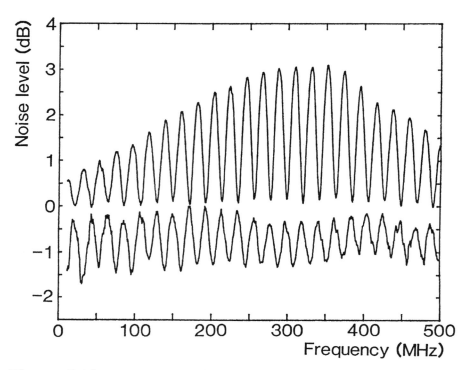

Figure 5.11: Current noise spectra normalized by the shot-noise level for bias levels $I/I_{th} = 1.03$ (upper curve) and $I/I_{th} = 13.6$ (lower curve). The amplifier thermal noise is subtracted in the normalization process.

difference is smaller than 0.1 dB. The current noise spectrum at $R_P = 0.03$ shows an enhanced noise peak at the relaxation-oscillation frequency. The current noise spectrum at $R_P = 12.6$ shows a noise level reduced to -1.3 dB below the shot-noise level. The observed 0.6-dB squeezing becomes 1.3-dB squeezing in Fig. 5.11 because the effect of the amplifier thermal noise is subtracted. This noise reduction is much larger than the error bar of the shot-noise-level calibration. If the increase in the noise level owing to the detection quantum efficiency of 60% is corrected, the measured noise level corresponds to the squeezing of -3 dB (50%) at the laser output.

Degree of squeezing versus optical loss

The photocurrent noise spectral density at 800 MHz at pump rate $R_P = 10.4$ versus optical loss L put in front of the balanced detectors is shown in Fig. 5.12. The ordinate is normalized by the corresponding shot-noise level. The noise level increases to the shot-noise level as the optical loss increases, and the amplitude squeezing is eventually lost in the limit of infinite loss. This is so because the optical loss process couples the vacuum field fluctuation to the laser light. In the limit of large optical loss, the original quantum noise of the laser is entirely replaced by the vacuum field fluctuation, and thus the shot-noise emerges. This apparent increase in normalized photocurrent noise is an unmistakable mark of amplitude squeezing.

The overall quantum efficiency from the laser injection-current increment to photodetector current increment is 22% in this experiment when artificial optical attenuation is eliminated. The overall detection quantum efficiency of 55% consists of a photodetector quantum efficiency of 0.93, focusing-lens loss of 0.90, isolator insertion loss of 0.81, mirror reflection loss of 0.95, cryostat window loss of 0.93, and laser collimating lens loss of 0.93. The amplitude-noise levels corrected for these factors are shown in Fig. 5.12. The observed noise level corresponds to amplitude squeezing of -1.7 dB (33%) below the SQL at the output of the laser facet, as shown in Fig. 5.12. This is the noise level that the laser actually produced at the output mirror. The laser output coupling efficiency is

$$\eta_c = \frac{L^{-1}\ln(R_1^{-1/2})}{\{\alpha + L^{-1}\ln\{(R_1 R_2)^{-1/2}]\}}$$

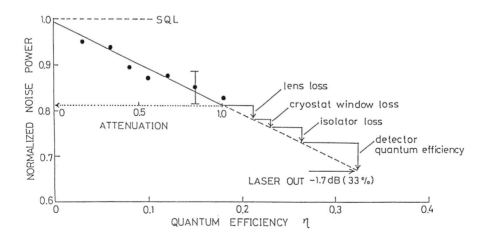

Figure 5.12: Degree of squeezing versus optical loss.

where $L = 250$ μm is the cavity length, $R_1 = 0.32$, and $R_2 = 0.6$ are power reflectivities of the front and rear facets, respectively, and $\alpha = 22$ cm^{-1} is the internal absorption loss. If the laser output-coupling efficiency due to nonideal real-facet reflectivity $\eta_M \simeq 0.57$ and that due to internal loss $\eta_A = 0.70$ are also corrected, the observed noise level corresponds to amplitude squeezing of -7 dB (80%) below the shot-noise level. This is the intrinsic noise level achievable if the rear-facet reflectivity is increased to 100% and internal absorption loss is eliminated.

10-dB Squeezing

The degree of squeezing is degraded by the Poissonian partition noise associated with optical loss. To increase the light-collection efficiency, a GaAs transverse junction stripe laser, with less than 3% front-facet reflectivity R_1 and more than 90% rear-facet reflectivity R_2, is operated at 66 K to minimize the free-carrier absorption loss and is directly coupled to an S_i photodetector to minimize the coupling loss. Figure 5.13 shows theoretical and experimental photon-number noise values normalized by the shot-noise value versus the normalized pump rate $I/I_{th} - 1$.[23] The experimental results are corrected for a detection quantum efficiency of about 89%, so the degree of squeezing shown in Fig. 5.13 corresponds to that of the laser output. The experimental maximum degree of squeezing, -14 dB ($\simeq 0.04$), is in reasonable agreement with the theoretical limit imposed by the output coupling efficiency

$$\eta = \frac{\ln(1/R_1)}{\ln(1/R_1R_2)} \simeq 0.97$$

of the laser.

5.2.4 Squeezed vacuum state generation

Squeezed vacuum states from amplitude-squeezed states

Figure 5.14 shows the setup for generating a squeezed vacuum state by destructively interfering with an amplitude-squeezed state from an injection-locked slave laser with strong coherent light from a master laser.[24] A constant-current-driven semiconductor laser, which is denoted a *slave laser*, is injection locked by an external *master laser*; therefore, the output signal from the injection-locked slave laser is phase coherent with the master laser signal. These two signals are combined at a high

5.2. Squeezing in semiconductor lasers

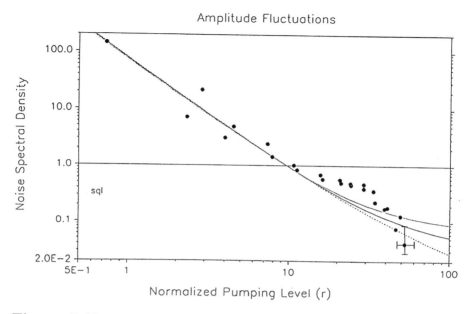

Figure 5.13: The theoretical (lines) and experimental (●) intensity-noise values normalized by the shot-noise value versus the normalized pump rate $R \equiv I/I_{th} - 1$. The experimental results are corrected for a detection quantum efficiency of 89%.

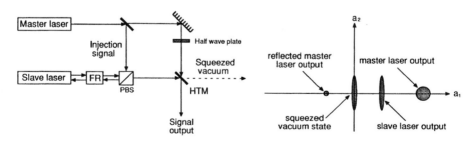

Figure 5.14: Squeezed-vacuum-state generation by mixing an amplitude-squeezed state with a coherent state. FR, Faraday-rotator; PBS, polarization beam splitter; HTM, high-transmission mirror.

transmission mirror, and the coherent excitation of the squeezed output signal from the injection-locked slave laser is canceled by the destructive interference with the master laser signal. By using the high transmission mirror, the squeezed output signal from the injection-locked slave laser is not degraded because the noise of the master laser signal is attenuated by the mirror.

Experimental setup

The experimental setup is shown in Fig. 5.15. The master laser is an AlGaAs single-mode high-power semiconductor laser. This master laser is mounted on a copper heat sink, which is attached to a temperature-controlled chamber. Two optical isolators are used in tandem to avoid the undesired feedback from the injected laser. The slave laser is a single-mode low-power GaAs transverse-junction strip semiconductor laser with antireflection coating (~10%) on the front facet and high-reflection coating (~90%) on the rear facet. This slave laser is mounted on a copper heat sink in a 77 K cryostat. Two polarization beam splitters, one half-wave

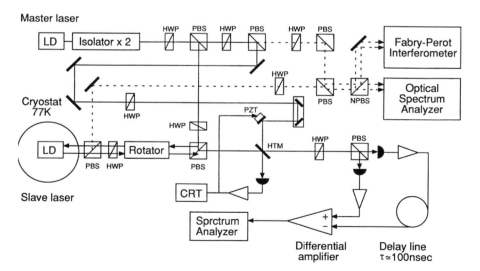

Figure 5.15: The experimental setup for the squeezed-vacuum-state generation by a semiconductor laser system. HWP, half-wave plate; PBS, polarization beam splitter, NPBS, nonpolarization beam splitter; HTM, high-transmission mirror; PZT, piezo translator.

5.2. Squeezing in semiconductor lasers

plate and one Faraday-rotator in front of the slave laser, are used to match the master laser polarization to the slave laser polarization and to avoid undesired feedback from the detection system. The master and slave laser outputs are collimated with antireflection-coated microlenses, respectively. The oscillation wavelengths of the master and slave lasers are monitored simultaneously by an optical spectrum analyzer and a Fabry-Perot interferometer. Longitudinal mode matching between the two lasers is carried out roughly by controlling the temperature of the master laser heat sink. The optical spectrum analyzer is used to identify the longitudinal modes roughly. Frequency fine-tuning is achieved by adjusting the master laser drive current. The Fabry-Perot interferometer is used to observe the locking phenomena. The master laser signal is divided into an injection signal and a reference signal for the interference. This master laser reference signal and the injection-locked slave laser signal are combined by a high-transmission mirror (~95%). The optical path length of the reference signal is adjusted by moving a corner reflector. To eliminate a minute optical reflection feedback to the slave laser, all optical elements are antireflection-coated and slanted with respect to the beam direction. The intensity-noise measurement of the injection-locked slave laser is performed by blocking the master laser reference signal and removing the high-transmission mirror. The detection system is a balanced direct detector with a delay line.[19] The signal wave is equally divided by a half-wave plate and a polarizing beam splitter. One detector output is delayed by about 100 ns, but the other detector output is not. The difference in these two outputs is produced by a differential amplifier.

Experimental results

Spectra of master, slave, and injection-locked laser. Figure 5.16(a,b) shows the spectra of the master and slave lasers. The oscillation wavelength of the slave laser is 788.6 nm at 77 K, and $I_d = 18$ mA. The oscillation wavelength of the master laser is matched to the same wavelength. Figure 5.16(c) shows the spectrum of an injection-locked laser. It is demonstrated that the side-mode intensity of the slave laser is suppressed by about 5 dB due to injection locking.

Observation of an injection-locking phenomenon by a Fabry-Perot interferometer. Figure 5.17(a) shows a Fabry-Perot interferometer output without injection locking. The linewidths of the master and slave lasers

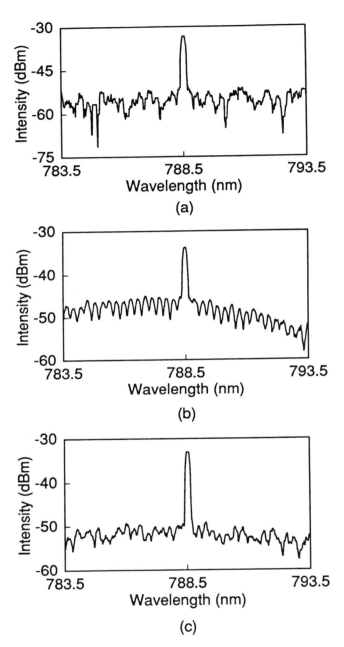

Figure 5.16: The spectra of master (a), slave (b), and injection-locked (c) semiconductor lasers.

5.2. Squeezing in semiconductor lasers

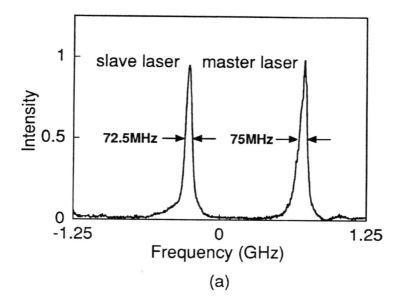

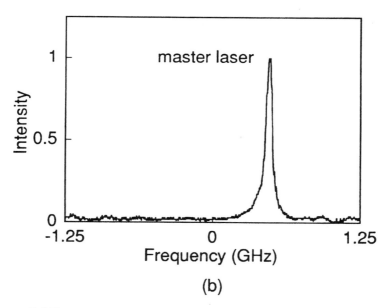

Figure 5.17: Fabry-Perot interferometer output without injection locking (a) and with injection locking (b).

are 75 and 72.5 MHz, respectively. Figure 5.17(b) shows a Fabry-Perot interferometer output with injection locking. It is demonstrated that the oscillation frequency of the slave laser is locked to that of the master laser by injection locking.

Intensity-noise spectra for the free-running master and slave lasers. Figure 5.18 shows the differential amplifier output spectra. Curves (a) and (b) show intensity noise spectra for the free-running master and slave lasers, respectively. These spectra are modulated in the frequency of 10 MHz because the output signal of one detector is delayed by 100 ns. Therefore, the laser intensity-noise value and the corresponding shot-noise value appear alternately with the frequency interval of 10 MHz. The intensity-noise spectrum for the master laser, which is shown by curve (a), features lower noise power at Ω_{in} than at Ω_{out}. This indicates that the master laser output field has excess amplitude noise. On the other hand, the intensity-noise spectrum for the slave laser, which is shown by curve (b), features higher noise power at Ω_{in} than at Ω_{out}. This

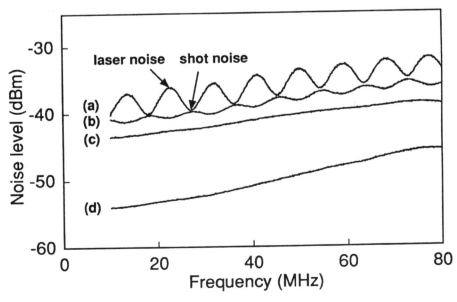

Figure 5.18: Intensity-noise spectra for the master laser (a), the free-running slave laser (b), the output of single detector (c), and the thermal noise (d).

5.2. Squeezing in semiconductor lasers

indicates that the slave laser output field has squeezed-amplitude noise. The slave laser drive current is 18 mA, which corresponds to 13.3 times the threshold current. The total dc photocurrent for this bias level is 5.14 mA, which is not the same as that for the master laser. Curve (c) shows the output spectrum of a single detector; i.e., one of the two signal beams is blocked. This curve shows the frequency characteristics of our detection system, which is used to correct the shot noise value at Ω_{out}. Curve (d) shows the thermal noise value, which is about 10 dB below the curve (c) in the low-frequency region.

Intensity-noise spectral density of an injection-locked laser. Figure 5.19 shows the intensity-noise spectral density of an injection-locked semiconductor laser at 23.9 MHz as a function of the master laser frequency detuning from the slave laser free-running frequency. The intensity-noise spectral density is normalized by the corresponding shot-noise value. The

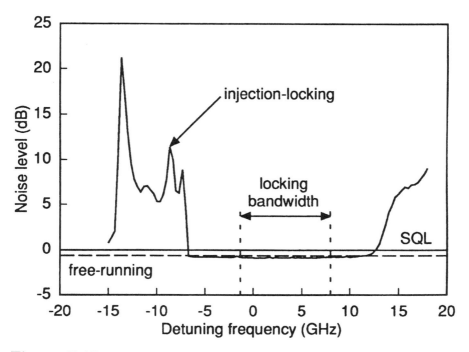

Figure 5.19: Intensity-noise spectral density of an injection-locked semiconductor laser as a function of the detuning.

salve laser drive current is 18 mA, which corresponds to 13.3 times the threshold current. The total dc photocurrent without injection locking is 5.14 mA, and that with injection locking is 5.06 mA. This slight decrease in the output power caused by injection locking is due to the cavity internal loss.[25] The dashed line shows the intensity-noise spectral density of a free-running slave laser, which is 0.72 dB below the standard quantum limit (SQL). The injection power into the slave laser, which is measured in front of the cryostat, is 7 mW. The locking bandwidth is measured to be 9.3 GHz. The minimum value of the measured noise level of the injection-locked semiconductor laser is 0.91 dB below the SQL. This corresponds to -2.56 dB squeezing when corrected for an optical loss from the laser to the detector. The linewidth enhancement factor α is estimated to be about 6 from the asymmetry of the locking bandwidth.

Comparing this experimental result with the numerical results presented in a later section [see Fig. 5.24(b)] and in ref. 25, we can find the following different points: In the experimental result, amplitude squeezing was enhanced by injection locking, although it is not enhanced but rather degraded in the numerical result. This enhancement of amplitude squeezing is due to the suppression of the side-mode intensity of the slave laser by injection locking [Fig. 5.16(c)] because the suppression of the side-mode intensity reduces the mode-partition noise caused by the multi-longitudinal mode oscillation.[26] Even in the case that the free-running slave laser is in a multilongitudinal mode oscillation and its intensity noise is excess noise, the squeezed intensity noise was observed by suppressing the side-mode intensity in terms of injection locking. In order to explain the above-mentioned phenomenon, we must analyze the injection locking to the slave laser, which is not in a single-mode oscillation but in a multilongitudinal mode oscillation. Amplitude squeezing is observed even outside of the negative edge of the locking bandwidth. Outside the locking bandwidth where amplitude squeezing was observed, the slave laser frequency changes with a constant frequency separation to the master laser frequency. The asymmetry of the locking bandwidth in the experimental result is opposite to that of the numerical result. These points require further theoretical study.

Intensity-noise spectral density with the interference. Figure 5.20 shows the differential amplifier output spectra. Curves (a) and (b) show the intensity noise spectra with the interference. Curve (c) shows the intensity-noise spectrum without the interference. The intensity-noise spec-

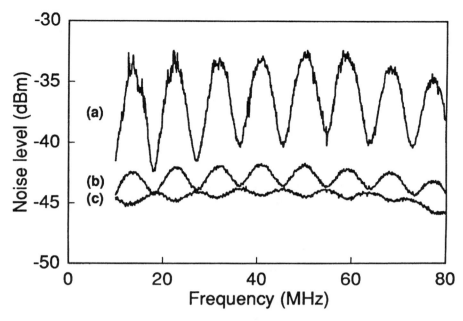

Figure 5.20: The intensity-noise spectra with and without the interference: (a) phase difference $(\pi - \delta) \simeq 90°$; (b) phase difference $(\pi - \delta) = 170 \pm 5°$; and (c) without the interference.

trum without the interference, which is shown by curve (c), features higher noise power at Ω_{in} than at Ω_{out}. This indicates that the output field from the injection-locked slave laser has a squeezed-amplitude noise. The slave laser drive current is 14 mA, which corresponds to 10.4 times the threshold current. The total dc photocurrent for this bias level is 3.5 mA. The injection power into the slave laser, which was measured in front of the cryostat, is 2.9 mW. The locking bandwidth is measured to be 9.6 GHz. On the other hand, the intensity noise spectra with the interference, which are shown by curves (a) and (b), feature lower noise power at Ω_{in} than at Ω_{out}. This indicates that the excess noise is produced by the interference. The measured intensity noise of the master laser signal was shot-noise level; therefore, the intensity noise of the master laser signal cannot contribute the excess noise with the interference. The amplitude cancellation rate, which is defined by the ratio of the average amplitude of the master laser signal to that of the injection-locked slave laser signal, is about 4%. The phase difference between the injection-locked slave laser signal and the master laser signal is about 90° in the case shown by curve (a). This phase difference was increased gradually, and the minimum intensity noise spectrum we could get is shown by curve (b). The phase difference between the master and injection-locked slave laser signals is estimated to be 170 ± 5° in the case of curve (b). Our phase-difference stabilization system is a simple electrical feedback circuit; therefore, the phase difference could not be stabilized exactly at 180°. This intensity excess noise is due to the residual phase error between the master and slave laser signals.

Figure 5.21 shows the intensity-noise spectral densities at 22.8 MHz with and without the interference as a function of the locking bandwidth. The different locking bandwidths can be obtained by changing the injection power. The measured locking bandwidths are 9.6, 6.7, and 6.0 GHz. The corresponding injection powers are 2.9, 1.65, and 1 mW, respectively. These intensity-noise spectral densities are normalized by the shot-noise

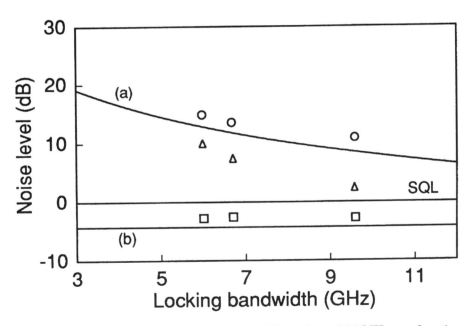

Figure 5.21: The intensity-noise spectral densities at 22.8 MHz as a function of the locking bandwidth. Circles: phase difference $(\pi - \delta) \simeq 90°$; triangles: phase difference $(\pi - \delta) = 170 \pm 5°$; squares: without the interference; solid lines: theory.

5.2. Squeezing in semiconductor lasers

value without the interference and corrected for the optical loss from the laser to the detector. The amplitude cancellation rate is about 4% for all locking bandwidths. The circles show the intensity-noise spectral densities when the phase difference is about 90°. The triangles show the minimum intensity-noise spectral densities we could get at 170 ± 5°. The squares show the intensity-noise spectral densities for the injection-locked slave laser without the interference. As the locking bandwidth becomes wider, the residual intensity noise spectral densities decrease. This is so because the phase noise of the injection-locked slave laser signal is reduced as the locking bandwidth becomes wider. Curves (a) and (b) show the intensity-noise spectral densities as a function of the locking bandwidth with the phase difference as a parameter. These theoretical curves were calculated numerically by using the results of the next section. Numerical parameters are the same as those in Fig. 5.24(b). Curve (a) shows the intensity-noise spectral density when the phase difference is 90°. The experimental results are in good agreement with this numerical result. Curve (b) shows the intensity-noise spectral density when the phase difference is 180°. If we can stabilize the phase difference exactly at 180°, then the intensity-noise spectral densities denoted by the triangles will be reduced to the noise level denoted by the corresponding squares. These experimental and numerical results demonstrate that the excess noise produced by the interference is the phase-to-amplitude conversion noise, and it can be reduced by increasing the locking bandwidth and/or suppressing the phase error by a more sophisticated phase-locked-loop circuit.

Phase-to-amplitude conversion noise in squeezed-vacuum-state generation

Squeezed states are usually detected by a balanced homodyne detector. In this detection system, we can always find the minimum quadrature noise by adjusting the local oscillator phase. However, if there is a residual phase error, the excess noise of the antisqueezed quadrature is partly added to the detection signal. In a similar way, we cannot avoid the phase-to-amplitude conversion noise when we cancel the coherent intensity of the slave laser output by the destructive interference with the master laser signal. Figure 5.22 shows the phase-to-amplitude conversion noise that appears in the interference experiment. The quasi-probability density of the master laser signal reflected by a high-transmission mirror is denoted by the open circle and that of the transmitted injection-locked slave

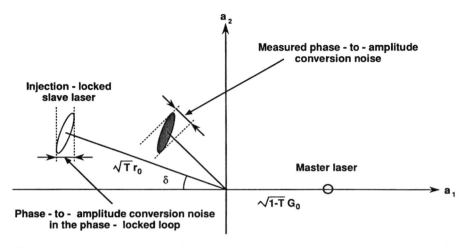

Figure 5.22: Phase-to-amplitude conversion noise with the interference.

laser signal is denoted by the open ellipse. The amplitude mean values of the master and injection-locked slave laser signals are $\sqrt{1 - T}G_0$ and $\sqrt{T}r_0$, respectively. Here r_0 and G_0 are the classic excitations of the master and injection-locked slave laser signals, and T is the transmissivity of the high-transmission mirror. The phase difference between these two signals is $\pi - \delta$, where δ is the phase error, which is equal to zero when the phase difference between them is exactly 180°. The quasi-probability density of the combined signal is denoted by the shaded ellipse. The intensity noise of the combined signal is measured with respect to its mean field. Therefore, the phase noise of the injection-locked slave laser signal is partly added to the detection signal if the phase error δ is not equal to zero. This kind of phase-to-amplitude conversion noise was measured in our experiment, and it depends on the phase error δ and the amplitude cancellation rate, which is defined by $\sqrt{1 - T}G_0/\sqrt{T}r_0$.

Next, we consider the phase-to-amplitude conversion noise that appears when the squeezed-vacuum state generated by our scheme is applied to the interferometric phase measurement. Figure 5.23 shows the setup for generating a squeezed-vacuum state and its application to the Mach-Zehender interferometer. In order to minimize the phase-difference noise between the two arms of the Mach-Zehnder interferometer, the squeezed-vacuum state must be injected into the open port of the 50/50 beam splitter with a 90° phase sift with respect to the master laser signal. For this phase-locked loop, the phase error from the 180° phase difference in the

5.2. Squeezing in semiconductor lasers

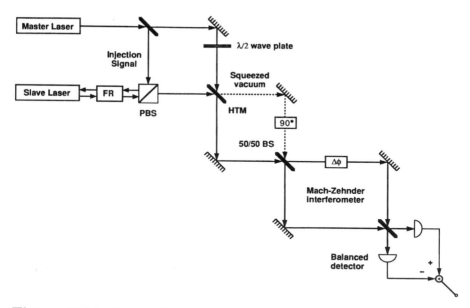

Figure 5.23: Squeezed-vacuum-state generation by mixing an amplitude-squeezed state with a coherent state and its application to the Mach-Zehnder interferometer. FR, Faraday-rotator; PBS, polarization beam splitter; HTM, high-transmission mirror; BS, beam splitter.

interferometric cancellation of the coherent intensity of the injection-locked slave laser signal causes excess phase-to-amplitude conversion noise in the detection of very small phase sifts $\Delta\phi$ by a balanced detector. In this section we calculate the phase-to-amplitude conversion noise that appears when the squeezed-vacuum state is applied to the interferometric phase measurement as a function of the phase error. Whether the squeezed-vacuum state can be generated or not should be estimated by the performance of the interferometric phase measurement.

Langevin equations for an injection-locked semiconductor laser. The operator Langevin equation for the cavity internal field $\hat{A}(t)$ is given by [17, 25]

$$\frac{d\hat{A}(t)}{dt} = -\frac{1}{2}\left[\frac{\omega}{Q} + i2(\omega_r - \omega) - \frac{\omega}{\mu^2}(\tilde{\chi}_i - i\tilde{\chi}_r)\right]\hat{A}(t) \\ + \left(\frac{\omega}{\mu^2}\langle\tilde{\chi}_i\rangle\right)^{1/2}\tilde{f}_G(t) + \left(\frac{\omega}{Q_0}\right)^{1/2}\hat{f}_L(t) + \left(\frac{\omega}{Q_e}\right)^{1/2}[F_0 + \hat{f}(t)] \quad (5.85)$$

where $F_0 + \hat{f}(t)$ is the amplitude of the injection signal, F_0 is the classic (c-number) excitation, and $\hat{f}(t)$ is the injection signal fluctuation operator. The optical angular frequency of the injection signal is denoted by ω. The total Q value of the laser cavity depends on the external (mirror) loss Q_e and the internal loss Q_0 according to

$$\frac{1}{Q} = \frac{1}{Q_e} + \frac{1}{Q_0} \tag{5.86}$$

The angular resonance frequency of an empty cavity is denoted ω_r, μ is the refractive index, and $\tilde{\chi}$ is the electronic susceptibility operator, whose imaginary part equals the stimulated emission gain:

$$\frac{\omega}{\mu^2}\tilde{\chi}_i = \tilde{E}_{cv} - \tilde{E}_{vc} \tag{5.87}$$

where $\tilde{E}_{cv}$ and $\tilde{E}_{vc}$ are the operators of the stimulated emission and absorption rates, respectively. The noise operators $\tilde{f}_G(t)$ and $\hat{f}_L(t)$ are associated with the gain fluctuation and the internal loss, respectively.

The operator Langevin equation for the total excited electron number $\tilde{N}_c(t)$ is given by

$$\frac{d\tilde{N}_c(t)}{dt} = p - \frac{\tilde{N}_c(t)}{\tau_{sp}} - (\tilde{E}_{cv} - \tilde{E}_{vc})\hat{n}(t) - \langle \tilde{E}_{cv}\rangle + \tilde{\Gamma}_p(t) + \tilde{\Gamma}_{sp}(t) + \tilde{\Gamma}(t) \tag{5.88}$$

where p is the pumping rate, τ_{sp} is the lifetime of the electrons due to spontaneous emission, and $\hat{n}(t) \equiv \hat{A}^\dagger(t)\hat{A}(t)$ is the photon-number operator. The three last terms in Eq. (5.88) are fluctuating noise operators: $\tilde{\Gamma}_p(t)$ is the pump noise, $\tilde{\Gamma}_{sp}(t)$ is the spontaneous emission noise, and $\tilde{\Gamma}(t)$ is the stimulated emission and absorption noise including spontaneous emission into a lasing mode. For the noise sources in Eqs. (5.85) and (5.88), the following shorter notations are used:

$$\hat{H}(t) = \left(\frac{\omega}{\mu^2}\langle \tilde{\chi}_i\rangle\right)^{1/2} \tilde{f}_G(t) + \left(\frac{\omega}{Q_0}\right)^{1/2} \hat{f}_L(t) + \left(\frac{\omega}{Q_e}\right)^{1/2} \hat{f}(t) \tag{5.89}$$

$$\tilde{F}_c(t) = \tilde{\Gamma}_p(t) + \tilde{\Gamma}_{sp}(t) + \tilde{\Gamma}(t) \tag{5.90}$$

Noise spectral densities. Let us consider a laser oscillator pumped at well above the threshold. For this region, the operator Langevin equations [Eq. (5.85) and (5.88)] can be solved by the quasi-linearization procedure.

5.2. Squeezing in semiconductor lasers

Operators are expanded into c-number mean values and fluctuating operators according to

$$\tilde{N}_c(t) = N_{c0} + \Delta\tilde{N}_c(t) \tag{5.91}$$

$$\hat{A}(t) = [A_0 + \Delta\hat{A}(t)]\exp\{-i[\phi_0 + \Delta\hat{\phi}(t)]\} \tag{5.92}$$

$$\hat{n}(t) = \hat{A}^\dagger(t)\hat{A}(t) \approx [A_0 + \Delta\hat{A}(t)]^2 \approx A_0^2 + 2A_0\Delta\hat{A}(t) \tag{5.93}$$

$$\tilde{\chi}_i = \langle\tilde{\chi}_i\rangle + \frac{d\langle\tilde{\chi}_i\rangle}{dN_{c0}}\Delta\tilde{N}_c \tag{5.94}$$

$$\tilde{\chi}_r = \langle\tilde{\chi}_r\rangle + \frac{d\langle\tilde{\chi}_r\rangle}{dN_{c0}}\Delta\tilde{N}_c \tag{5.95}$$

Here, N_{c0}, A_0, and ϕ_0 are the average excited electron number, field amplitude, and phase (c-numbers), and $\Delta\tilde{N}_c, \Delta\hat{A}$, and $\Delta\hat{\phi}$ are the Hermitian excited electron number, field amplitude, and phase operators.

The following notations are also introduced:

$$\frac{1}{\tau_{st}} = \frac{\omega}{\mu^2}A_0^2\frac{d\langle\tilde{\chi}_i\rangle}{dN_{c0}} \tag{5.96}$$

$$\omega_0 = \omega_r + \frac{\omega}{2\mu^2}\langle\tilde{\chi}_r\rangle \tag{5.97}$$

$$\alpha = \frac{d\langle\tilde{\chi}_r\rangle/dN_{c0}}{d\langle\tilde{\chi}_i\rangle/dN_{c0}} \tag{5.98}$$

$$\beta = \frac{\langle\tilde{E}_{cv}\rangle}{N_{c0}/\tau_{sp}} \tag{5.99}$$

$$n_{sp} = \frac{\langle\tilde{E}_{cv}\rangle}{\langle\tilde{E}_{cv}\rangle - \langle\tilde{E}_{vc}\rangle} \tag{5.100}$$

and

$$G = \frac{\omega}{\mu^2}\langle\tilde{\chi}_i\rangle \tag{5.101}$$

where τ_{st} is the electron lifetime due to stimulated emission, ω_0 is the resonance angular frequency of the pumped cavity including the frequency shift introduced by the injection, α is the linewidth enhancement factor, β is the spontaneous emission factor, n_{sp} is the population inversion factor, and G is the average stimulated emission gain. The Q values are replaced by photon lifetimes according to

$$\frac{1}{\tau_{p0}} = \frac{\omega}{Q_0}, \quad \frac{1}{\tau_{pe}} = \frac{\omega}{Q_e}, \quad \frac{1}{\tau_p} = \frac{\omega}{Q} \tag{5.102}$$

Substituting Eqs. (5.91) to (5.95) into Eqs. (5.85) and (5.88), and neglecting the terms in the order of Δ^2, the following linearized operator equations are obtained:

$$\frac{d\Delta\tilde{N}_c(t)}{dt} = A_1\Delta\tilde{N}_c(t) + A_2\Delta\hat{A}(t) + \tilde{F}_c(t) \tag{5.103}$$

$$\frac{d\Delta\hat{A}(t)}{dt} = A_3\Delta\tilde{N}_c(t) - A_5\Delta\hat{A}(t) - A_0A_6\Delta\hat{\phi}(t) + \hat{H}_r(t) \tag{5.104}$$

$$\frac{d\Delta\hat{\phi}(t)}{dt} = A_4\Delta\tilde{N}_c(t) + \frac{A_6}{A_0}\Delta\hat{A}(t) - A_5\Delta\hat{\phi}(t) - \frac{1}{A_0}\hat{H}_i(t) \tag{5.105}$$

where

$$A_1 = -\frac{1}{\tau_{sp}} - \frac{1}{\tau_{st}} \tag{5.106}$$

$$A_2 = -2A_0 G \tag{5.107}$$

$$A_3 = \frac{1}{2A_0\tau_{st}} \tag{5.108}$$

$$A_4 = \frac{\alpha}{2A_0^2\tau_{st}} \tag{5.109}$$

$$A_5 = \frac{F_0}{A_0}\left(\frac{1}{\tau_{pe}}\right)^{1/2}\cos\phi_0 \tag{5.110}$$

$$A_6 = \frac{F_0}{A_0}\left(\frac{1}{\tau_{pe}}\right)^{1/2}\sin\phi_0 \tag{5.111}$$

The operators $\hat{H}_r(t)$ and $\hat{H}_i(t)$ are the Hermitian quadrature noise operators:

$$\hat{H}_r = \frac{1}{2}(\hat{H}\exp\{i[\phi_0 + \Delta\hat{\phi}(t)]\} + \hat{H}^\dagger\exp\{-i[\phi_0 + \Delta\hat{\phi}(t)]\}) \tag{5.112}$$

$$\hat{H}_i = \frac{1}{2i}(\hat{H}\exp\{i[\phi_0 + \Delta\hat{\phi}(t)]\} - \hat{H}^\dagger\exp\{-i[\phi_0 + \Delta\hat{\phi}(t)]\}) \tag{5.113}$$

5.2. Squeezing in semiconductor lasers

Fourier transforms of Eqs. (5.103) to (5.105) are expressed as

$$i\Omega\Delta\tilde{N}_c(\Omega) = A_1\Delta\tilde{N}_c(\Omega) + A_2\Delta\hat{A}(\Omega) + \hat{F}_c(\Omega) \tag{5.114}$$

$$i\Omega\Delta\hat{A}(\Omega) = A_3\Delta\tilde{N}_c(\Omega) - A_5\Delta\hat{A}(\Omega) - A_0A_6\Delta\hat{\phi}(\Omega) + \hat{H}_r(\Omega) \tag{5.115}$$

and

$$i\Omega\Delta\hat{\phi}(\Omega) = A_4\Delta\tilde{N}_c(\Omega) + \frac{A_6}{A_0}\Delta\hat{A}(\Omega) - A_5\Delta\hat{\phi}(\Omega) - \frac{1}{A_0}\hat{H}_i(\Omega) \tag{5.116}$$

These equations are solved in the limit of $\Omega \to 0$. The actual observation frequency is well below the cavity cutoff and relaxation peak frequency, so this condition is justified for explaining the experimental results. From these equations, $\tilde{N}_c(0)$ is eliminated, and the expressions for the cavity internal amplitude and phase noise $\Delta\hat{A}(0)$ and $\Delta\hat{\phi}(0)$ are obtained as

$$\Delta\hat{A}(0) = \frac{1}{B_1}[B_2\tilde{F}_c(0) + B_3\hat{H}_i(0) + B_4\hat{H}_r(0)] \tag{5.117}$$

and

$$\Delta\hat{\phi}(0) = \frac{1}{B_5}[B_6\tilde{F}_c(0) + B_7\hat{H}_i(0) + B_8\hat{H}_r(0)] \tag{5.118}$$

The expressions for the coefficients B_i are given in the Table 5.1.

The output field $\hat{r}(t)$ is expressed in terms of the cavity internal field $\hat{A}(t)$ and the injection signal $F_0 + \hat{f}(t)$ as

$$\hat{r}(t) = -[F_0 + \hat{f}(t)] + \left(\frac{\omega}{Q_e}\right)^{1/2}\hat{A}(t) \tag{5.119}$$

As in the case of the cavity internal field calculations, the operator $\hat{r}(t)$ is expanded as

$$\hat{r}(t) = [r_0 + \Delta\hat{r}(t)]\exp\{-i[\psi_0 + \Delta\hat{\psi}(t)]\} \tag{5.120}$$

where r_0 and ψ_0 are the average output field amplitude and phase. $\Delta\hat{r}(t)$ and $\Delta\hat{\psi}(t)$ are the Hermitian output field amplitude and phase operator. Substituting Eqs. (5.92 and (5.120) into Eq. (5.119) and neglecting the terms in the order of Δ^2, the fluctuating parts of Eq. (5.119) are obtained as

$$\frac{C_4}{r_0}\Delta\hat{r}(t) + C_3[\Delta\hat{\phi}(t) - \Delta\hat{\psi}(t)] = -\hat{f}_r(t) + C_1\Delta\hat{A}(t) \tag{5.121}$$

and

$$-\frac{C_3}{r_0}\Delta\hat{r}(t) + C_4[\Delta\hat{\phi}(t) - \Delta\hat{\psi}(t)] = -\hat{f}_i(t) - C_2\Delta\hat{A}(t) - F_0\Delta\hat{\phi}(t) \tag{5.122}$$

Symbol	Expression
B_1	$A_2 A_3 A_5 + A_1(A_5^2 + A_6^2) - A_0 A_2 A_4 A_6$
B_2	$A_0 A_4 A_6 - A_3 A_5$
B_3	$A_1 A_6$
B_4	$A_1 A_5$
B_5	$-A_0 B_1$
B_6	$A_0 A_4 A_5 + A_3 A_6$
B_7	$A_2 A_3 + A_1 A_5$
B_8	$A_0 A_2 A_4 - A_1 A_6$
C_5	$\frac{r_0}{C_4}$
C_6	$C_1 + C_3 C_{10} C_{11}$
C_7	$C_3(C_{10} C_{12} - 1)$
C_8	$C_3^2 C_{10} - 1$
C_9	$C_3 C_4 C_{10}$
C_{10}	$\frac{1}{C_3^2 + C_4^2}$
C_{11}	$-C_2 F_0$
C_{12}	$C_3^2 + C_4(C_4 + F_0)$
D_1	$\frac{B_2}{B_1} C_6 + \frac{B_6}{B_5} C_7$
D_2	$\frac{B_3}{B_1} C_6 + \frac{B_7}{B_5} C_7$
D_3	$\frac{B_4}{B_1} C_6 + \frac{B_8}{B_5} C_7$
D_4	$\frac{B_2}{B_1} C_{11} + \frac{B_6}{B_5} C_{12}$
D_5	$\frac{B_3}{B_1} C_{11} + \frac{B_7}{B_5} C_{12}$
D_6	$\frac{B_4}{B_1} C_{11} + \frac{B_8}{B_6} C_{12}$

Table 5.1: Expressions for coefficients B_i, C_i, and D_i defined in the text.

where

$$C_1 = \left(\frac{1}{\tau_{pe}}\right)^{1/2} \cos\phi_0 \qquad (5.123)$$

$$C_2 = \left(\frac{1}{\tau_{pe}}\right)^{1/2} \sin\phi_0 \qquad (5.124)$$

$$C_3 = A_0 C_2 \qquad (5.125)$$

5.2. Squeezing in semiconductor lasers

$$C_4 = A_0 C_1 - F_0 \tag{5.126}$$

$$\hat{f}_r = \frac{1}{2}\{\hat{f}\exp[i\Delta\hat{\phi}(t)] + \hat{f}^\dagger \exp[-i\Delta\hat{\phi}(t)]\} \tag{5.127}$$

$$\hat{f}_i = \frac{1}{2i}\{\hat{f}\exp[i\Delta\hat{\phi}(t)] - \hat{f}^\dagger \exp[-i\Delta\hat{\phi}(t)]\} \tag{5.128}$$

These equations are solved and Fourier transformed as

$$\Delta\hat{r}(0) = C_5[C_6\Delta\hat{A}(0) + C_7\Delta\hat{\phi}(0) + C_8\hat{f}_r(0) + C_9\hat{f}_i(0)] \tag{5.129}$$

and

$$\Delta\hat{\psi}(0) = C_{10}[C_{11}\Delta\hat{A}(0) + C_{12}\Delta\hat{\phi}(0) + C_3\hat{f}_r(0) + C_4\hat{f}_i(0)] \tag{5.130}$$

Substituting Eqs. (5.117) and (5.118) into Eqs. (5.129) and (5.130), the final expressions for the output amplitude and phase noise are obtained as

$$\Delta\hat{r}(0) = C_5[D_1\tilde{F}_c(0) + D_2\hat{H}_i(0) + D_3\hat{H}_r(0) + C_8\hat{f}_r(0) + C_9\hat{f}_i(0)] \tag{5.131}$$

and

$$\Delta\hat{\psi}(0) = C_{10}[D_4\tilde{F}_c(0) + D_5\tilde{H}_i(0) + D_6\hat{H}_r(0) + C_3\hat{f}_r(0) + C_4\hat{f}_i(0)] \tag{5.132}$$

The expressions for the coefficients C_i and D_i are given in the Table 5.1.

In order to obtain the noise spectral densities, the correlation functions between the noise operators have to be known. They are given in the Table 5.2. Finally the resulting expressions for the output field amplitude and phase-noise spectral densities are given by

$$\begin{aligned}\langle\Delta\hat{r}(0)\Delta\hat{r}(0)\rangle = C_5^2[&D_1^2\langle\tilde{F}_c(0)\tilde{F}_c(0)\rangle + D_2^2\langle\hat{H}_i(0)\hat{H}_i(0)\rangle + D_3^2\langle\hat{H}_r(0)\hat{H}_r(0)\rangle \\ &+ C_8^2\langle\hat{f}_r(0)\hat{f}_r(0)\rangle + C_9^2\langle\hat{f}_i(0)\hat{f}_i(0)\rangle] \\ + 2C_5^2[&D_1D_3\langle\tilde{F}_c(0)\hat{H}_r(0)\rangle + D_2C_8\langle\hat{H}_i(0)\hat{f}_r(0)\rangle + D_2C_9\langle\hat{H}_i(0)\hat{f}_i(0)\rangle \\ &+ D_3C_8\langle\hat{H}_r(0)\hat{f}_r(0)\rangle + D_3C_9\langle\hat{H}_r(0)\hat{f}_i(0)\rangle],\end{aligned} \tag{5.133}$$

$$\begin{aligned}\langle\Delta\hat{\psi}(0)\Delta\hat{\psi}(0)\rangle = C_{10}^2[&D_4^2\langle\tilde{F}_c(0)\tilde{F}_c(0)\rangle + D_5^2\langle\hat{H}_i(0)\hat{H}_i(0)\rangle + D_6^2\langle\hat{H}_r(0)\hat{H}_r(0)\rangle \\ &+ C_3^2\langle\hat{f}_r(0)\hat{f}_r(0)\rangle + C_4^2\langle\hat{f}_i(0)\hat{f}_i(0)\rangle] \\ + 2C_{10}^2[&D_4D_6\langle\tilde{F}_c(0)\hat{H}_r(0)\rangle + D_5C_3\langle\hat{H}_i(0)\hat{f}_r(0)\rangle + D_5C_4\langle\hat{H}_i(0)\hat{f}_i(0)\rangle \\ &+ D_6C_3\langle\hat{H}_r(0)\hat{f}_r(0)\rangle + D_6C_4\langle\hat{H}_r(0)\hat{f}_r(0)\rangle]\end{aligned} \tag{5.134}$$

Correlation Function	Expression
$\langle \hat{H}_r(t)\hat{H}_r(u)\rangle$	$\frac{1}{4}\left[\frac{1}{\tau_p} + \langle \tilde{E}_{cv}\rangle + \langle \tilde{E}_{vc}\rangle\right]\delta(t-u)$
$\langle \hat{H}_i(t)\hat{H}_i(u)\rangle$	$\frac{1}{4}\left[\frac{1}{\tau_p} + \langle \tilde{E}_{cv}\rangle + \langle \tilde{E}_{vc}\rangle\right]\delta(t-u)$
$\langle \tilde{F}_c(t)\tilde{F}_c(u)\rangle$	$\left[p + \frac{N_{c0}}{\tau_{sp}} + \langle \tilde{E}_{cv}\rangle(A_0^2+1) + \langle \tilde{E}_{vc}\rangle A_0^2\right]\delta(t-u)$
$\langle \tilde{F}_c(t)\hat{H}_r(u)\rangle$	$-\frac{A_0}{2}\left[\langle \tilde{E}_{cv}\rangle + \langle \tilde{E}_{vc}\rangle\right]\delta(t-u)$
$\langle \tilde{F}_c(t)\hat{H}_i(u)\rangle$	0
$\langle \hat{H}_r(t)\hat{H}_i(u)\rangle + \langle \hat{H}_i(t)\hat{H}_r(u)\rangle$	0
$\langle \hat{f}_r(t)\hat{f}_r(u)\rangle$	$\frac{1}{4}\delta(t-u)$
$\langle \hat{f}_i(t)\hat{f}_i(u)\rangle$	$\frac{1}{4}\delta(t-u)$
$\langle \hat{f}_r(t)\hat{f}_i(u)\rangle + \langle \hat{f}_i(t)\hat{f}_r(u)\rangle$	0
$\langle \hat{H}_r(t)\hat{f}_r(u)\rangle = \langle \hat{f}_r(t)\hat{H}_r(u)\rangle$	$\frac{1}{4}\left[\frac{1}{\tau_{pe}}\right]^{1/2}(\cos\phi_0)\delta(t-u)$
$\langle \hat{H}_i(t)\hat{f}_r(u)\rangle = \langle \hat{f}_r(t)\hat{H}_i(u)\rangle$	$\frac{1}{4}\left[\frac{1}{\tau_{pe}}\right]^{1/2}(\sin\phi_0)\delta(t-u)$
$\langle \hat{H}_r(t)\hat{f}_i(u)\rangle = \langle \hat{f}_i(t)\hat{H}_r(u)\rangle$	$-\frac{1}{4}\left[\frac{1}{\tau_{pe}}\right]^{1/2}(\sin\phi_0)\delta(t-u)$
$\langle \hat{H}_i(t)\hat{f}_i(u)\rangle = \langle \hat{f}_i(t)\hat{H}_i(u)\rangle$	$\frac{1}{4}\left[\frac{1}{\tau_{pe}}\right]^{1/2}(\cos\phi_0)\delta(t-u)$

Table 5.2: Expressions for correlation functions of noise operators.

The cross-correlation spectral density is given by

$$\begin{aligned}\langle \Delta\hat{r}(0)\Delta\hat{\psi}(0)\rangle &= C_5C_{10}[D_1D_4\langle \tilde{F}_c(0)\tilde{F}_c(0)\rangle + D_2D_5\langle \hat{H}_i(0)\hat{H}_i(0)\rangle + D_3D_6\langle \hat{H}_r(0)\hat{H}_r(0)\rangle \\
&+ C_8C_3\langle \hat{f}_r(0)\hat{f}_r(0)\rangle + C_9C_4\langle \hat{f}_i(0)\hat{f}_i(0)\rangle] \\
&+ C_5C_{10}[(D_1D_6 + D_3D_4)\langle \tilde{F}_c(0)\hat{H}_r(0)\rangle + (D_2C_3 + C_8D_5)\langle \hat{H}_i(0)\hat{f}_r(0)\rangle \\
&+ (D_2C_4 + C_9D_5)\langle \hat{H}_i(0)\hat{f}_i(0)\rangle + (D_3C_3 + C_8D_6)\langle \hat{H}_r(0)\hat{f}_r(0)\rangle \\
&+ (D_3C_4 + C_9D_6)\langle \hat{H}_r(0)\hat{f}_i(0)\rangle]\end{aligned}$$

(5.135)

5.2. Squeezing in semiconductor lasers

Phase-to-amplitude conversion noise. The amplitude operator of the combined signal by a high-transmission mirror, whose transmissivity is denoted by T, is expressed as

$$\hat{I}(t) = \sqrt{T}[r_0 + \Delta\hat{r}(t)]\exp\{-i[\pi - \delta + \Delta\hat{\psi}(t)]\} + \sqrt{1-T}[G_0 + \hat{g}(t)] \quad (5.136)$$

where δ is the phase error, which is zero when the slave and master laser signals are completely out of phase (see Fig. 5.22). The first term in Eq. (5.136) expresses the transmitted injection-locked slave laser signal, and the second term expresses the reflected master laser signal. Here G_0 and $\hat{g}(t)$ are the classic (c-number) excitation and fluctuation operators of the master laser signal, respectively. One quadrature component, which is the a_1 component in Fig. 5.22, is calculated because the local oscillator phase is supposed to fixed to a_1 component. The a_1 component of $\hat{I}(t)$ is given by

$$\hat{I}_{a_1}(t) = -\sqrt{T}r_0\cos\delta + \sqrt{1-T}G_0 - \sqrt{T}\Delta\hat{r}(t)\cos\delta \\ - \sqrt{T}r_0\Delta\hat{\psi}(t)\sin\delta + \sqrt{1-T}\Delta\hat{g}_r(t) \quad (5.137)$$

where $\Delta\hat{g}_r(t)$ is defined by

$$\Delta\hat{g}_r(t) = \tfrac{1}{2}[\hat{g}(t) + \hat{g}^\dagger(t)] \quad (5.138)$$

$\hat{I}_{a_1}(t)$ is divided into the average amplitude I_{a_10} and amplitude fluctuation operator $\Delta\hat{I}_{a_1}(t)$:

$$I_{a_10} = -\sqrt{T}r_0\cos\delta + \sqrt{1-T}G_0 \quad (5.139)$$

$$\Delta\hat{I}_{a_1}(t) = -\sqrt{T}\Delta\hat{r}(t)\cos\delta - \sqrt{T}r_0\Delta\hat{\psi}(t)\sin\delta + \sqrt{1-T}\Delta\hat{g}_r(t) \quad (5.140)$$

The second term in Eq. (5.140) is the phase-to-amplitude conversion noise, and this term disappears when the phase error δ is equal to zero. Finally, the resulting expression for the intensity-noise spectral density is given by

$$\langle\Delta\hat{I}_{a_1}(0)\Delta\hat{I}_{a_1}(0)\rangle = T\cos^2\delta\langle\Delta\hat{r}(0)\Delta\hat{r}(0)\rangle + 2Tr_0\cos\delta\sin\delta\langle\Delta\hat{r}(0)\Delta\hat{\psi}(0)\rangle \\ + Tr_0^2\sin^2\delta\langle\Delta\hat{\psi}(0)\Delta\hat{\psi}(0)\rangle + (1-T)\langle\Delta\hat{g}_r(0)\Delta\hat{g}_r(0)\rangle \quad (5.141)$$

where

$$\langle\Delta\hat{g}_r(0)\Delta\hat{g}_r(0)\rangle = \tfrac{1}{2} \quad (5.142)$$

Substituting Eqs. (5.133) to (5.135) into Eq. (5.141), the intensity-noise spectral density $\langle\Delta\hat{I}_{a_1}(0)\Delta\hat{I}_{a_1}(0)\rangle$ is calculated.

When the phase difference between the master and injection-locked slave laser signals is 90°, the intensity noise of the combined signal can be calculated by setting the phase error δ in Eq. (5.141) to $\tan^{-1}(\sqrt{1-TG_0}/\sqrt{Tr_0})$. The numerical results are shown in Fig. 5.21.

Numerical results

The noise spectral densities $\langle \Delta \hat{r}(0) \Delta \hat{r}(0) \rangle$, $r_0^2 \langle \Delta \hat{\psi}(0) \Delta \hat{\psi}(0) \rangle$, and $\langle \Delta \hat{I}_{a_1}(0) \Delta \hat{I}_{a_1}(0) \rangle$ are calculated numerically. The following numerical parameters are assumed for a diode laser: $\tau_{sp} = 2 \times 10^{-9}$ s, $\tau_p = 2 \times 10^{-12}$ s, $\tau_{pe} = 2.5 \times 10^{-12}$ s, $\beta = 2 \times 10^{-5}$, and $n_{sp} = 1$. The pump rate $R(\equiv I/I_{th} - 1) = 10$, and the pump noise is completely suppressed, i.e., $\langle \tilde{\Gamma}_p(t) \tilde{\Gamma}_p(u) \rangle = 0$.

Figure 5.24(a,b) shows the external amplitude-noise spectral densities as a function of the frequency detuning with the linewidth enhancement factor $\alpha = 0$ and $\alpha = 6$, respectively. In both cases, the locking bandwidth is 9.3 GHz, which is determined by the experimental result. These numerical results are valid within the locking bandwidth because it is assumed in Eq. (5.85) that the angular frequency of the injection signal is identical to the resonance frequency of the cavity. In the case of $\alpha = 0$ [Fig. 5.24(a)], the amplitude-noise spectral density and locking bandwidth are symmetric, and the minimum noise value is seen at detuning $\Delta f = 0$. In this case, the amplitude squeezing is seen only within the locking bandwith. In the case of $\alpha = 6$ [Fig. 5.24(b)], the amplitude-noise spectral density and the locking bandwith are asymetric, and the minimum value of the amplitude noise is seen close to the negative edge of the locking bandwidth.

Figure 5.25 shows the intensity-noise spectral density with the interference, which is normalized by the shot-noise value as a function of the phase error δ. The frequency detuning is assumed to be zero. The transmissivity of the high-transmission mirror is 0.95. The locking bandwidth is 9.6 GHz, and the α parameter is 6. These parameters are determined experimentally. The dashed line shows the noise level that is obtained when the phase error δ is equal zero. This numerical result demonstrates that the phase error must be reduced to below 0.01^0 to generate a squeezed-vacuum state under the present experimental conditions.

The allowable values of the phase error are not fixed but depend on the residual phase-noise level that cannot be suppressed by injection

5.2. Squeezing in semiconductor lasers

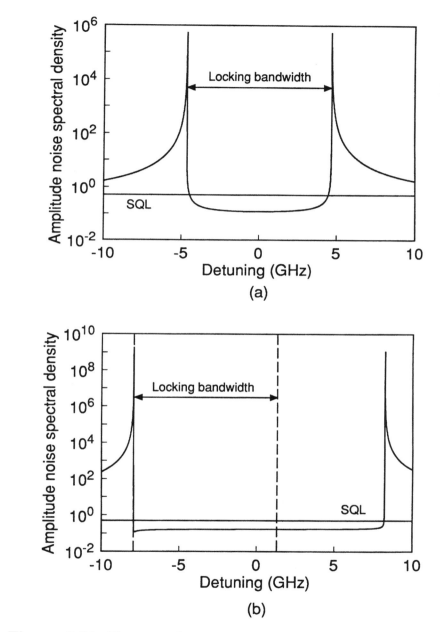

Figure 5.24: The external amplitude-noise spectral density at $\Omega = 0$ as a function of the detuning: (a) $\alpha = 0$; (b) $\alpha = 6$.

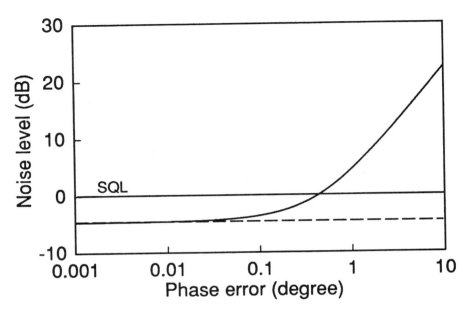

Figure 5.25: The intensity-noise spectral density with the interference as a function of the phase error δ.

locking. Figure 5.26 shows the output-field amplitude and phase-noise spectral densities as a function of the locking bandwidth. Numerical parameters are the same as those in Fig. 5.25. If the locking bandwidth increases, then the phase noise decreases [curve (a)]. However, the amplitude noise then increases [curve (b)]. Therefore, a larger phase error is allowable if the locking bandwidth is increased, but there exists a limit beyond which the amplitude squeezing is degraded.

5.3 Controlled spontaneous emission in semiconductor lasers

5.3.1 Brief review of cavity quantum electrodynamics

Spontaneous emission is not a fixed property of an atom but a consequence of atom-vacuum field coupling. It has been known for some time that the decay rate, transition energy, and radiation pattern of spontaneous

5.3. Controlled spontaneous emission in semiconductor lasers

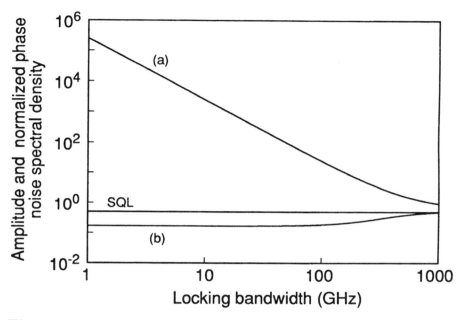

Figure 5.26: The amplitude and normalized phase-noise spectral densities as a function of the locking bandwidth: (a) normalized phase-noise spectral density; (b) amplitude-noise spectral density.

emission can be altered by modifying vacuum field fluctuations by a cavity wall. The principle, often referred to as *cavity quantum electrodynamics (cavity QED)*, is a classic theoretical problem. Cavity-enhanced and -suppressed spontaneous decay rates were predicted by Purcell in 1946.[27] A cavity-induced radiative energy shift was pointed out by Casimir in 1948.[28] The nonlinear coupling of atom and vacuum fields was formulated by Jaynes and Cummings in 1963,[29] in which it was predicted that spontaneous emission becomes even reversible if an atom is put in a high-Q single-mode cavity. However, these effects have been observed only recently using either Rydberg atoms in microwave superconductor cavities or atoms in optical microcavities.[30] The only exception is the pioneering work by Drexhage in 1974,[31] in which he demonstrated that the decay rate and radiation pattern of spontaneous emission are modified by a metal mirror.

Since spontaneous emission is a major source of energy loss, speed limitation, and noise in semiconductor optical devices, the possibility of employing cavity QED to control spontaneous emission is expected to improve device performances fundamentally.

In this section we review the principle of cavity QED, the performance of various microcavity structures, and applications to semiconductor optical devices such as light-emitting diodes and semiconductor lasers.

There are two distinct regimes in cavity QED, i.e., a low-Q regime and high-Q regime. A decisive parameter separating the two regimes is the vacuum Rabi frequency,

$$\Omega_R = \frac{P_{12}}{\hbar} \mathscr{E}_{\text{vac}} = \frac{P_{12}}{\hbar} \sqrt{\frac{\hbar\omega}{2\varepsilon_0 V_0}} \qquad (5.143)$$

where P_{12} is the atomic dipole moment, $\mathscr{E}_{\text{vac}}$ is the vacuum field amplitude, and V_0 is the optical cavity volume. When Ω_R is much smaller than the decay rates of the photon and atomic dipole moment, i.e., $\Omega_R \ll \gamma_{\text{photon}}, \gamma_{\text{dipole}}$, the atom-vacuum field coupling is only weakly perturbed by the cavity, and spontaneous emission remains an irreversible and incoherent process. Nevertheless, a cavity can modify the radiation pattern and decay rate.

On the other hand, when $\Omega_R \gg \gamma_{\text{photon}}, \gamma_{\text{dipole}}$, the atom-vacuum field coupling is strongly perturbed by the cavity, and spontaneous emission becomes a reversible and coherent process. A coupled atom-field system, often referred to as a *dressed state,* has two split eigenfrequencies. This vacuum Rabi splitting corresponds to a periodic energy exchange between the atom and field.

When N atoms collectively couple with the cavity field, the atom-field coupling strength is enhanced in proportion to $\sqrt{N}$. In the low-Q regime, the cavity enhances a so-called cooperative superradiance decay.[32] The decay rate is enhanced, and the emission pattern is highly concentrated due to the formation of a giant dipole moment. In the high-Q regime, it corresponds to a collective Rabi splitting.[33] The difference between the two eigenfrequencies is increased to $\Omega_R\sqrt{N}$ from Ω_R for the single-atom case.

Enhanced and inhibited spontaneous emission rates at microwave frequencies have been demonstrated with either Rydberg atoms[34,35] or cyclotron electrons.[36] Enhanced and inhibited spontaneous emission at optical frequencies also has been demonstrated with either dye molecules[37] or Rb atoms.[38] Collective Rabi splitting has been demonstrated both at microwave[39] and optical frequencies.[40,41] Cavity QED experiments

5.3. Controlled spontaneous emission in semiconductor lasers

in semiconductors require a microcavity structure with an optical wavelength size. This is so because the spontaneous emission spectral width in semiconductors is much broader than that of an atom in vacuum due to a very short dephasing time and a very large inhomogeneous broadening of the electron–hole pair dipole. The cavity structures used for cavity-QED experiments in semiconductors are quite similar to those utilized in vertical cavity surface-emitting lasers.[42–44] In fact, a major thrust in the study of cavity-QED effects in semiconductors is the development of small-sized, ultra-low-threshold, high-speed lasers and light-emitting diodes for applications in optical communications, parallel optical signal processing, and optical computing.

5.3.2 Rate-equation analysis of microcavity lasers

As was mentioned in the very beginning of this section, the spontaneous emission into nonlasing modes is a major source of energy loss in macroscopic lasers. The energy loss leads to a low quantum efficiency below threshold and to a high pump power threshold. As soon as lasing begins, however, the quantum efficiency increases in all lasers due to the very high rate of stimulated emission into the lasing mode and the correspondingly short free carrier lifetime.

The power dissipation due to spontaneous emission into nonlasing modes is so severe, in fact, that reducing the number of cavity modes by QED, the laser threshold of a semiconductor laser, can be reduced by several orders of magnitude[45–47] from present threshold current levels, which are in the range of 1 mA. A prerequisite for threshold current decrease in microcavity lasers is that the photon lifetime can still be of the same order of magnitude as that of a macroscopic laser. This has become possible with the advent of monolayer-precision, lattice-matched semiconductor growth techniques, since the mirror reflectivity of planar, epitaxially grown Bragg mirrors can exceed 99%.[44] This should be compared with a cleaved semiconductor facet mirror, whose reflectivity is only about 30%.

Below threshold, in any good macroscopic laser, the largest radiative loss mechanism is spontaneous emission into nonlasing modes (including radiation modes). In a macroscopic laser, the rate of spontaneous emission radiated into every mode is approximately the same, meaning that only a small fraction, the inverse of the number of modes within the gain bandwidth, of the dissipated power reaches the (eventually) lasing mode and acts as a seed for stimulated emission. To reduce the number of modes, and thus increase the efficiency with which the pump power is coupled to the lasing mode, one can either reduce the mode density (number of modes per unit energy) or reduce the gain bandwidth. The former is usually easier to accomplish and is more effective than the latter. Reducing the cavity length from, say, 300 μm to 300 nm (roughly the length of a one wavelength cavity), the spontaneous emission coupling efficiency will increase by a factor of 10^3, and the threshold current will drop by the same factor (if the cavity lifetime remains constant).

The spontaneous emission coupling factor β is defined as the spontaneous emission rate into the lasing mode divided by the total rate of spontaneous emission.[47] It is a key parameter in the analysis of semiconductor laser threshold current reduction. It is important to understand that β depends on *both* the cavity characteristics and the characteristics of the excited electron–hole pairs or excitons.

The total rate of spontaneous emission is usually expressed as the inverse of the spontaneous emission lifetime τ_{sp}. Again, it is important to bear in mind that the spontaneous emission lifetime may be different in a bulk sample than in a particular microcavity configuration. This effect has been confirmed experimentally by several groups.[37,38,48]

Using β, it is possible to express the spontaneous emission rate into the lasing mode as $\beta N/\tau_{sp}$, where N is the number of free carriers. Using a linear gain model, $g = g_0(N - N_0)$, where N_0 is the number of free carriers in the active material that must be injected to achieve transparency, one can write the stimulated emission into the mode $g_0 N p$, where p is the mean photon number in the lasing mode. It is well known that at a mean photon number of unity in the mode, the rates of stimulated emission and spontaneous emission are equal,[47] allowing one to eliminate g_0 by the identity $g_0 \equiv \beta/\tau_{sp}$. From this point it is relatively straightforward to write down the rate equations for the mean number of free carriers and the mean cavity photon number in the lasing mode:

$$\frac{d}{dt}N = \frac{I}{q} - \frac{N}{\tau_{sp}} - \frac{N}{\tau_{nr}} - \frac{\beta(N - N_0)p}{\tau_{sp}} \qquad (5.144)$$

$$\frac{d}{dt}p = -\left(\gamma - \frac{\beta(N - N_0)}{\tau_{sp}}\right)p + \frac{\beta N}{\tau_{sp}} \qquad (5.145)$$

where I is the injection current, q is the elementary charge, τ_{nr} is the nonradiative recombination lifetime, and γ is the cavity loss rate per unit time.

5.3. Controlled spontaneous emission in semiconductor lasers

In a macroscopic laser, the characteristics are abruptly changed when the laser is driven above threshold. Both the output power versus pump power curve and the gain versus pump power curve have sharp "knees" defining the threshold pump power (or threshold current). In a microcavity laser, these transitions are usually much smoother, causing $\beta \approx 1$ lasers to be dubbed "thresholdless" lasers.[45] In our opinion, a threshold can always be defined, the laser characteristics (such as its temporal coherence or general noise behavior) under the defined threshold being markedly different from the characteristics above. The correct definition, in our opinion, is that threshold is reached when the mean photon number in the cavity reaches unity. Our experience is that this definition better corresponds to the onset of linewidth narrowing, gain clamping, and intensity-noise behaviour for a larger class of lasers than the conventional definition.[47,49] When the laser is macroscopic, this definition coincides roughly with the conventional definition that the gain should equal the loss.

Solving the steady-state behavior of the rate equations, it is relatively easy to find that the threshold injection current I_{th} needed to bring the photon number to unity is

$$I_{th} = \frac{q\gamma}{2\beta}\left[1 + \beta + \frac{\tau_{sp}}{\tau_{nr}} + \xi\left(1 - \beta + \frac{\tau_{sp}}{\tau_{nr}}\right)\right] \tag{5.146}$$

where the dimensionless parameter $\xi = \beta N_0/\gamma\tau_{sp}$ has been introduced. ξ has a direct physical interpretation: It corresponds to the mean photon number at material transparency. The fact that ξ can be larger than unity indicates that the laser threshold can be reached even before the active material is inverted in a microcavity.[49]

In a good laser, nonradiative recombination should play a minor role, and the factor τ_{sp}/τ_{nr} in Eq. (5.146) can be neglected. (In practice, unfortunately, this is not always the case.) Furthermore, in microcavity quantum well lasers, the active volume is usually so small that $\xi \ll 1$. In this regime, Eq. (5.146) can be simplified to

$$I_{th} = \frac{q\gamma(1 + \beta)}{2\beta} \tag{5.147}$$

From this equation it is clear that the figure of merit for threshold current reduction is the ratio γ/β.

In Fig. 5.27 the threshold current as a function of β is plotted for two different values of ξ. The parameter values are given in the plot. The larger N_0 corresponds to a transparency free carrier density of 10^{18} cm^{-3}

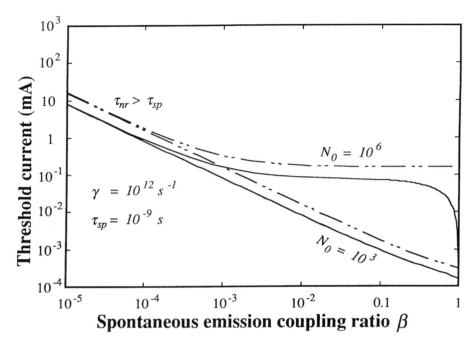

Figure 5.27: Threshold current versus spontaneous emission coupling ratio. The solid lines correspond to the definition $p_{th} \equiv 1$; the dashed line corresponds to the conventional definition gain equals loss.

and an active material volume of 1 μm^3, i.e., a microcavity completely filled with active material. The smaller N_0 value corresponds to an active volume of 10^{-3} μm^3. This is roughly the volume of a single quantum well in a microcavity laser. The dashed line represents the conventional definition, that the gain equals the loss at threshold. It is clear that the remarkable reduction in threshold current with increasing β has nothing to do with our definition of threshold; almost everywhere the two definitions coincide to within a factor of 2. The only exception is when both ξ and β are large. In this region we feel that the conventional definition fails because it neglects spontaneous emission, which is important in this regime.

5.3.3 Semiconductor microcavity lasers: experiments

We have carried out experiments with three different types of microcavity lasers, which exhibit a high spontaneous emission factor and ultralow

5.3. Controlled spontaneous emission in semiconductor lasers

threshold and allow comparison with the theoretical predictions for threshold and/or spontaneous emission factor: First, unmodified planar microcavities[72]; since the pumped spot size for this case is larger than the lasing mode diameter, several transverse modes contribute in the threshold regime. In order to get single-mode operation, a three-dimensional structure was produced by etching the planar sample, which led to a further threshold reduction.[73] A comparison of experimental data with theoretical predictions gives a reasonable agreement. The third type of microcavity uses a curved mirror (hemispherical cavity),[52] which is an alternative structure for obtaining transversal confinement.

Planar Bragg mirror microcavity lasers

Two planar microcavity structures (A and B) were grown by metal-organic vapor-phase epitaxy (MOVPE). The structure consists of a GaAs quantum well as active medium, embedded between two distributed Bragg reflectors made from alternating layers of $Al_{0.15}Ga_{0.85}As$ and AlAs, grown on a GaAs substrate. Figure 5.28 shows the case of sample B, which has a 200-Å GaAs quantum well. The quantum well is located at the antinode

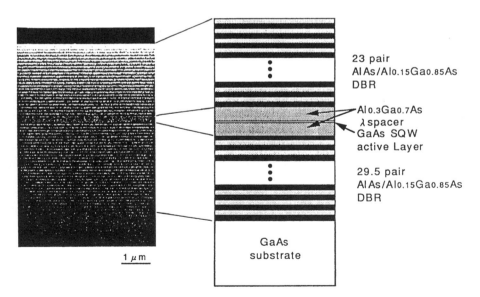

Figure 5.28: Vertical structure of the planar and etched microlasers. Active medium is a GaAs quantum well; the $Al_{0.15}Ga_{0.85}As$/AlAs Bragg reflectors are spaced by a $Al_{0.30}Ga_{0.70}As$ layer such that a one-wavelength cavity is formed.

position of the standing-wave field, which ensures optimal coupling between the active medium and the laser mode. Structure A has a 70-Å quantum well but is otherwise identical. By a variation of the layer thicknesses over the wafer (tapering), different wavelengths are resonant at different positions; thus the cavity can be tuned easily to the quantum well emission wavelength by translation of the sample. The optical characteristics of the samples were investigated by illuminating with white light and observing the spectrum of the reflected light with a monochromator (upper trace in Fig. 5.29 for the case of structure A). The Bragg mirrors are highly reflective over a wavelength range of 80 nm (stop band of

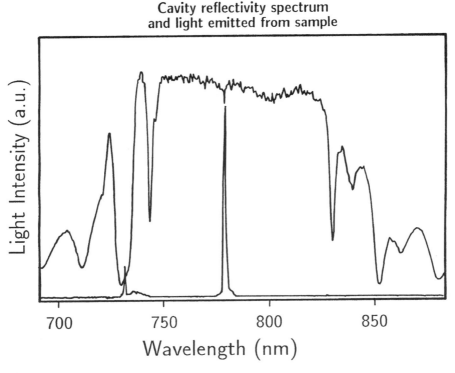

Figure 5.29: Upper trace: Spectrum obtained by reflection of white light from sample A. Lower trace: Fluorescence spectrum obtained from sample A when pumping with the Ti:sapphire laser. The small peak at 732 nm is the residual reflected pump light. The curves are somewhat distorted by filtering effects of the optical elements in the beam.

5.3. Controlled spontaneous emission in semiconductor lasers

the cavity transmission), centered at the cavity resonance (dip in the reflectivity curve). The lower trace in Fig. 5.29 shows the fluorescence from the quantum well, filtered by the cavity.

The lasing characteristics were investigated after placing the microcavity in a liquid helium dewar and cooling to 4 K, which substantially reduced the fluorescence linewidth. This reduction was confirmed with samples containing only a quantum well embedded in an AlGaAs layer: For a bare 200-Å quantum well, the linewidth could be reduced from 33 to 2 nm when cooling from room temperature to 4 K. In structure B, the observed emission linewidth was 2.5 nm. The reduced fluorescence linewidth allows a much better match with the cavity-resonance linewidth and an improved coupling of the spontaneous emission to the lasing mode. The narrower quantum well in structure A has a bigger inhomogeneous broadening, which leads to a bigger linewidth of 5.3 nm. The samples are pumped optically by a Ti:sapphire laser tuned to the point of maximum cavity transmission (740 nm for sample B), thereby reducing reflection losses to about 15%. The pump photon energy at this wavelength is below the band-gap energy of the layers constituting the mirrors; therefore, no serious absorption occurs. The pump light is chopped at 330 Hz with a duty cycle of 1:17 in order to avoid the thermal problems observed at higher powers. The incident beam is focused to a spot of 25-μm diameter. The focusing lens is also used for collimating the outgoing light, which is then analyzed by a monochromator and detected by a CCD camera.

The results of input versus output power measurements for the two microcavity structures are shown in Fig. 5.30. The output power was measured in an approximately 0.75-nm-wide band centered around the lasing wavelength (786 and 811 nm for samples A and B, respectively). The threshold in sample A occurs at an incident power of 55 mW or an absorbed power of 330 μW. Here we used as a threshold definition that the average number of photons in the lasing mode is unity at threshold; this definition corresponds to the initial bending point of the curves in the figure. The height of the step as it appears in the logarithmic plot is equal to the spontaneous emission factor of the cavity. This is so because in the absence of nonradiative processes, the quantum efficiency above threshold is close to unity, and below threshold, by the definition of the spontaneous emission coefficient, it is equal to β. For the actual determination of β, we used the method of fitting the theoretical curve obtained by solving the laser rate equations in the steady state to the experimental data, with the threshold power and β as fitting parameters. This method

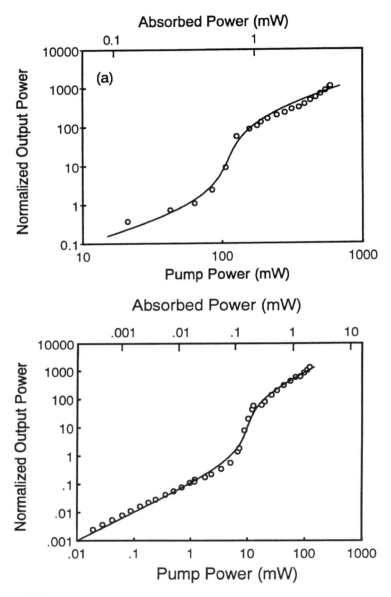

Figure 5.30: Output power versus input power for the two planar structures. The solid lines are the theoretical curves. (a) Sample A (70-Å quantum well). (b) Sample B (200-Å quantum well). The vertical scales are normalized to the theoretical number of photons in the lasing mode, being unity at threshold.

5.3. Controlled spontaneous emission in semiconductor lasers

yields a β of about 6×10^{-3} for sample A. The estimation of the absorbed threshold power took into account 15% reflection losses at the sample surface and an absorption coefficient of 10^4 cm^{-1} at 4 K.[76] If the pump light forms a standing wave inside the cavity, the absorption efficiency may be altered; however, a model calculation showed that this effect is at most 50%.

The reduced inhomogeneous broadening due to the thicker quantum well of sample B leads to a better match with the cavity resonance width and therefore to a reduced absorbed threshold power of 85 μW. Because of the increased absorption, the incident power threshold is reduced by an even bigger factor to 5 mW. The absorbed power density at threshold is 18 W/cm^2. β is about 10^{-2}, being slightly higher than in sample A, although the difference is within the measurement uncertainty. A higher value is what one would expect because of the narrower fluorescence linewidth.

The linewidth of the radiation emitted from sample B was measured after replacing the monochromator with a plane-parallel scanning Fabry-Perot resonator with a mirror reflectivity of 97.5% and various mirror spacings. The obtained values below threshold were additionally verified using the monochromator. The result is shown in Fig. 5.31 (solid circles). The linewidth of microcavity lasers is treated theoretically in refs. 47 and 77; a drop in linewidth by a factor of β from the cold cavity value at threshold is predicted [47] for the case of a linewidth enhancement factor $\alpha = 0$. In order to allow a comparison, the experimental values must be corrected by $(1 + \alpha^2)$ (open circles, for a typical value of $\alpha = 3$). The height of the step from the cold cavity bandwidth to the value above threshold in Fig. 5.31 confirms a β on the order of 10^{-2}.

It is desirable to check the estimate of the spontaneous emission coupling factors presented in the preceding paragraphs by independent measurements. Unfortunately, a direct measurement of β is impossible with the present sample because most of the spontaneous emission goes into the absorbing substrate. However, the theory developed in ref. 78 allows us to estimate β from the sample structure and the quantum well fluorescence linewidth, given in the preceding paragraphs, and the cavity resonance width.

Theory predicts that a perfect (and optimal) cavity made from the given material composition would have a linewidth of 0.065 nm. This theoretical minimum cavity decay rate implies mirror reflectivities of 99.91% and 99.94%. A slight imperfection in layer thickness or scattering

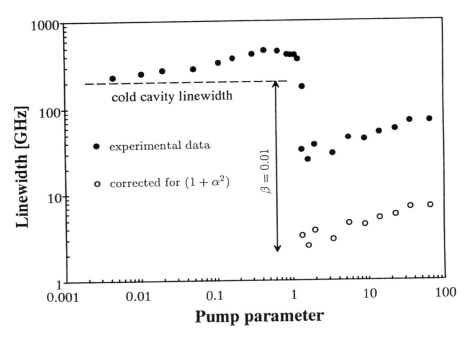

Figure 5.31: Laser linewidth of the planar structure versus pump power (normalized to unity at threshold).

would inevitably lead to a shorter photon lifetime and broader linewidth. This fact is confirmed by the direct measurement of the cold cavity linewidth $\Delta\nu_{\text{FWHM 0}} = 200$ GHz, or $\Delta\lambda_{\text{FWHM 0}} = 0.43$ nm (Fig. 5.31, left-most points). The optimal β achievable with the structure of Fig. 5.28 can be calculated[47] to be 0.05. However, since the cold cavity linewidth is smaller than the fluorescence linewidth, the optimal match in the frequency regime is not realized, and the β value is degraded approximately by the ratio of the two linewidths, yielding $\beta_{\text{the}} \approx 5 \times 10^{-3}$ and 9×10^{-3} for the two samples. A justification of this argument and a more rigorous calculation can be found in ref. 78. These values are in quite good agreement with the experimental observation.

Etched microcavity

In order to determine the number of transverse modes being simultaneously excited by the pump laser in the planar sample, the lasing mode diameter should be compared with the pump spot size. A measurement of the divergence angle of the radiation below threshold yielded a $1/e^2$ angle of 8°, corresponding to a lasing mode diameter of about 7 μm. Comparing with the pump focus diameter of 25 μm, one sees that at threshold several independent lasers can start to oscillate in the pumped area, which is equivalent to multiple-transverse-mode operation. In order to get a more well-defined situation, a three-dimensional structure was produced by etching, which made single-transverse-mode operation possible and further reduced the threshold power by one order of magnitude.

An array of three-dimensional microlasers was made by chemical wet etching of sample B, yielding circular mesa-like structures of various diameters (Fig. 5.32). The etch depth was chosen such as to remove the upper mirror, the active region, and half the lower mirror in the unmasked parts. The resulting microlasers have a small top mirror diameter, which acts like a pinhole for the cavity field, leading to increased losses for non-TEM_{00} modes and therefore encouraging single-mode operation. On the other hand, the diameter at the quantum well is 4 to 5 μm bigger; this reduces losses due to carriers diffusing to the etched surface and recombining there. Although the etched lasers have a three-dimensional structure, the larger lasers must still be considered planar microcavities, since the diameters involved are larger than the planar cavity mode size and substantially larger than the wavelength. This means, for example, that waveguide effects in these lasers are negligible.

The measurement setup was similar to the planar case, with a somewhat smaller pump focus diameter of 20 μm. Aiming the focused pump beam to a particular microlaser was done by shifting the sample while observing the pumped spot with a TV imaging system.

The etched sample showed lasing at 805 nm. The result of the input-output measurements for a 3.4-μm microlaser is shown in Fig. 5.33. In terms of input power, threshold occurs at 13.5 mW, which is equivalent to about 7 μW absorbed power. The calculation of the absorbed power takes into account that most of the incident light does not hit the top mirror. The penetrating power is calculated geometrically from the ratio of the mirror and pump spot areas; the part of the light actually absorbed is then determined as in the planar case. The solid line in Fig. 5.33 is the theoretical curve for the $\beta = 9 \times 10^{-3}$ estimated in the preceding section with a fitted threshold. The absorbed threshold power observed corresponds to a threshold current of 4 μA for the case of electrical pumping, which, compared with the approximately 10 mA of commercial laser di-

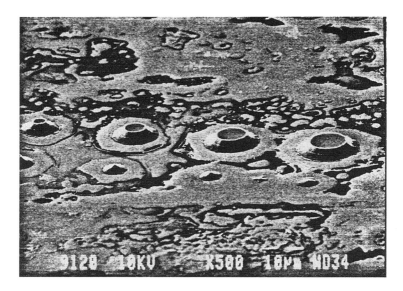

Figure 5.32: Scanning electron microscopic picture of a family of microlasers, as obtained from the planar sample B by etching, and an enlarged view of the 3.4-μm microlaser. The top surface acts as a pinhole, determining both the incident light beam diameter and the transverse mode structure of the laser. The fact that the quantum well diameter is bigger than the top mirror diameter reduces the surface recombination.

5.3. Controlled spontaneous emission in semiconductor lasers

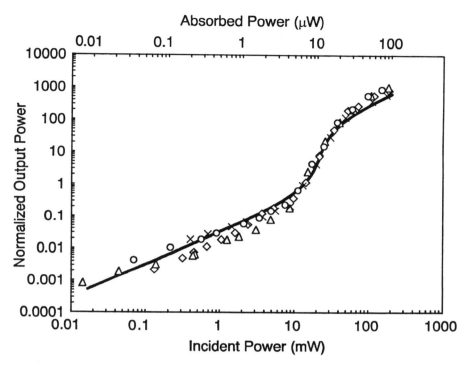

Figure 5.33: Some overlaid 3.4-μm microlaser input-output data and the theoretical curve for $\beta = 9 \times 10^{-3}$ fitted to the threshold power. The ordinate is normalized to unity at threshold and thus gives the theoretical number of photons in the lasing mode. The threshold values of individual lasers scatter by $\pm 40\%$.

odes, indicates a three order of magnitude threshold reduction. This is in agreement with the three orders of magitude increase from $\beta \approx 10^{-5}$ for a ordinary laser diode to $\beta \approx 10^{-2}$ for the microlaser.

Table 5.3 shows the threshold values obtained for microlasers of different diameters, together with the data of the planar sample B for comparison. The incident threshold power density is approximately the same for all diameters except the smallest one; the increased pump power transmitted by the top mirror for the bigger structures is apparently consumed by an increase in the number of independent oscillators. Conversely, the enhanced power density required for the 3.4-μm laser confirms that this structure actually lases single mode, as we could already conclude from the measured single-mode diameter.

	d	ρ	P_{abs}	P_{sing}
Single mode	3.4	4300	6.7	6.7
	8.6	960	9.4	6.2
Multimode	13.8	1150	29.1	7.5
	19.1	1110	54.3	7.3
	24.3	1750	93.5	11.4
Planar	25	1020	85	6.7

Note: d: top mirror diameter (μm), ρ: incident pump power density (W/cm^2), P_{abs}: absorbed power (μW), P_{sing}: threshold power for the TEM$_{00}$ mode, calculated using a 7-μm TEM$_{00}$ mode diameter for the multimode cases (μW).

Table 5.3: Threshold data for different microlasers.

As mentioned earlier, a large-diameter micropost cavity behaves in most respects as a planar cavity. The theoretical spontaneous emission coupling coefficient corresponding to the linewidth of 0.43 nm is calculated above to be 9×10^{-3}. This is within a factor of 2 or 3 from the experimental data. Neglecting nonradiative decay, the threshold power can be written[47] as $P_{th} = h\nu/2\beta\tau_{ph}$. Inserting $\tau_{ph} = 0.8$ ps as derived from the cold cavity linewidth measurement, the threshold power is found to be about 18 μW, a factor of 3 from the power estimated from the input versus output power measurement.

The collection of micropost cavities with varying diameters enabled us to independently estimate the mode radius of the planar cavity. As expected, the required pump power density remained relatively constant with micropost diameter as long as the post diameter was larger or equal to 8.6 μm. A best fit of the data assuming a mode diameter of 7 μm yielded a single-mode threshold power of 6.7 μW. This value is in very good agreement with the value estimated from the input versus output power measurement for the real single-mode case. However, considering the discrepancy between the theoretical and experimental values for β and P_{th}, we must emphasize that all numbers are not known better than to within a factor of 2. What is clear from Table 5.3 is that the assumption of a well-defined mode radius in a planar structure is valid.

Hemispheric cavities

In a planar microcavity, the mode diameter increases, and consequently, the radiation lobe width decreases, when the cavity Q is enhanced. There-

5.3. Controlled spontaneous emission in semiconductor lasers

fore, while the higher photon lifetime tends to decrease the lasing threshold, the spatial coupling efficiency of spontaneous emission to the lasing mode is reduced, and no net threshold reduction can be gained by an increased mirror reflectivity. This is different in a hemispheric microcavity, as illustrated schematically in Fig. 5.34. Here, the laser mode geometry is defined by the planar and curved mirrors, and adding more layers to the Bragg mirrors only increases the photon lifetime without changing β. Moreover, a big lobe angle collects a high percentage of the spontaneous emission and thus gives rise to a high β. Therefore, this cavity type has the potential for a very low threshold.

For our experiments, an initially planar microcavity structure was grown by MOVPE on top of a GaAs substrate, which consisted of a 29.5-period GaAs/AlAs Bragg reflector followed by an $Al_{0.2}Ga_{0.8}As$ spacer layer, the active medium, in this case a 100-Å $In_{0.2}Ga_{0.8}As$ single quantum well, and another $Al_{0.2}Ga_{0.8}As$ spacer layer. Resonators with $In_xGa_{1-x}As$ as active material are interesting because the operating wavelengths of 900 to 1000 nm are close to those of GaAs, yet GaAs is transparent and can

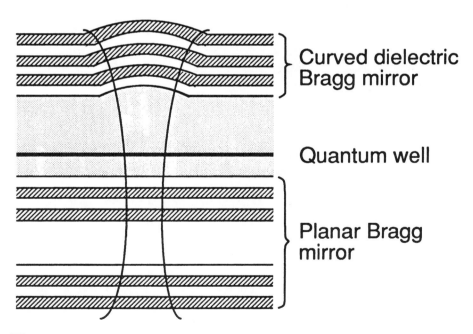

Figure 5.34: Hemispherical cavity configuration. A thin quantum well is sandwiched between the upper curved and the lower planar Bragg reflectors.

be used in mirrors and substrates. Room-temperature lasing has been reported previously in $In_{0.2}Ga_{0.8}As$ vertical cavities.[79–81]

To fabricate the curved mirrors, a method previously used in the fabrication of microlenses for light-emitting diodes was employed.[82,83] Arrays of photoresist disks were first deposited on the upper spacer layer. The disks were subsequently melted at 300°C, which resulted in a pattern of half lentil shape [Fig. 5.35(a)]. The curvature and diameter of the microlenses were controlled by varying the diameter and the thickness of the resist disk, resulting in curvatures ranging from roughly 5 to 20 μm. The microlens shape was thereafter transferred to the semiconductor by dry etching using an Ar^+ O_2 (5%) ion beam [Fig. 5.35(b)]. The etch rate and smoothness of the etched surface were controlled by the O_2 concentration and by the incidence angle of the ion beam to the sample (65°). Finally, to create a high-reflecting curved mirror, a 13.5-period SiO_2/TiO_2 dielectric Bragg reflector was deposited on the hemispheres.

The investigated cavity had a length of approximately 2.5 μm and a curvature around 15 ± 5 μm. The cavity length was determined from the white light reflectance characteristics of the cavity before deposition of the dielectric mirror. However, the white light reflected over an area much larger than the curved mirror does not give any direct information of the cavity length at the location of the curved mirror. This we obtained by adding to the inferred planar cavity length the height of the curved mirror as estimated from test samples etched under similar conditions. The samples were grown with a tapered cavity length, and a point where a cavity resonance through a curved mirror coincided with the emission wavelength was found experimentally by scanning the pump beam across the sample.

Our experiments were made at 4 K by placing the samples in a helium cold finger cryostat. This reduced the nonlinear (Auger) recombination and resulted in a narrow gain medium emission linewidth, giving, as explained before, a better atom–cavity coupling. The linewidth observed by photoluminescence in cavity samples without the dielectric mirror was roughly 40 nm at room temperature and was reduced to 7.9 nm at 4 K. The sample was investigated by optical pumping from a Ti:sapphire laser at 780 nm focused to a spotsize of 13 μm. The pump light was chopped in order to get a lock-in signal. No effects of thermal heating were observed. The emitted light was collected from the output side of the curved mirror by the same lens used to focus the pump light, passed through a monochromator (0.4-nm resolution, to spectrally select only the emission of one mode) and detected by a photomultiplier.

5.3. *Controlled spontaneous emission in semiconductor lasers* 433

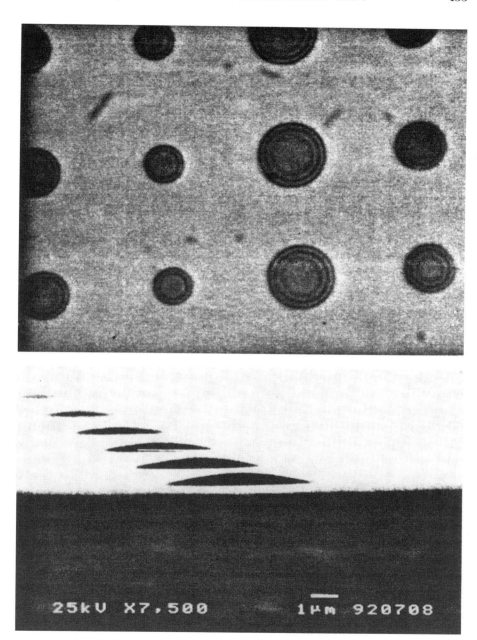

Figure 5.35: (a) The white-light reflection from the melted photoresist disks. The Newton ring pattern indicates the thickness variations. (b) An SEM image of an array of microlenses transferred to the semiconductor.

In Fig. 5.36, the measured output power (at a wavelength of ~ 930 nm) is shown as a function of incident optical pump power. In order to ensure that the lasing was a result of the optical feedback from the curved mirror and not from the flat surface, the pump beam was scanned over the sample. Only at the locations of the curved mirrors could strong emitted light be observed. The threshold incident pump power, corresponding to the point of the "jump" in Fig. 5.36 is 14 mW (11 kW/cm^2 power density). By estimating that approximately 3% is absorbed in the quantum well at this wavelength, [84] an absorbed threshold power density of 0.32 kW/cm^2 results, corresponding to an absorbed pump power of 0.4 mW in the (large) pumped area of 1.3×10^2 μm^2. The value of the spontaneous emission factor can be estimated from the input-output curve[47] to be about 10^{-2}. The solid line in Fig. 5.36 is a theoretical prediction of the input-output power characteristics calculated assuming a 15-μm mirror curvature, a 3% pump absorption, a carrier density required for transparency $N_o = 0.1 \times 10^{24}$ m^{-3}, and a bulk spontaneous lifetime $\tau_{sp} = 2$ ns. However, we never succeeded in resolving the cavity linewidth experimentally. Therefore, in the theoretical prediction, we assumed reasonable internal losses of 50 cm^{-1}, which, together with the calculated Bragg reflectances, gave a photon lifetime of 2.2 ps. The theoretical input-output characteristics are within order of magnitude agreement with the experiment. When operating the sample at room temperature, we also observed lasing, but the threshold was increased by about 1 order of magnitude; this is consistent with the broadening of the emission linewidth by about 1 order of magnitude.

Figure 5.37 shows the experimentally measured polarization dependence below and above threshold. Curve A is for pump powers below threshold. The output light was unpolarized (the residual modulation is caused by the dielectric beamsplitters). Curve B is for a pump power above threshold. The output power transmitted through the polarizer was modulated more than 95% when rotating the λ/2-plate, showing linearly polarized output. Geometrically, there is no preferred polarization direction in surface-emitting lasers, but several studies have found the output light above threshold to be linearly polarized.[85,86] In the present case the polarization direction was the same as the crystal direction [011], and the selection may perhaps be related to strain induced during the epitaxial growth.

The input-output power characteristics as predicted from rate-equation analysis agreed well with the experimental results. However, the far-

5.3. Controlled spontaneous emission in semiconductor lasers

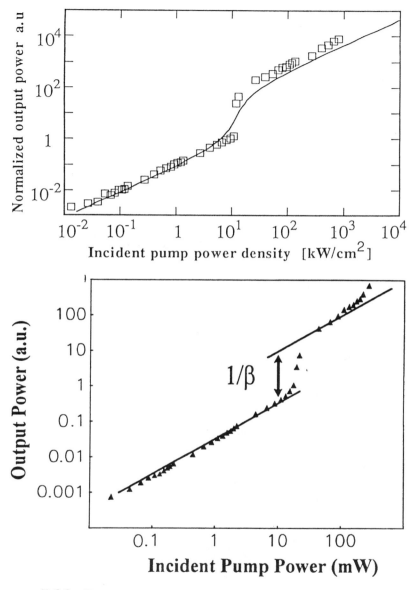

Figure 5.36: Experimental (points) and theoretical (solid line) input-output power curves for a hemispherical microcavity laser. The ordinate is normalized to the theoretical photon number per mode and is unity at threshold. (a) 2.5-μm cavity; (b) 0.5-μm cavity.

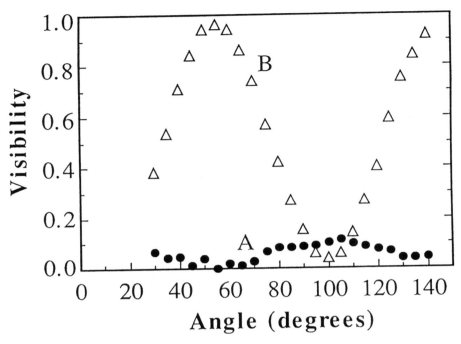

Figure 5.37: Experimentally measured polarization dependence; light power as a function of rotation angle of the Babinet-Soleil compensator. Curve A is for low pump powers, curve B for high pump powers.

field exit angle in the threshold region, experimentally measured to be about 2.5°, was substantially smaller than the theoretical estimate of far-field angle from the fundamental mode (of 0.7-μm spot size) of about 20°, taking into account the lens effects of the curved mirror. We believe that this could be due to the effective radius of curvature being larger than expected. If this is so, the theoretically estimated β will be lower. But this also makes the area of the lasing mode larger,[78] implying that the theoretically predicted threshold power density remains virtually the same. The mode selection will be the subject of further experimental and theoretical studies.

The 2.5-μm sample is not quite optimal for achieving a high β because, due to the long cavity, the resonance linewidth is small, and the match with the fluorescence linewidth is deteriorated. Therefore, in order to obtain a better coupling efficiency, we have made a second hemispheric microcavity with a 2λ (0.5 μm) cavity spacing. The basic structure is the same as in the previous case [Fig. 5.36(a)], whereas the individual parameters are somewhat different. The gain medium here is a 100-Å $In_{0.1}Ga_{0.9}As$ quantum well having a fluorescence linewidth of 5.6 nm at 4 K. The upper mirror has a curvature of about 20 μm. Due to the different quantum well material, the lasing wavelength is 828 nm at 4 K. The measured input-output data are shown in Fig. 5.36(b). From the height of the step, we can infer a β value of roughly 0.05, a factor of 5 higher than in the case of the 2.5-μm cavity. A measurement of the laser linewidth above and below threshold, similar to the planar case, confirmed this value. It is also in reasonable agreement with the theoretical prediction of $\beta = 0.1$. The theoretical prediction of the threshold power is 16 μW. We could not yet directly verify this value because the pumped area is much bigger than the mode diameter.

5.4 Conclusion

Squeezing and control of spontaneous emission in semiconductor lasers have been reviewed. A squeezed-vacuum-state generation by a semiconductor laser requires a phase-stabilized cancellation of a coherent amplitude excitation. An injection-locking technique turns out to be useful for suppressing spurious side modes and obtaining more squeezing from a semiconductor laser. A high-quantum-efficiency and low-threshold semiconductor laser would produce a phase-coherent-squeezed state with a small coherent amplitide excitation. A surface-emitting microlaser is a promising device in this direction. A monolithic device may improve the stability of such a system.

A low-threshold semiconductor laser by control of spontaneous emission requires low-temperature operation at present. An improved microcavity configuration is needed to cope with a rather broad emission from a room-temperature quantum well, and such efforts seem to be promising. A microcavity semiconductor laser or light-emitting diode with a high quantum efficiency and low noise would have many potential applications including optical interconnects in VLSI.

References

1. D. Bohm, *Quantum Theory,* Prentice-Hall, Englewood Cliffs, NJ, 1951.
2. A. Messiah, *Quantum Mechanics,* McGraw-Hill, New York, 1961.

3. R. J. Glauber, *Phys. Rev.* **131**, 2766 (1963).
4. R. J Glauber, in *Quantum Optics and Electronics,* (C. DeWitt, A. Blandin, and C. Cohen-Tannoudji, eds.), Gordon & Breach, New York, 1965, p. 65.
5. H. Takahashi, *Adv. Commun. Syst.* **1**, 227 (1965).
6. D. Stoler, *Phys. Rev.* **D4**, 1925 (1971).
7. H. P. Yuen, *Phys. Rev.* **A13**, 2226 (1976).
8. R. Jackiw, *J. Math. Phys.* **9**, 339 (1968).
9. R. E. Slusher, L. W. Hollberg, B. Yurke, J. C. Mertz, and J. F. Valley, *Phys. Rev. Lett.* **55**, 2409 (1985).
10. R. M. Shelby, M. D. Levenson, S. H. Perlmutter, R. G. DeVoe, and D. F. Walls. *Phys. Rev. Lett.* **57**, 691 (1986).
11. L. A. Wu, H. J. Kimble, J. L. Hall, and H. Wu, *Phys. Rev. Lett.* **57**, 2520 (1986).
12. M. W. Maeda, P. Kumar, and J. H. Shapiro, *Opt. Lett.* **12**, 161 (1987).
13. H. Haken, *Light and Matter,* Vol. 25 in *Handbuch der Physics,* Springer-Verlag, Berlin, 1970.
14. M. Sargent III, M. O. Scully, and W. E. Lamb, Jr., *Laser Physics,* Addison-Wesley, Reading, MA, 1974.
15. M. Lax and W. H. Louisell, *Phys. Rev.* **185**, 568 (1969).
16. P. Carruthers and M. M. Neito, *Rev. Mod. Phys.* **40**, 411 (1968).
17. H. A. Haus and Y. Yamamoto, *Phys. Rev.* **A29**, 1261 (1984).
18. Y. Yamamoto and N. Imoto, *IEEE J. Quantum Electron.* **QE-22**, 2032 (1986).
19. S. Machida and Y. Yamamoto, *Opt. Lett.* **14**, 1045 (1989).
20. H. P. Yuen and V. W. S. Chan, *Opt. Lett.* **8**, 177 (1983).
21. G. L. Abbas, V. W. S. Chan, and T. K. Yee, *IEEE J. Lightwave Technol.* **LT-3**, 1110 (1985).
22. S. Machida and Y. Yamamamoto, *IEEE J. Quantum Electron.* **QE-22**, 6017 (1986).
23. W. H. Richardson, S. Machida, and Y. Yamamoto, *Phys. Rev. Lett.* **66**, 2867 (1991).
24. Y. Lai, H. A. Haus, and Y. Yamamoto, *Opt. Lett.* **16**, 1517 (1991).
25. L. Gillner, G. Björk, and Y. Yamamoto, *Phys. Rev.* **A41**, 5053 (1990).
26. S. Inoue, H. Ohzu, S. Machida, and Y. Yamamoto, *Phys. Rev.* **A46**, 2757 (1992).
27. E. M. Purcell, *Phys. Rev.* **69**, 681 (1946).
28. H. Casimir and D. Polder, *Phys. Rev.* **73**, 360 (1948).
29. E. T. Jaynes and F. W. Cummings, *Proc. IEEE* **51**, 89 (1963).

30. S. Haroche and D. Kleppner, *Physics Today* **42** (1989).
31. K. H. Drexhage, *Progress in Optics,* Vol. 12, (E. Wolf, ed.), North-Holland, New York, 1974, p. 165.
32. R. H. Dicke, *Phys. Rev.* **93**, 99 (1954).
33. S. Haroche and J. M. Raimond, *Adv. Atomic Mole. Phys.* **20**, 347 (1984).
34. R. G. Hulet, E. S. Hilfer, and D. Kleppner, *Phys. Rev. Lett.* **55**, 2137 (1985).
35. W. Jhe, A. Anderson, E. A. Hinds, D. Meschede, L. Moi, and S. Haroche, *Phys. Rev. Lett.* **58**, 666 (1987).
36. G. Gabrielse and H. Dehmelt, *Phys. Rev. Lett.* **55**, 67 (1985).
37. F. DeMartini, G. Innocenti, G. R. Jacobovitz, and P. Mataloni, *Phys. Rev. Lett.* **59**, 2955 (1987).
38. D. Heinzen, J. J. Childs, J. E. Thomas, and M. S. Feld, *Phys. Rev. Lett.* **58**, 1320 (1987).
39. M. Brune, J. M. Raymond, and S. Haroche, to be published.
40. M. G. Raizen, R. J. Thomason, R. J. Brecha, H. J. Kimble, and H. J. Carmichael, *Phys. Rev. Lett.* **63**, 240 (1989).
41. Y. Zhu, D. J. Gauthier, S. E. Morin, Q. Wu, H. J. Carmichael, and T. W. Mossberg, *Phys. Rev. Lett.* **64**, 2499 (1990).
42. H. Soda, K. Iga, C. Kitahara, and Y. Suematsu, *Jpn. J. Appl. Phys.* **18**, 2329 (1979).
43. K. Iga, F. Koyama, and S. Kinoshita, *IEEE J. Quantum Electron.,* **24**, 1845 (1988).
44. J. L. Jewell, J. P. Harbison, A. Scherer, Y. H. Lee, and L. T. Florez, *IEEE J. Quantum Electron.* **27**, 1332 (1991).
45. F. DeMartini and G. R. Jacobovitz, *Phys. Rev. Lett.* **60**, 1711 (1988).
46. H. Yokoyama and S. D. Brorson, *J. Appl. Phys.* **66**, 4801 (1989).
47. G. Björk and Y. Yamamoto, *IEEE J. Quantum Electron.* **27**, 2386 (1991).
48. P. Goy, J. M. Raimond, M. Gross, and S. Haroche, *Phys. Rev. Lett.* **50**, 1903 (1983).
49. Y. Yamamoto and G. Björk, *Jpn. J. Appl. Phys.* **30**, 2039 (1991).
50. S. L. McCall, A. F. J. Levi, R. E. Slusher, S. J. Pearton, and R. A. Logan, *Appl. Phys. Lett.* **60**, 289 (1992).
51. K. H. Lin and W. F. Hsieh, *Opt. Lett.* **16**, 1608 (1991).
52. F. M. Matinaga, A. Karlsson, S. Machida, Y. Yamamoto, T. Suzuki, Y. Kadota, and M. Ikeda, *Appl. Phys. Lett.* **62**, 443 (1993).
53. G. Björk, Y. Yamamoto, S. Machida, and K. Igeta, *Phys. Rev.* **A44**, 669 (1991).

54. P. W. Milonni, *Phys. Rev.* **A25**, 1315 (1982).
55. P. W. Milonni and P. L. Knight, *Optics Commun.* **9**, 119 (1973).
56. J. P. Dowling, M. O. Scully, and F. DeMartini, *Optics Commun.* **82**, 415 (1991).
57. L. Yang et al., *Appl. Phys. Lett.* **56**, 889 (1990).
58. C. Lei, T. J. Rogers, D. P. Deppe, and B. G. Streetman, *Appl. Phys. Lett.* **58**, 1122 (1991).
59. Y. Yamamoto et al., *Optics Commun.* **80**, 337 (1991).
60. H. Yokoyama et al., *Optics Quantum Electron.* **24**, S245 (1992).
61. D. G. Deppe and C. Lei, *J. Appl. Phys.* **70**, 3443 (1991).
62. F. DeMartini et al., *Phys. Rev.* **A43**, 2480 (1991).
63. G. Björk, H. Heitmann, and Y. Yamamoto, to be published.
64. K. Ujihara, *Jpn. J. Appl. Phys.* **30**, L901 (1991).
65. F. DeMartini, M. Marrocco, and D. Murra, *Phys. Rev. Lett.* **65**, 1853 (1990).
66. T. Baba, T. Hamano, F. Koyama, and K. Iga, *IEEE J. Quantum Electron.* **27**, 1347 (1991).
67. H. Haus, *Waves and Fields in Optoelectronics,* Prentice-Hall, Englewood Cliffs, NJ, 1990.
68. T. Baba, T. Hamano, F. Koyama, and K. Iga, *IEEE J. Quantum Electron.*, **27**, 1310 (1992).
69. E. Yablonovitch et al., *Optics Quantum Electron.* **24**, S276 (1992).
70. S. D. Brorson, H. Yokoyama, and E. Ippen, *IEEE J. Quantum Electron.*, **26**, 1492 (1990).
71. E. Yablonovitch, *Phys. Rev. Lett.* **58**, 2059 (1987).
72. R. J. Horowicz, H. Heitmann, Y. Kadota, and Y. Yamamoto, *Appl. Phys. Lett.* **61**, 393 (1992).
73. H. Heitmann et al., *QELS '92 Conference Proceedings, QPD11-1/19, Anaheim, California,* May 10–15, 1992; and H. Heitmann, Y. Kadota, T. Kawakami, and Y. Yamamoto, *Jap. J. Appl. Phys.* **32**, L1147 (1993).
74. T. Kobayashi, T. Segawa, Y. Morimoto, and T. Sueta, (in Japanese), presented at the 46th fall meeting of the Japan Appl. Phys. Soc, 1982, paper 29a-B-6.
75. T. Kobayashi, Y. Morimoto, and T. Sueta, (in Japanese), presented at Nat. Top. Meet. Rad. Sci., 1985, paper RS85-06.
76. Landolt-Börnstein, *Numerical Data and Functional Relationships in Science and Technology,* Vol. 17, Springer-Verlag, Berlin, 1984.
77. G. Björk, A. Karlsson, and Y. Yamamoto, *Appl. Phys. Lett.* **60**, 304 (1992).
78. G. Björk, H. Heitmann, and Y. Yamamoto, *Phys. Rev.* **A47**, 4451 (1993).

79. J. L. Jewell, K. F. Huang, K. Tai, Y. H. Tai, Y. H. Lee, R. J. Fischer, S. L. McCall, and A. Y. Cho, *Appl. Phys. Lett.* **55**, 424 (1989).
80. R. S. Geels, S. W. Corzine, J. W. Scott, D. B. Young, and L. A. Coldren, *IEEE Photon. Technol. Lett.* **2**, 234 (1990).
81. R. S. Geels, S. W. Corzine, and L. A. Coldren, *IEEE J. Quantum Electron.* **27**, 1359 (1991).
82. O. Wada, H. Hamaguchi, Y. Nishitani, and T. Sakurai, *IEEE Trans. Electron Dev.* **29**, 1454 (1982).
83. O. Wada, *J. Electrochem. Soc.* **131**, 2373 (1984).
84. K. F. Huang, K. Tai, Y. H. Tai, S. N. G. Chu, and A. Y. Cho, *Appl. Phys. Lett.* **54**, 2026 (1989).
85. C. J. Chang-Hasnian, J. P. Harbison, G. Hasnian, A. C. Von Lehmen, L. T. Florez, and N. G. Stoffel, *IEEE J. Quantum Electron.* **27**, 1402 (1991).
86. F. Koyama, K. Morito, and K. Iga, *IEEE J. Quantum Electron.* **27**, 1410 (1991).
87. J. Eberly, N. B. Narozhny, and J. J. Sanchez-Mondragon, *Phys. Rev. Lett.* **44**, 1323 (1980).
88. G. Rempe, H. Walter, and N. Klein, *Phys. Rev. Lett.* **58**, 353 (1987).
89. Y. Yamamoto, M. Machida, and O. Nilsson, *Phys. Rev.* **A34**, 4025 (1986).
90. S. Machida, Y. Yamamoto, and Y. Itaya, *Phys. Rev. Lett.* **58**, 1000 (1987).
91. P. R. Tapster, J. G. Rarity, and J. S. Satchell, *Europhys. Lett.* **4**, 293 (1987).
92. Y. Yamamoto, S. Machida, and G. Björk, *Phys. Rev.* **A44**, 669 (1991).
93. N. Ochi, T. Shiotani, M. Yamanishi, Y. Honda, and I. Suemune, *Appl. Phys. Lett.* **58**, 2735 (1991).
94. M. Yamanishi, Y. Yamamoto, and T. Shiotane, *IEEE Photon. Technol. Lett.* **2**, 889 (1991).
95. K. Wakita, I. Kotaka, O. Mitomi, H. Asai, Y. Kawamura, and M. Naganuma, presented at CLEO '90, paper CTuC6, Anaheim CA, 1990.

Index

Active region, 2
  photon density in, 37, 38
  reduction in dimensions of, 4, 6
Alpha cutoff frequency, 198
Ambipolar transport, 196, 235
Amplitude modulation (AM), 63
Amplitude-phase coupling factor, 63–67, 98–103
Amplitude squeezing
  experimental results, 383–390
  in pump-noise-suppressed laser, 379–382
  squeezed vacuum states from, 390–414
Atomic layer epitaxy (ALE), QD grown by strained-induced, 347
Auger recombination and intervalence band absorption, 152–155
  influence of strain on, 155–161

Band filling. *See* State filling
Bandgap offset at heterojunctions, reducing state filling with, 81–84
Band structure, strained-layer, 133–136
Band tail model, 204

Barrier transport time, 233
Bernard-Duraffourg condition, 124
Bipolar junction transistor (BJT), 197
Bloch function, 8–9
  for bulk semiconductor structure under unitary transformation, 25–26
  at conduction band edge, 22
  at valence band edge, 23
Bookkeeping analysis, 46
Bragg mirror microcavity lasers, 421–426
Bragg scattering, 306
Bulk quantum wells
  See also Three-dimensional (3D) bulk quantum wells
  conduction band and valence band structure, 24–25
  differential gain and state filling, 56–60
  transition matrix elements for TE and TM modes, 27

Capture time, 199–202
Carrier capture time, 52–53
Carrier density, injected
  rate equations for, 37–40, 179, 180–182

relative intensity noise, 187–189
Carrier distribution functions and induced polarization, 12–16
Carrier escape time, designing high-speed lasers and, 225–233
Carrier population, 49, 50
Carrier relaxation time, 38, 198
  slow, 41, 53
Carriers and optical field, interaction between injected, 8–21
Carrier transport theory, 53
  rolloff, 185, 186, 218, 220, 221
Carrier transport times
  ambipolar, 196
  capture time, 199–202
  designing high-speed lasers and, 218–225
  SCH, 193–194, 195–198
  thermionic emission time, 194, 198
  tunneling time, 194, 198–199
Cavity current, 302
Cavity external ouput field, 372–374
  quantum noise properties of, 377–379
Cavity internal field
  and external ouput field, 372–374
  noise power spectra of, 368–372
  quantum noise properties of, 374–376
Cavity quantum electrodynamics (cavity QED)
  microcavity lasers, experiments, 420–437
  overview of, 414–417
  rate-equation analysis of microcavity lasers, 417–420

CHCC Auger recombination process, 154, 157, 158
Chirping
  high-speed lasers and wavelength, 190–193
  for QWR and QD lasers, 306–307
Chirp width, 168
CHSH Auger recombination process, 154, 157, 158
Cleaved-edge overgrowth (CEO), QWR, 312–315
Conduction band structure
  Bloch function at, 22
  for bulk semiconductor structure, 24–25
  envelope function, 23
  zinc blende crystal, 22
Correlation functions of noise operators, 367–368
Coupling factors for, 18–20
Critical layer thickness
  designing high-speed lasers and, 212
  strained layer structures and, 129–131

Dampening rate, 42–43
Density matrix analysis, 13–14, 204, 207
  nonlinear gain and, 240–241
Density of states (DOS)
  carrier, 4, 5, 6
  for QWR and QD lasers, 295–299
  for QW structures, 34–37
  reduced, 37
Dephasing time, 207
Differential gain, 21, 44
  effective, 221

INDEX 445

effect of state filling and thermionic emission on, 227
function, 204
Differential gain, designing high-speed lasers and, 204–206
  p-doping, 213–216
  quantum size effects and strain, 207–213
  wavelength detuning, 216–217
Differential gain, state filling and
  compared with bulk lasers, 56–60
  compared with experiments in modulation bandwidth, 62
  multiple quantum wells and, 60–62
  simplified model, 54–56
Differential index, 64
Dimensional carrier density, 21
Dimensional gain, 20–21
Dimensional volume, 16, 18
Distributed Bragg reflector (DBR), 103
Distributed feedback (DFB), 103, 180, 216, 248
Doping offset, 193
Double heterostructure (DH) lasers, 2, 6, 7
Double quantum well (DQW) lasers, submilliampere threshold current, 88
Dressed state, 416

Effective index, 191–192
Elastic properties, of strained layer structures, 127–129
Electronic states, in semiconductor structures, 8–11
Envelope function, 9, 10

at conduction band structure, 23, 28, 30
for QWR and QD lasers, 295
Etching and regrowth
  QD, 341, 343
  QWR, 309, 312–314

Fabry-Perot cavity semiconductor lasers, 40, 105, 180
Fabry-Perot interferometer, 393–396
Fermi function, 205, 206
Fluctuation-dissipation theorem, 367
Fourier-series analysis, 370
Frequency deviation, 191
Frequency modulation (FM)
  high-speed lasers and, 190–183
  undesired, 63

Gain
  nonlinear, 240–247
  strained layer structures and, 141–145
Gain–carrier density relationships, 209
Gain coefficients
  See also under type of
  differential, 21, 44
  dimensional, 20–21
  linear, 41
  modal/exponential, 19–20, 41–44
  optical transitions and, 16–21
  threshold modal gain, 41–44
Gain spectrum and sublinear gain relationship, 45–48
Gain switching, 262, 263–266

Hamiltonian, 12
Harmonic distortion, 183
Heaviside function, 35, 205, 296
Heisenberg uncertainty principle, 362
Hemispheric microcavity lasers, 430–437
Hermitian amplitude, phase, and excited electron fluctuating operators, 368–369
High-speed modulation, 50–54
　at low operation current, 95–98
　for QWR and QD lasers, 303–305
　strain and, 168–169
High-speed semiconductor lasers
　advances in, 178
　advantages of, 177
　carrier transport times, 193–202
　frequency modulation and chirping, 190–193
　future of, 278–279
　large-signal modulation, 262–278
　rate equations, 179, 180–182
　relative intensity noise, 187–189
　small-signal amplitude modulation, 182–186
High-speed semiconductor lasers, designing
　carrier escape time, 225–233
　carrier transport parameters, optimization of, 217–240
　carrier transport time, 218–225
　device operating conditions, 248–249
　device size and microwave propagation effects, 256–262
　device structures with low parasitics, 250–256
　differential gain, 204–217
　multiple quantum well structures, 233–240
　nonlinear gain, 240–247
　p-doping, 213–216
　photon density, 247–248
　quantum size effects and strain, 207–213
　steps in, 202–203
　wavelength detuning, 216–217
Hot carrier effects, 53–54
Hydrostatic pressure techniques, 156

Index of refraction, 190, 236
Intervalence band absorption (IVBA), Auger recombination and, 152–155
　influence of strain on, 155–161
Intrinsic region, 2
　reduction in dimensions of, 4, 6
Inversion carrier density, 4

Jitter, 268, 277–278

k conservation principle, 206
K factor, 185–186, 189, 226, 228–231, 244
Kramers-Kronig relationship, 216–217
k-selection rule, 13, 14

Large-signal modulation
　bit-rate performance, 262
　gain switching, 262, 263–266
　mode locking, 262, 269–278
　without prebias, 266–269
Lateral patterning, 292, 296

Latticed-matched [111] lasers, 145–146
Linewidth enhancement factor, 63, 98–103, 167–168, 190–191, 193
Long-wavelength lasers
 influence of strain on loss mechanisms, 155–161
 influence of strain on temperature sensitivity, 161–167
 loss mechanisms of Auger recombination and intervalence band absorption, 152–155
Loss mechanisms
 of Auger recombination and intervalence band absorption, 152–155
 influence of strain on, 155–161
Low parasitic devices, structure of, 250–256
Luminescence up-conversion, capture time and, 200
Luttinger-Kohn (LK) Hamiltonian, 134, 136–141

Magnetic field confinement, QWR, 307, 309
Maxwell equations, 12, 17
Metal organic chemical vapor deposition (MOCVD), 2, 178
Metal organic vapor phase epitaxy (MOVPE), 421
Microcavity lasers
 etched, 426–430
 experiments, 420–437
 hemispheric, 430–437
 planar Bragg mirror, 421–426
 rate-equation analysis of, 417–420

Microwave propagation effects, 256–262
Minimum-uncertainty state, 362–363, 364
Modal gain coefficients, 19–20, 41–44
Mode locking, 262
 principles of, 269–274
 pulsewidth, energy, and spectral width, 274–276
 techniques, types of, 272–274
 timing jitter, 277–278
Modulation
 See also High-speed modulation; Large-signal modulation
 small-signal amplitude, 182–186
 without prebias, 266–269
Modulation current efficiency factor (MCEF), 96–98
Molecular beam epitaxy (MBE), 2, 178, 292, 345–347, 348
Multiple quantum well (MQW) lasers
 capture time, 199–200
 carrier transportation complications, 233–240
 differential gain enhancement in, 60–62
 modulation bandwidth in, 51
 reducing state filling in, 69–70, 101–102
 submilliampere threshold current, 88
 thermionic emission time, 194, 198
 tunneling time, 194, 198–199

Noise operators, correlation functions of, 367–368

Noise power spectra of cavity internal field, 368–372
Noise properties
  of cavity external output field, 377–379
  of cavity internal field, 374–376
Nonclassic photon states, 365
Nonlinear gain, 240–247
Nonplanar substrates, QWRs grown on, 325–340

One-dimensional (1D) quantum wire. *See* Quantum wire (QWR), one-dimensional lasers
Operator Langevin equations, 366–367, 368, 403–404
Optical confining layers (OCL), 49–50
  quantum well barrier height, state filling, and, 70–74
Optical field, interaction between injected carriers and, 8–21
Optical gain theory
  gain spectrum and sublinear gain relationship, 45–48
  for QWR and QD lasers, 299–301
  universal, 6
Optical transitions and gain coefficients, 16–21
Organometallic chemical vapor deposition (OMCVD), 292, 312, 313, 318

Parabolic band approximation, 11
Patterned substrate (PS) approach, 89–90
p-doping
  designing high-speed lasers and, 213–216
  strain and, 169

Phonon bottleneck effect, 304
Photon density, 37, 38
  designing high-speed lasers and, 247–248
  rate equations, 179, 180–182
  saturation, 40
  at steady state, 41
Piezoelectric fields, 146
Planar Bragg mirror microcavity lasers, 421–426
Polarization
  anisotropy, 146–147
  carrier distribution functions and induced, 12–16
Prebias, modulation without, 266–269

Q-switched laser, 266
Quadrature amplitude squeezed state, 363
Quantum dot (QD) lasers
  advantages of, 291–292
  approaches to, 293–294
  characteristics of, 342
  configurations, 292, 293
  coordinates, dimensions, wavefunctions for, 10
  density of states, 295–299
  etching and regrowth, 341, 343
  future of, 352–353
  high-speed modulation, 303–305
  optical gain, 299–301
  principles of, 294–307
  self-organized growth, 343–352
  spectral control, 305–307
  threshold current, 301–303
Quantum size effects and strain, designing high-speed lasers and, 207–213

INDEX

Quantum well barrier height, state filling and, 70–74
Quantum well (QW) semiconductor lasers
  applications of, 1–2
  basics of, 21–44
  carrier distribution functions and induced polarization, 12–16
  coordinates, dimensions, wave-functions for, 10
  coupling factors for, 19
  density of states for, 34–37
  electronic states in, 8–11
  interaction between injected carriers and optical field, 8–21
  optical transitions and gain coefficients, 16–21
  rate equations for, 37–40
  reducing threshold current, 4
  research on, 6
  schematic description, 2, 3, 4, 5
  state filling in, 44–84
  static and dynamic properties, 41–44
  transition matrix elements, 22–34
Quantum well (QW) semiconductor lasers, performance characteristics of
  amplitude-phase coupling and spectral linewidth, 98–103
  high-speed modulation at low operation current, 95–98
  submilliampere threshold current, 84–95
  wavelength tunability and switching, 103–108
Quantum wire (QWR), one-dimensional lasers, 4, 5
  advantages of, 291–292
  approaches to, 293–294
  characteristics of, 310–311
  cleaved-edge overgrowth, 314–317
  configurations, 292, 293, 307, 308
  coordinates, dimensions, wavefunctions for, 10
  coupling factors for, 18–19
  density of states, 295–299
  etching and regrowth, 309, 312–314
  future of, 352–353
  grown on nonplanar substrates, 325–340
  grown on vicinal substrates, 317–321
  high-speed modulation, 303–305
  magnetic field confinement, 307, 309
  optical gain, 299–301
  principles of, 294–307
  spectral control, 305–307
  strained-induced self-ordering, 321–325
  threshold current, 301–303
Quasi-equilibrium relaxation times, 14–15
Quasi-probability density, 365

Rate equations
  analysis of microcavity lasers, 417–420
  high-speed lasers, 179, 180–182
  limitations of, 40
  for quantum well laser structures, 37–40
  small-signal solution, 182–186
  static and dynamic properties and, 41–44

Recombination, reduction in carriers' spontaneous, 4
Reduced density of states, 37
Refraction index, 190, 236, 306
Relative intensity noise (RIN), 187–189
Relaxation broadening model, 204–207, 241
Relaxation oscillations, 303–305
Relaxation resonance frequency, 42
Residual phase noise, 277
Resonance frequency, 248

Schawlow-Townes formula, 63, 188, 306
Schrödinger equations
 electronic states and, 8, 9–10, 11, 12
 for quantum well semiconductor structures, 31–32
 for quantum wire and quantum dot lasers, 295
 transition matrix elements and, 22, 23–24
Schwartz inequality, 362
Self-organized growth, QD, 343–352
Separate confinement heterostructure (SCH) quantum well lasers, 6, 7
 capture time, 199–202
 carrier transport time, 193–194, 195–198
 graded-index (GRIN), 49–50, 71
 linearly graded index (L-GRINSCH), 201–202
 parabolically graded index (P-GRINSCH), 201–202
 rate equations for, 38–40, 179, 180–182
 reducing state filling in, 74–77
 step-index (STIN), 49–50
Serpentine SL (SSL) structures, 318, 320
Single quantum well (SQW) lasers
 modulation bandwidth in, 51
 submilliampere threshold current, 88
Small-signal amplitude modulation, 182–186
Spatial confinement, 291–292
Spectral confinement, 291–292, 296, 298
Spectral control, for QWR and QD lasers, 305–307
Spectral dynamics, state filling and, 63–67
Spectral hole burning, 15, 241, 243–246
Spectral linewidth, 98–103
Spontaneous emission
 microcavity lasers, experiments, 420–437
 overview of cavity quantum electrodynamics, 414–417
 rate-equation analysis of microcavity lasers, 417–420
Squeezed vacuum states
 from amplitude-squeezed states, 390–392
 experimental results, 393–401
 experimental setup, 392–383
 numerical results, 412–414
 phase-to-amplitude conversion noise in, 401–412
Squeezing/squeezed states
 amplitude, in pump-noise-suppressed laser, 379–382

INDEX 451

cavity internal field and external ouput field, 372–374
correlation functions of noise operators, 367–368
degree of squeezing versus optical loss, 388–390
experimental results, 383–390
noise power spectra of cavity internal field, 368–372
operator Langevin equations, 366–367
overview of, 362–366
quantum noise properties of cavity external output field, 377–379
quantum noise properties of cavity internal field, 374–376
uncertainty product, 382–383
Standard quantum limit (SQL), 363, 364
State filling, in quantum well lasers, 44
   bandgap offset at heterojunctions, 81–84
   defined, 45
   differential gain and, 54–62
   effect on, differential and threshold gain, 227
   gain spectrum and sublinear gain relationship, 45–48
   high-speed modulation, 50–54
   quantum well barrier height, impact of, 70–74
   reduction of, 67–84
   spectral dynamics and, 63–67
   substrate orientation, 81
   threshold current and, 49–50
Steady-state carrier distribution, 196

Strain, designing high-speed lasers and, 207–213
Strained-induced self-ordering QWRs, 321–325
Strained layer quantum well structures, 123–127
   amplifiers, 169–170
   band structure, 133–136
   critical layer thickness, 129–131
   elastic properties, 127–129
   electronic structure and gain, 131–147
   equation for net strain, 127
   laser gain, 141–145
   linewidth, chirp, and high-speed modulation, 167–169
   long-wavelength lasers, 152–167
   Luttinger-Kohn (LK) Hamiltonian, 134, 136–141
   on non-[001] substrates, 145–147
   reducing state filling in, 78–81
   requirements for efficient lasers, 131–133
   visible lasers, 147–152
Stranski-Krastanow (SK) self-organized growth, 345–352

TE mode
   laser gain, 141–145
   strained laser amplifiers, 169–170
TE mode, transition matrix elements
   for bulk semiconductor structures, 27
   polarization modification factors for, under decoupled valence band approximation, 32, 33

polarization modification factors for, in quantum well semiconductor structures, 31
Temperature sensitivity, influences of strain on, 161–167
Thermionic emission time, 194, 198, 227
Third-order perturbation theory, 240, 241
Three-dimensional (3D) bulk quantum wells, 4, 5
  coordinates, dimensions, wavefunctions for, 10
  coupling factors for, 18–19
  density of states, 35
Threshold current
  for QWR and QD lasers, 301–303
  reducing, 4
  state filling on, 49–50
  submilliampere, 84–95
Threshold modal gain, 41–44
  effect of static filling and thermionic emission on, 227
TM mode
  laser gain, 141–145
  strained laser amplifiers, 169–170
TM mode, transition matrix elements
  for bulk semiconductor structures, 27
  polarization modification factors for, under decoupled valence band approximation, 32, 33
  polarization modification factors for, in quantum well semiconductor structures, 31
Transient carrier heating, 241, 246–247

Transition matrix elements, 22–23
  for bulk semiconductor structures, 24–27
  for quantum well semiconductor structures, 28–34
Transparency carrier density equation, 2, 4
  reduction in active layer and reduction in, 4, 6
Transparency current, 302
Triple quantum well (TQW) lasers, submilliampere threshold current, 88
Tunneling injection (TI) laser, 76–77
Tunneling transport time, 194, 198–199
Two-dimensional (2D) quantum wells, 4, 5
  coordinates, dimensions, wavefunctions for, 10
  coupling factors for, 18–19
  density of states, 35

Uncertainty product, 382–383

Vacuum Rabi frequency, 416–417
Valence band approximation, decoupled, 32
  density of states for QW structures using, 34–37
Valence band edge
  Bloch function at, 23
  for bulk semiconductor structure, 24–25
  density of states for holes in, 35
  for zinc blende crystals, 23
Valance band structure

Luttinger-Kohn (LK) Hamiltonian, 134, 136–141
  strained-layer, 133–136
Vertical cavity surface emitting lasers (VCSEL), 92–94, 98
V-grooved QWRs, grown on nonplanar substrates, 325–340
Vicinal substrates, QWRs grown on, 317–321
Visible lasers, 147–152

Wavefunctions, 10, 11
  for electrons in conduction band, 22, 28
  for holes in valence band, 24, 30
Wavelength chirping, 190–193
Wavelength detuning, 216–217
Wavelength tunability and switching, 103–108
Well-barrier hole burning, 51–52, 185
Wiener-Khintchin's theorem, 371

Zero-dimensional (0D) quantum dot lasers, 4, 5
Zinc blende crystal
  conduction band structure, 22
  valence bands, 23

Optics and Photonics
(Formerly Quantum Electronics)

Edited by Paul F. Liao, *Bell Communications Research, Inc., Red Bank, New Jersey*
Paul L. Kelley, *Tufts University, Medford, Massachusetts*
lvan P. Kaminow, *AT&T Bell Laboratories, Holmdel, New Jersey*
Gorvind P. Agrawal, *University of Rochester, Rochester, New York*

N. S. Kapany and J. J. Burke, *Optical Waveguides*
Dietrich Marcuse, *Theory of Dielectric Optical Waveguides*
Benjamin Chu, *Laser Light Scattering*
Bruno Crosignani, Paolo DiPorto and Mario Bertolotti, *Statistical Properties of Scattered Light*
John D. Anderson, Ir, *Gasdynamic Lasers: An Introduction*
W. W. Duly, CO_2 *Lasers: Effects and Applications*
Henry Kressel and J. K. Butler, *Semiconductor Lasers and Heterofunction LEDs*
H. C. Casey and M. B. Panish, *Heterostructure Lasers: Part A. Fundamental Principles; Part B. Materials and Operating Characteristics*
Robert K. Erf, editor, *Speckle Metrology*
Marc D. Levenson, *Introduction to Nonlinear Laser Spectroscopy*
David S. Kliger, editor, *Ultrasensitive Laser Spectroscopy*
Robert A. Fisher, editor, *Optical Phase Conjugation*
John F. Reintjes, *Nonlinear Optical Parametric Processes in Liquids and Gases*
S. H. Lin, Y. Fujimura, H. J. Neusser and E. W. Schlag, *Multiphoton Spectroscopy of Molecules*
Hyatt M. Gibbs, *Optical Bistability: Controlling Light with Light*
D. S. Chemla and J. Zyss, editors, *Nonlinear Optical Properties of Organic Molecules and Crystals, Volume 1, Volume 2*
Marc D. Levenson and Saturo Kano, *Introduction to Nonlinear Laser Spectroscopy, Revised Edition*
Govind P. Agrawal, *Nonlinear Fiber Optics*
F. J. Duarte and Lloyd W. Hillman, editors, *Dye Laser Principles: With Applications*
Dietrich Marcuse, *Theory of Dielectric Optical Waveguides, 2nd Edition*
Govind P. Agrawal and Robert W. Boyd, editors, *Contemporary Nonlinear Optics*
Peter S. Zory, Jr., editor, *Quantum Well Lasers*
Gary A. Evans and Jacob M. Hammer, editors, *Surface Emitting Semiconductor Lasers and Arrays*
John E. Midwinter, editor, *Photonics in Switching, Volume I, Background and Components*
John E. Midwinter, editor, *Photonics in Switching, Volume II, Systems*
Joseph Zyss, editor, *Molecular Nonlinear Optics: Materials, Physics, and Devices*
F. J. Duarte, editor, *Tunable Lasers Handbook*
Jean-Claude Diels and Wolfgang Rudolph, *Ultrashort Laser Pulse Phenomena: Fundamentals, Techniques, and Applications on a Femtosecond Time Scale*
Eli Kapon, editor, *Semiconductor Lasers I: Fundamentals*
Eli Kapon, editor, *Semiconductor Lasers II: Materials and Structures*

Yoh-Han Pao, Case Western Reserve University, Cleveland, Ohio, Founding Editor 1972–1979